Cities and Development

T0144598

For the first time in human history more people now live in towns and cities than in rural areas. In the wealthier countries of the world, the transition from predominantly rural to urban habitation is more or less complete. But in many parts of Africa, Asia and Latin America, urban populations are expanding rapidly. Current UN projections indicate that virtually all population growth in the world over the next 30 years will be absorbed by towns and cities in developing countries. These simple demographic facts have profound implications for all those concerned with understanding and addressing the pressing global development challenges of reducing poverty, promoting economic growth, improving human security and confronting environmental change.

This revised and expanded second edition of *Cities and Development* explores the dynamic relationship between urbanism and development from a global perspective. The book surveys a wide range of topics, including: the historical origins of world urbanisation; the role cities play in the process of economic development; the nature of urban poverty and the challenge of promoting sustainable livelihoods; the complexities of planning and managing urban land, housing, infrastructure and urban services; and the spectres of endemic crime, conflict and violence in urban areas. The second edition also benefits from two new chapters: one that examines the links between urbanisation and environmental change, and a second that focuses on urban governance and politics.

Adopting a multidisciplinary perspective, *Cities and Development* critically engages with debates in urban studies, geography and international development studies. Each chapter includes supplements in the form of case studies, chapter summaries, questions for discussion and suggested further readings. Written for upper-level undergraduate and graduate students interested in geography, urban studies and international development studies, the book will also serve as a useful reference guide for policymakers, urban planners and development practitioners.

Sean Fox is Lecturer in Urban Geography and Global Development at the University of Bristol. His research has been published in journals such as *Environment and Planning C: Government and Policy*, *Population and Development Review* and *World Development*. He has also served as an advisor on urban development issues for a variety of international organisations, including DfID, CARE International and UN-Habitat.

Tom Goodfellow is Lecturer in Urban Studies and International Development at the University of Sheffield. His work has been published in a range of journals including *Urban Studies*, *Comparative Politics* and *Development and Change*. He has acted as an advisor to Oxfam on urban development issues and provided policy analysis for the UK Parliamentary Select Committee on International Development, as well as several government authorities in sub-Saharan Africa.

Routledge Perspectives on Development

Series Editor: Professor Tony Binns, *University of Otago*

Since it was established in 2000, the same year as the Millennium Development Goals were set by the United Nations, the *Routledge Perspectives on Development series* has become the pre-eminent international textbook series on key development issues. Written by leading authors in their fields, the books have been popular with academics and students working in disciplines such as anthropology, economics, geography, international relations, politics and sociology. The series has also proved to be of particular interest to those working in interdisciplinary fields, such as area studies (African, Asian and Latin American studies), development studies, environmental studies, peace and conflict studies, rural and urban studies, travel and tourism.

If you would like to submit a book proposal for the series, please contact the Series Editor, Tony Binns, on: jab@geography.otago.ac.nz

Published:

Third World Cities, 2nd edn
David W. Drakakis-Smith

Rural–Urban Interactions in the Developing World
Kenneth Lynch

Environmental Management & Development
Chris Barrow

Southeast Asian Development
Andrew McGregor

Postcolonialism and Development
Cheryl McEwan

Disaster and Development
Andrew Collins

Non-Governmental Organisations and Development
David Lewis and Nazneen Kanji

Gender and Development, 2nd edn
Janet Momsen

Economics and Development Studies
Michael Tribe, Frederick Nixson and Andrew Sumner

Water Resources and Development
Clive Agnew and Philip Woodhouse

Theories and Practices of Development, 2nd edn
Katie Willis

Food and Development
E. M. Young

An Introduction to Sustainable Development, 4th edn
Jennifer Elliott

Latin American Development
Julie Cupples

Religion and Development
Emma Tomalin

Development Organizations
Rebecca Shaaf

Cities and Development

Second Edition

Sean Fox and
Tom Goodfellow

Routledge
Taylor & Francis Group

LONDON AND NEW YORK

Second edition published 2016

by Routledge

2 Park Square, Milton Park, Abingdon, Oxon OX14 4RN

and by Routledge

711 Third Avenue, New York, NY 10017

Routledge is an imprint of the Taylor & Francis Group, an informa business

© 2016 Sean Fox and Tom Goodfellow

The right of Sean Fox and Tom Goodfellow to be identified as authors of this work has been asserted by them in accordance with sections 77 and 78 of the Copyright, Designs and Patents Act 1988.

First edition published by Routledge 2009

British Library Cataloguing in Publication Data
A catalogue record for this book is available from the British Library

Library of Congress Cataloging in Publication Data
Names: Fox, Sean, 1979– | Goodfellow, Tom. | Beall, Jo, 1952– Cities and development.Title: Cities and development / Sean Fox and Tom Goodfellow.Description: Second Edition. | New York : Routledge, 2016. | Series: Routledge perspectives on development series | Revised edition of Cities and development, 2009. | Includes bibliographical references and index.Identifiers: LCCN 2015032847| ISBN 9780415740715 (hardback) | ISBN 9780415740722 (pbk.) | ISBN 9781315815527 (ebook)Subjects: LCSH: Urbanization--Economic aspects. | Urbanization--History--21st century. | Cities and towns--Growth--History--21st century. | Urban economics--History--21st century. | Poverty--History--21st century.Classification: LCC HT371 .F69 2015 | DDC 307.7609/05--dc23LC record available at http://lccn.loc.gov/2015032847

ISBN: 978-0-415-74071-5 (hbk)
ISBN: 978-0-415-74072-2 (pbk)
ISBN: 978-1-315-81552-7 (ebk)

Typeset in Times New Roman
by Saxon Graphics Ltd, Derby

To Su Lin, Juna, Sally, Hal and Tess

Contents

Plates

Figures

Tables

Boxes

Foreword

More than half of the world's population today lives in cities, making urbanisation a defining characteristic of the twenty-first century and a factor in international development that can no longer be ignored. As this volume demonstrates, throughout history cities have played a critical role in economic, political and social development. Yet until quite recently, in low- and middle-income countries national development policies and international development interventions have paid scant attention to cities, towns and urban dwellers.

Today, this neglect is no longer tenable. Economists have demonstrated that economic innovation and growth is closely associated with benefits deriving from urban agglomeration. And it is in cities that we see more vibrant civic engagement and the exercise of political voice and choice. People move to cities to improve their life chances and generally do so. Nevertheless, urbanisation also presents enormous challenges for development. There are severe infrastructure and services deficits in many cities and towns in low- and middle-income countries, leading to severe congestion and other negative outcomes of urban agglomeration. The weight on people living in poorly planned and managed urban centres is considerable, particularly those living in burgeoning slums where poverty and its effects remain widespread, or in cities affected by conflict and environmental changes. This second edition of *Cities and Development* marks a significant step in our understanding of the ways in which processes of urbanisation and international development variously intersect, impact upon and bypass one another. It does so because of the multidisciplinary insights and methodologies on which it draws.

The first volume on cities in the Routledge development series was entitled *The Third World City*, with the second edition being called *Third World Cities*, recognising differences among the wide variety of cities in the developing world. Their author, the late David Drakakis-Smith, was a geographer and these early volumes reflect the perspectives and preoccupations of the disciplines of geography and urban planning. David Drakakis-Smith was one of my doctoral examiners and I was consequently very pleased to be asked to write a third edition. Still, it was also clear to me that to encompass and comprehend the complex and multidimensional ways in which urbanisation and development processes are and have been entwined was beyond the insights and tools of a single discipline. So it was that I asked Sean Fox to co-author with me the first edition of *Cities and Development*, with us each bringing to the task different approaches and methodologies in a manner consistent with urban studies and development studies as areas of interdisciplinary scholarly pursuit.

The first edition of *Cities and Development* made an important contribution towards how we look at cities in the context of development and why it is important that we do so. Tom Goodfellow provided invaluable research assistance at the time of writing the book and I am therefore delighted that he and Sean Fox have now taken on the baton to produce this second edition, which advances our understanding even further, offering as it does accumulated and enriching perspectives from economics, politics, international relations and social policy, alongside the more conventional urban disciplines of geography and planning.

Jo Beall
Director Education and Society
British Council
London, August 2015

Acknowledgements

We would like to thank our Editor Andrew Mould for giving us an opportunity to undertake this unusual revision for the second edition, as well as Sarah Gilkes at Routledge for her patience throughout the process.

We are grateful for the financial support received by the Sheffield Institute for International Development (SIID) for the preparation of this edition, and to SIID directors Paula Meth and Chasca Twyman in particular. We would also like to thank Nick Clare, Mary Davies, Steph Gamauf, Will Wright and Jessica Swinburne-Cloke for valuable research assistance.

Many friends and colleagues have influenced our research and thinking at the intersection of urban studies and development studies, but none more so than Jo Beall, who initiated the first edition of this book and encouraged us to carry forward this intellectual project.

Finally, we would like to thank Su Lin Lewis and Sally Lavelle for their unwavering support.

 # Development in the first urban century

- Defining 'cities' and 'urbanisation'
- Defining 'development'
- A (very) brief history of development theory, policy and practice
- Cities as problems, solutions and contexts

Introduction

For the first time in human history more people now live in towns and cities than in rural areas, and over the next few decades this trend towards an increasingly urban world is set to continue. However, the process of world urbanisation has been geographically uneven. All economically advanced nations are predominantly urban societies, having undergone this spatial-demographic transition in the nineteenth and twentieth centuries. By contrast, most low- and middle-income countries remained predominantly rural societies until the mid-twentieth century when a global population boom set in motion a historically unprecedented expansion of urban populations in Africa, Asia and Latin America. Today cities of unprecedented size are emerging; slums, shanty towns, and squatter settlements are growing; urban poverty is rising; violence and armed conflict threaten the security of urban dwellers in rich and poor countries alike; and the stability of the global environment is under threat, with particular implications for urban centres. At the same time, cities have historically played an important role as drivers of social, political and economic transformations; they are social melting pots, nodes of regional and international communication and transportation, engines of economic growth, seats of political power and iconic cultural spaces. As such, the relationship between cities and development is complex and intimately tied to both the history of human development and contemporary efforts to improve the human condition through development policy and practice.

In this book we provide a global and interdisciplinary perspective on the relationship between cities and development. Drawing on a wide range of literature from the broader social sciences, we aim in particular to synthesise insights from the fields of Urban Studies and Development Studies – both interdisciplinary subject areas that emerged in the twentieth century as social scientists sought to grapple with the causes and consequences of industrialisation, urbanisation and globalisation. Through this approach we address the contributions that cities have made to development processes, as well as the unique economic, political and social development challenges posed by complex urban environments. We begin by defining some key terms and concepts and providing a rough outline of broad trends in the history of development research and practice.

Defining 'cities' and 'urbanisation'

What is a city? Most people recognise a city when they see one, and implicitly or explicitly recognise a difference between rural and urban ways of living, but under closer scrutiny these categories can be problematic. What is the fundamental difference between a large village and a small town, or a large town and a small city? Social scientists have been pondering these questions for more than a century. While there is no consensus in theory or in practice, the perspectives of two eminent urban scholars writing in the early twentieth century remain influential among urban scholars and policymakers alike.

The first comes from Chicago School sociologist Louis Wirth, whose essay 'Urbanism as a way of life' is a classic in the field of urban studies. Wirth defined a city as a 'relatively large, dense, and permanent settlement of socially heterogeneous individuals' (1938: 8). This remains the most concise and enduring definition of an urban settlement. Importantly, Wirth argued that these conditions – size, density and heterogeneity – create a distinctly 'urban way of life' and an identifiable 'urban personality'. It is the unique nature of the social, political, economic and cultural life of cities – or *urbanism* – that lies at the heart of urban scholarship.

Wirth's contemporary, Lewis Mumford, went further by defining the urban condition not only in relation to demographic and physical characteristics, but also the socioeconomic relationships typical of human settlements with these characteristics:

> The essential physical means of a city's existence are the fixed site, the durable shelter, the permanent facilities for assembly, interchange, and storage; the essential social means are the social division of labour, which serves not merely the economic life but the cultural processes.

> (Mumford 1937: 93–94)

This definition highlights an important aspect of urban: the social division of labour. For many social scientists this is a defining aspect of urban life and forms an essential component of the definition of what is 'urban'. However, Mumford recognises the fundamental role of population size and density in creating this dynamic:

> [W]ithout the social drama that comes into existence through the focusing and intensification of group activity there is not a single function performed in the city that cannot be performed – and has not in fact been performed – in the open country'.

> (ibid.)

In other words, it is ultimately the concentration of human energies and activities that brings a place to life and gives it a distinctly urban character. Many authors before and since have sought to define or redefine the city, but the definitions offered by Wirth and Mumford remain fundamental to our conceptualisation of cities.

However, translating these definitions into practice for the purposes of research and policymaking highlights the ambiguities inherent in all efforts to create categories for the purposes of quantification and analysis. What is the appropriate size or population density threshold for classifying a settlement as distinctly 'urban'? What degree of labour specialisation characterises an 'urban' way of life in a community? In practice, human settlements are classified very differently across countries, generally with reference to factors such as administrative status, population size, 'urban characteristics' or economic function (see Table 1.1). Of the 233 countries and territories tracked by the United Nations, 30 per cent use strictly administrative criteria for classifying cities, 21 per cent use only population size, and 4 per cent have no definitive criteria (United Nations, Department of Economic and Social Affairs, Population Division 2014).

Table 1.1 *Defining 'urban': settlement classification criteria for selected countries*

Country	Definition of an 'urban' settlement
Angola	Localities with 2,000 inhabitants or more.
Australia	Urban centres with 1,000 inhabitants or more.
Botswana	Agglomerations of 5,000 inhabitants or more where 75 per cent of the economic activity is non-agricultural.
Burkina Faso	Localities with 10,000 inhabitants or more and with sufficient socio-economic and administrative infrastructures.
Canada	Since the 1981 census, areas with 1,000 inhabitants or more and a population density of at least 400 inhabitants per square kilometre.
Cuba	Places with 2,000 inhabitants or more, and places with fewer inhabitants but with paved streets, street lighting, piped water, sewage, a medical centre and educational facilities.
Germany	Communes ('kreisfreie Staedte' and 'Kreise') with population density equal or greater than 150 inhabitants per square kilometre.
India	Statutory places with a municipality, corporation, cantonment board or notified town area committee; and places satisfying the following three criteria simultaneously: (a) a minimum population of 5,000 inhabitants, (b) at least 75 per cent of male working population engaged in non-agricultural pursuits, and (c) a density of population of at least 400 per square kilometre.
Iran	Districts with a municipality. In censuses before 1986, all county centres regardless of size and places with a population of 5,000 inhabitants and more.
Kenya	Municipalities, town councils, and other urban centres with 2,000 inhabitants or more.
Mexico	Localities with 2,500 inhabitants or more.
Nigeria	Towns with 20,000 inhabitants or more.
Pakistan	Places with municipal corporation, town committee or cantonment.
United Kingdom	Settlements with 10,000 inhabitants or more. For the censuses up to 1971, administrative boundaries were used.
Vietnam	Places with 4,000 inhabitants or more.
Zimbabwe	Places officially designated as urban, as well as places with 2,500 inhabitants or more whose population resides in a compact settlement pattern and where more than 50 per cent of the employed persons are engaged in non-agricultural occupations.

Source: United Nations, Department of Economic and Social Affairs, Population Division (2014). World Urbanization Prospects: The 2014 Revision, CD-ROM Edition.

These differences in settlement classification make direct comparisons between countries somewhat tricky. Throughout this book we rely primarily on population statistics from the United Nations, which are generally considered to be the most reliable and comparable. But given national-level variation in settlement classification, even these are considered problematic and should be used with caution (Cohen 2004; Satterthwaite 2007).

Further confusion arises when discussing and analysing the process of urbanisation more broadly. The word 'urbanisation' is used in the English language in at least three ways: a) to refer to the process of rural–urban migration, b) to refer to the expansion of built-up areas, and c) to refer to an increase in proportion of a nation's population living in settlements classified as urban (see Parnell and Walawege 2011). In other languages the word is sometimes used to denote the development of physical infrastructure in a town or city, such as paved roads, water mains or public buildings. For the purposes of conceptual and analytical clarity, we employ three distinct terms to describe the kinds of changes that occur as countries transition from predominantly rural to predominantly urban societies: urbanisation, urban growth and urban expansion.

Within this framework the word *urbanisation* is defined specifically in relation to the proportion of a country or region's population living in urban (as opposed to rural) areas. Thus a country can be said to be 'urbanising' if the proportion of its population living in urban areas is increasing (e.g. from 35 per cent to 40 per cent). The speed of this change is known as the 'rate' of urbanisation (e.g. 2 per cent per annum), while the 'level' of urbanisation refers to the percentage of the population living in urban areas at any given time (e.g. as in 'China's level of urbanisation increased from 36% to 56% between 2000 and 2015'). This is a far more narrow use of the word than that found in general public discourse, in which the word urbanisation is used in a much broader manner.

We contrast urbanisation with *urban growth*, which refers specifically to an increase in the *absolute number* of people living in urban settlements rather than the proportion of people living in urban settlements. For example, if a country's urban population increased from 1 million to 1.1 million people in a single year, we might say that it has an urban growth rate of 10 per cent per annum. Importantly, this is distinct from the rate of urbanisation: while urbanisation and urban growth often go together, it is possible for urban growth to occur without urbanisation (see Chapter 2).

Finally, *urban expansion* refers to an increase in the physical built-up area of a settlement or collection of settlements. Urban growth is often accompanied by urban expansion as families, developers or governments build on empty land to accommodate increasing numbers. But it is also possible for urban growth to be accommodated without expansion by increasing population densities in existing built-up areas (e.g. by constructing taller buildings), or urban expansion to occur without urban growth (e.g. de-densification through suburbanisation).

Box 1.1 Urban vocabulary

City

A relatively large, dense and permanent human settlement. Cities are also referred to as towns, urban agglomerations or urban settlements. There is no standard set of criteria for identifying a settlement as urban, although population size, economic function and political/administrative status are frequently employed.

Urbanism

The way of life associated with habitation in a city. Generally speaking, urbanism refers to the unique social, cultural, economic and political dynamics that result from the close habitation of people in space. However, urbanism is not necessarily confined to cities. Some social scientists argue that urban ways of living, although endemic to cities, are exported to rural areas – particularly in advanced economies. For example, urban culture and consumption patterns are frequently found in rural and urban areas alike.

Urbanisation

A demographic process involving a shift in the proportion of a population living in settlements defined as 'urban' as opposed to 'rural'. A country's *rate* of urbanisation refers to the pace at which the relative proportion of the urban population is changing in a defined period of time, while the *level* of urbanisation indicates the proportion of the population residing in urban areas at any given moment of time.

Urban growth

An increase in the absolute size of an urban population. This could be at the level of an individual settlement or a collection of settlements (e.g. at the national level). Urban growth and urbanisation often occur together, but not always. A nation's urban population can grow in absolute terms without increasing in relative terms.

Urban expansion

An increase in the built-up area of a settlement or collection of settlements (e.g. at the national level). This often accompanies an increase in urban population size (i.e. urban growth), but not necessarily. Urban growth can happen without expansion in contexts of increasing habitation density; conversely, urban expansion can occur without urban growth where de-densification happens (e.g. suburbanisation).

Maintaining a clear distinction between these processes is analytically important when seeking to understand the dynamics of urban change. This is particularly true in 'developing countries' – a term we define and justify our use of below – where urban change is happening fast.

Defining development

The word 'development' denotes change, and for most people it implies *good* change (Chambers 2004). However, in the field of development studies it can take on a variety of meanings. Generally speaking, development is understood as a historical process, a project or practice, a social objective or a discourse.

As a historical process, development has traditionally been equated with industrialisation and improvements in material living standards. These changes involve qualitative and quantitative shifts in the economic, social and political organisation of human societies. For example, the transition from a low-income economy dominated by subsistence agriculture to a relatively high-income one where most labour is employed in industry and services can be understood as 'development' within this process-oriented perspective. Indeed, one of the most enduring questions in the field of development studies is: why and how have some countries become rich while others have remained poor?

As a project or practice, development can be seen as a 'conscious and collective enterprise' to improve the human condition (Brett 2009). Development in this sense may involve a wide range of actors, including individual activists, community groups, governments, bilateral agencies (such as USAID and the UK's Department for International Development [DFID]), multilateral agencies (such as the United Nations Development Programme [UNDP], World Bank and International Monetary Fund [IMF]) and non-governmental organisations (such as Oxfam, CARE International, ActionAid, Christian Aid and Save the Children). These actors engage in deliberate interventions at various scales (i.e. from local to global) to promote 'good change' as they define it.

Development can also be seen as a goal or objective for a community or society – something desirable to be achieved. Clear examples of this at the international level are the Millennium Development Goals (MDGs), which were recently extended in the form of new Sustainable Development Goals (SDGs). The MDGs were agreed and adopted by 189 countries at a United Nations summit in September 2000. These were a set of targets to be achieved by 2015, which included reducing poverty, hunger, disease, illiteracy, environmental degradation and discrimination against women. In 2015 the MDGs were replaced by a new set of international targets known as the

Sustainable Development Goals (SDGs). These international goal-setting exercises serve to focus the minds and energies (and financial resources) of governments, donors and development practitioners seeking to achieve positive change.

As a discourse, 'development' can be understood as a vocabulary and set of concepts that shape the way we think about different communities and societies around the world. To speak of 'developed' and 'developing' countries assumes a set of values about what is desirable and casts some communities or countries as 'latecomers' that are seeking to 'catch up' with the wealthy, industrialised countries of the world. While the discourse of development is pervasive in the international public realm it is also contested. Indeed, there is a substantial body of work that views development as a highly problematic discourse. Some authors suggest that it serves to mask the political nature of development intervention, misguidedly presenting it as technical (Ferguson 1990, Li 2007), or that it constrains academic theorisation by introducing a dichotomy that privileges the ideas and theories of the so-called 'developed' world (Robinson 2006). At the extreme, some claim that the development discourse has been an instrument of social and psychological control, serving to perpetuate structures of power that reinforce privilege for some and penury for others (e.g. Escobar 1995; Appadurai 2006).

These critics make an important contribution by highlighting the embedded assumptions in the linguistic and conceptual architecture underpinning development scholarship and practice. Increasing numbers of scholars therefore avoid the idea of development and its counterpart – the term 'developing countries' – by writing instead about the 'global North' and 'global South', and the specific challenges facing the latter as a region that has been marginalised discursively as well as economically. Yet there is a risk of throwing out the baby with the bathwater by rejecting the discourse entirely, and relying on the idea of global North and South elides the question of what exactly is supposed to significantly differ between these two regions thereby rendering this geographical distinction relevant. We argue that the basis for this distinction is something that can be termed 'development', despite the problems with this term. This book is therefore predicated on the idea that the discourse of development still has relevance today. As Thomas has pointed out, abandoning the study and practice of development will not alleviate the very real problems of 'poverty and powerlessness, environmental degradation and social disorder' (2000a: 21) that continue to affect billions of people across the world.

Perhaps the most philosophically consistent and influential approach to thinking about development as a process and objective, with implications for discourse and practice, has emerged from the work of Nobel Prize-winning economist Amartya Sen and American philosopher Martha Nussbaum. Together these scholars have formulated the 'capabilities' approach to conceptualising development, which is outlined in Sen's seminal book *Development as Freedom* (1999). Sen argues that freedom is both the principal means and ultimate goal of development, with freedom defined as having the ability to live the kind of life that one has reason to value. According to Sen,

> Development requires the removal of major sources of unfreedom: poverty as well as tyranny, poor economic opportunities as well as systematic deprivation, neglect of public facilities as well as intolerance or overactivity of repressive states.
>
> (1999: 3)

From this perspective, development involves the expansion of real, substantive freedoms to communities and societies across the globe. While this may involve conscious interventions, the capabilities perspective clearly champions human agency:

> With adequate social opportunities, individuals can effectively shape their own destiny and help each other. They need not be seen primarily as passive recipients of the benefits of cunning development programs. There is indeed a strong rationale for recognizing the positive role of free and sustainable agency—and even of constructive impatience.
>
> (1999: 11)

This view has been very influential in shaping the way development is conceptualised and even measured in recent decades. By far the most common metric of 'development' since the end of the Second World War has been GDP per capita (i.e. income per person). However, in recognition that development is not just about income but also other capabilities that influence individuals' abilities to flourish, the Human Development Index (HDI) was introduced by the United Nations Development Programme in 1990. The HDI is a composite index that ranks a nation's level of development based upon income, literacy rates and life expectancy at birth. These later two measures are included to capture differences in levels of education and health, which are seen as essential capabilities underpinning individual freedom. While income level and HDI level are closely linked, a comparison of where countries rank on each metric reveals some significant discrepancies. As Table 1.2 illustrates, for some countries the HDI ranking is significantly higher than the income rank (e.g. Australia, Netherlands and Cuba) while in others the HDI ranking is far below the income rank (e.g. Qatar, Malaysia and South Africa). This highlights just how important concepts are in shaping our interpretation of 'development' and who is 'developing'.

In this book we engage with development as a multifaceted process involving the expansion of substantive capabilities, as a project aimed at improving the human condition and promoting inter-regional equality, and as a discourse. Although problematic, we follow the convention of classifying countries as 'developing' based solely on income level, following current World Bank categories: countries that fall into the low- and middle-income categories are referred to as developing, while those in the high-income group are referred to as developed. We avoid the anachronistic phrase Third World to indicate developing countries. We also do not use the currently popular term 'global South', for reasons indicated above (and because it is geographically flawed) except where discussing the work of scholars who use the term to make specific points.

Table 1.2 *Comparison of income and HDI rankings for selected countries*

Country	HDI rank	GDP per capita rank	GDP rank – HDI rank
Norway	1	6	5
Australia	2	20	18
Switzerland	3	9	6
Netherlands	4	17	13
United States	5	11	6
Qatar	31	1	−30
Cuba	44	56	12
Malaysia	62	49	−13
Mexico	71	70	−1
China	91	88	−3
South Africa	118	84	−34
India	135	130	−5
Kenya	147	160	13
Afghanistan	169	161	−8
Liberia	175	183	8

Source: UNDP (2014).

A (very) brief history of development theory, policy and practice

The conceptual foundations of development, and contemporary debates surrounding it, can be traced back at least to the Age of Enlightenment and the emergence of industrial capitalism in Europe in the eighteenth and nineteenth centuries. Adam Smith, in his *Enquiry into the Nature and Causes of the Wealth of Nations* (1776), famously celebrated the productivity gains that could be realised through the division of labour and the power of the 'invisible hand' of the market to channel individual self-interest into an orderly system of production and exchange. He was writing at a time when technological innovation was accelerating, inter-regional trade was intensifying, and manufacturing and industry were beginning to emerge as pillars of a new, more bountiful economic system. However, by the nineteenth century it had become clear that industrial capitalism had a dark side.

In 1842 the German social scientist Friedrich Engels was sent by his father to work in a Manchester textile firm. It was here that he observed the dreadful conditions under which English mill workers laboured, conditions he subsequently exposed and questioned along with Karl Marx. Marx, of course, went on to write some of the most influential and enduring critiques of capitalism and the social dislocations associated with industrialisation. In doing so he posited a theory of history influenced by the work of German philosopher Georg Hegel. Hegel had theorised history as progress; a 'progression to the better' in the realm of metaphysics and the human spirit. Marx famously turned Hegel on his head and concentrated his formidable analytic prowess on the nature of progress in the material world (Leys 1996). He theorized history as a progression through various modes of production – communal, feudal, capitalist, and eventually communist/socialist – with each step along the way representing progress towards the full realisation of human potential. Capitalism, according to Marx, would give way to socialism or communism due to the inherent contradictions of the system, which was clearly generating great wealth and great inequality simultaneously on a historically unprecedented scale. In this way, capitalism would produce its own 'grave-diggers' (Marx and Engels 1848).

This materialist theory of history echoed the dialectical logic of Hegel's philosophy, but also the evolutionary logic of another influential thinker of the nineteenth century – Charles Darwin. The idea that human societies evolve was further developed in the late nineteenth and early twentieth centuries by the founders of the sociological tradition, Emile Durkeim and Max Weber. These scholars explored the ways in which social structures change in response to industrialisation and urbanisation, introducing notions of transition from 'tradition' to 'modernity' and from the 'rule of man' to the 'rule of law' – i.e. a de-personalisation of the foundations of social order. Taken together, the intellectual legacy of these thinkers was an evolutionary understanding of social change and an explicit concern with the consequences of such change for human welfare. In sum, these thinkers laid the foundations for subsequent research and debates in development theory and practice.

Despite eighteenth- and nineteenth-century roots, it was not until the middle of the twentieth century that the idea of development was popularised to the extent that it inspired an international political agenda and distinct field of study. In the wake of the trauma of World War II, a new international order was negotiated that sought to improve international economic collaboration, eliminate conflicts between nations and maintain world peace. A new family of multilateral organisations were established to serve these ends, including the United Nations (UN), the International Monetary Fund (IMF), the World Bank Group (including the International Bank for Reconstruction and Development) and the General Agreement on Tariffs and Trade (GATT), set up in 1947 to promote free trade between nations – a precursor to the present World Trade Organisation (WTO). Initially, these organisations focused on facilitating the reconstruction of Europe in concert with the European Recovery Plan (otherwise known as the Marshall Plan), which was developed and financed by the United States and funnelled billions of dollars into post-war reconstruction efforts over the course of four years. However, there was also a push for the decolonisation of Africa, the Middle East, Asia and some remaining European possessions in the Americas. President Harry S. Truman's inaugural address on 20 January 1949 illustrates well the sentiments of the time:

> We must embark on a bold new program for making the benefits of our scientific advances and industrial progress available for the improvement and growth of underdeveloped areas. The old imperialism – exploitation for foreign profit – has no place in our plans. What we envisage is a program of development based on the concepts of democratic fair dealing.
>
> (cited in Cowen and Shenton 1996: 7)

Truman's speech ushered in the first development decade. The fact that European nations (and Japan) had been successfully reindustrialised after the war with international assistance fed optimism that this approach could be extended elsewhere in the world, giving rise to the field of development studies and an explicit global project to promote improvements in human welfare across the globe. Roughly speaking, we can identify five post-World War II periods in development theory, discourse and practice. These are summarised in Table 1.3 and explored briefly in the following pages.

The 1950s and 1960s were decades of optimism dominated by 'modernisation' theory, which emphasised state-driven industrial expansion to accelerate economic growth and development, modelled, to some extent, on the Marshall Plan. Modernisation theorists, who were largely based in the United States of America (US) and the United Kingdom (UK), conceptualised development as a dynamic process of economic and social transformation driven by a class of enlightened political elites. It involved a shift from agricultural to industrial production, from rural to urban habitation, and from 'traditional' to 'rational' socio-political values (Thomas 2000b; Willis 2005). Countries that remained largely agricultural, or that were deemed 'traditional' in their forms of social organisation and values, were referred to by modernisation theorists as 'undeveloped' or 'backward'. Development was understood as a linear process of cumulative change and the key problematic addressed by scholars in the post-World War II era was how to set the wheels of development in motion – or to use the terminology of American economist and modernisation theorist Walter Rostow (1960) – how to achieve 'take-off'. In his book *The Stages of Economic Growth: A Non-Communist Manifesto*, Rostow articulated a process of development not dissimilar in direction to that proposed by Marx although, as his title suggests, with very different goals in mind. In his view, all societies could advance from a traditional, low-productivity state to one of capitalist high mass-consumption, such as had been achieved by the US, so long as successive steps were followed to ensure the 'preconditions for take-off', then economic 'take-off', and finally a 'drive to maturity'.

In order to achieve this, it was believed that poorer countries would have to mobilise savings to finance the investments in infrastructure and technology required to support industrialisation. International aid was seen as a way to accelerate the process by making up for the shortfalls of capital limiting the scale of investment in poorer

Table 1.3 Key trends in development theory, discourse and practice

1950s–1960s	1970s	1980s–1990s	2000s	2010–
Modernisation theory	*Dependency theory*	*Structural adjustment/ neoliberalism*	*Governance/institutions*	*Evidence-based aid*
Structuralism/ developmentalism	*Poverty & inequality*	*Sustainability*	*State-building/state-failure*	*South–South cooperation/ global development goals*
	Participation & empowerment	*Rights-based approaches*		*Urbanisation*
Development as social transformation involving industrialisation, urbanisation, shift from 'traditional' to 'modern' social values	Wealthy nations achieve & maintain status through inequitable capitalist world system	Privatisation, liberalisation & deregulation to minimise distortions of state intervention in economy & unleash market forces	Focus on improving accountability, transparency, rule of law, security of property rights, bureaucratic efficiency and state effectiveness	Use of RCTs & means-tested targeting of aid to maximise impact & 'value for money'
State intervention to promote structural transformation of economy	Shift in focus from economic development to poverty alleviation & growth with redistribution	Environmental protection & intergenerational equity	Linking of security & development agendas, focus on weak states/ state-building	Emergence of South–South cooperation frameworks & reframing of development goals as global
	Community scale intervention, participatory methods	Protection of individual civil, political and 'social' rights is foundational		Growing attention to the challenges of managing & harnessing urban population change

countries. National governments were seen as playing a pivotal, active role in cultivating industrial development by investing directly in 'infant industries' and indirectly encouraging expansion and diversification through policies to restrict trade in certain sectors of the economy – a perspective known as structuralism or developmentalism. Investment and industrial policy were accompanied by large-scale infrastructure projects to build roads, housing, dams and energy infrastructure. The World Bank, in particular, promoted and financed such projects through grants and loans. The success of state-directed development is most closely associated with the 'developmental states' of East Asia, such as Japan, Taiwan and South Korea, which successfully facilitated a dramatic economic transformation over the course of the latter half of the twentieth century. This East Asian 'economic miracle' has been attributed to effectively 'governed markets' where there was a 'synergistic connection between a public system and a mostly market system' (Wade 2004: 5). In other words, government officials and politicians were able to construct and apply economic rules that advanced technological development and long-term capital growth.

Countries across Latin America, South East Asia, South Asia and sub-Saharan Africa also experienced post-war economic booms underpinned by state-directed investments and activist policies, but here developmentalism had a more chequered fate. Although there were some notable successes, by the 1970s it had become clear that the fruits of progress did not necessarily 'trickle down' from the wealthy to the poor as modernisation theory had predicted, and that industrial policies were facilitating corruption and graft at the highest levels of government. Poverty continued to rise, urban informal settlements were growing and state-driven industrialisation was failing to generate sufficient jobs to absorb a growing non-agricultural labour force. The limits of developmentalism were thrown into sharp relief in the 1970s when an oil price shock sent devastating ripples through the world economy. It was also in the 1970s that a wide range of critiques – many emerging from those parts of the world where developmentalism had failed – were levelled against the international development agenda more generally.

The first and most scathing critique of modernisation theory, which emerged in the late 1960s and heavily influenced development discourse and policy throughout the 1970s, came in the form of dependency theory. Proponents of this school of thought argued that countries were not 'backward' or 'undeveloped' but that they were deliberately 'underdeveloped' by the international capitalist system in a process that helped the advanced economies extend and maintain their prosperity at the expense of weaker economies. According to the dependency perspective, developing countries had been incorporated into this system on terms beyond their control and against their general interests during the colonial era and post-independence period. So-called 'modernising' elites were dubbed a 'comprador' class that worked to advance their own interests and those of international capital, rather than the general interests of national economies and citizens.

Dependency theory is most closely associated with the writings of the German-American economist Andre Gunder Frank, who developed his ideas while working with the Economic Commission for Latin America (CEPAL). Through detailed historical studies of Chile and Brazil, and drawing on both Latin American development debates and the writings of North American Marxian Paul Baran (1957), Frank argued in *Capitalism and Underdevelopment in Latin America* (1967) that development in low- and middle-income countries was not possible because economic surplus was extracted from the global periphery (i.e. less industrialised economies) by or on behalf of the metropolitan centres of wealthy countries. Dependency theorists also saw the whole development project as a grand design aimed at masking the underlying extractive intentions of rich and powerful nations and firms (Cardoso 1972). As an analytical framework, dependency theory was very compelling, particularly to policymakers in developing countries who had to answer to domestic constituencies for economic failures. However, it was problematic in that it was essentially deterministic and never really offered a pragmatic alternative vision for development aside from a global revolution (Palma 1981). Without a realistic programme of action, dependency theory was confined to the arena of intellectual debate and never really established an operational framework.

However, other perspectives began to emerge in the 1970s that changed the way the practice of development was understood and pursued. Macro-level concerns about economic transformation were augmented by more micro-level concerns about the lived experience of the poor. Increased attention was paid to the notion that the ultimate goal of development is improvement in the lives of people, and the expansion of production and exchange in national economies was increasingly viewed as the means rather than the end goal of development. The World Bank, which was a major investor in and advisor to less economically advanced countries, explicitly shifted its focus away from promoting industrialisation and towards poverty alleviation and the satisfaction of 'basic needs' (see Chapter 4) by financing investments in social services such as health and education.

At the same time, a new generation of scholar-practitioners challenged the top-down, technocratic approach to development policy which had prevailed in the two decades following the end of the Second World War. One of the leading thinkers in this movement was Robert Chambers, who argued that development practice should be turned on its head: development initiatives should be more participatory, more attentive to local voices, and respectful of indigenous knowledge and the expertise of people on the ground (1983). In other words, the practice of development should seek to empower the most vulnerable by giving them greater control over development initiatives. This gave rise to the participatory development paradigm: the theory that participation by the poor in designing and implementing initiatives aimed at improving their lives is both intrinsically and instrumentally desirable. This rapidly evolved into an interest in 'participatory development' more broadly, with the concept widely adopted and promoted by mainstream development agencies such as the World Bank by the 1990s. Meanwhile, the critiques of early 'modernisation' approaches to development that inspired this interest in participation also led to a keener focus on poverty, given the rising disillusionment with the idea that wealth would 'trickle down' in countries that managed to achieve growth.

Amid this growing concern with the alleviation of poverty and empowerment of the marginalised, in the 1970s development researchers and professionals also began to recognize the importance of gender and development. Initially, the focus was very much on women's relationship to development, leading to the first UN Conference on Women in Mexico City in 1975. Feminists pointed out the differential impact of development interventions on women and men, arguing that development policies had variously ignored, harmed or used women instrumentally and this had led to outcomes that were biased towards the interests of men (Boserup 1970; Elson 1991; Kabeer 1994; Moser 1993). Over time, the focus shifted from a concern with 'women and development' towards a broader focus on gender relations and gender-sensitive programming in development practice (Young 1997; Momsen 2004). As a result of sustained critique and activism, gender issues have become increasingly embedded in mainstream development policy and practice at the international level.

In retrospect, the most significant shift of all in the 1970s was the growing influence in certain circles of the economists Friedrich von Hayek and Milton Friedman, which sowed the seeds of the 'neoliberal turn' of the 1980s. A series of economic shocks and crises in the 1970s rocked rich and poor countries alike and set in motion new political and intellectual currents that shaped development in subsequent decades. In a context of rising unemployment, inflation and social unrest, Ronald Reagan and Margaret Thatcher were elected to power in the USA and UK respectively. These leaders shared a distinct worldview and championed a new economic ideology, influenced by Hayek and Friedman, which portrayed state intervention in the economy as a fundamental obstacle to the efficient functioning of markets, and advocated market-based approaches as the most effective mechanism for allocating resources in society. This body of ideas – which came to be known as neoliberalism – amounted to a total rejection of the state-driven modernisation paradigm. Indeed, state intervention became the prevailing explanation for the economic turmoil of the 1970s and 'rolling-back' the state was the prescription that Reagan and Thatcher pursued and promoted.

Neoliberalism was not, however, confined to the US and UK; it became a global intellectual movement that led to a seismic shift in development theory in the 1980s (Toye 1987). Development economists such as Deepak Lal actively promoted *laissez-faire* economic policies, arguing that economic development should be left to the market (Lal 1985). State intervention – or developmentalism – came to be seen as the very source of underdevelopment. By the end of the 1980s international development policy and practice had been thoroughly revamped based on these neoliberal ideas. International financial institutions (IFIs) such as the World Bank and IMF actively promoted a package of reforms known as 'the Washington Consensus', involving the privatisation of state enterprises, fiscal austerity (e.g. through reducing public sector employment), liberalisation of trade (e.g. reducing tariffs on imported goods), liberalisation of financial markets to facilitate foreign investment, tax reform, strengthening of private property rights and deregulation of domestic markets. These reforms, designed to create more competitive, internationally integrated markets that would attract private sector investment, were pushed onto developing countries through the World Bank and IMF's Structural Adjustment

Programmes (SAPs). These SAPs included 'conditionalities' under which countries had to agree to undertake some combination of these reforms in order to qualify for desperately needed grants and soft loans to fend off fiscal crisis and economic implosion. The Washington Consensus recipe was applied to countries across Africa, Asia and Latin America, as well as to those of Eastern Europe and the former Soviet Union, with little regard to the variations in context and circumstance.

The rise of neoliberalism coincided with the emergence of two other paradigms that have indelibly shaped development discourse and practice: rights-based approaches and sustainability. Towards the end of the 1980s, a renewed focus on human rights was growing in importance, evidenced for example by the adoption of the UN Declaration on the Right to Development in 1986 and UN Convention on the Rights of the Child in 1989. This focus on rights contrasted with the earlier focus on basic 'needs', and gathered momentum in the 1990s with the increasing salience of normative conceptions to development such as Sen's capabilities approach. A focus on rights can also be seen as part of a broader move to repoliticise development discourse, because while needs may (or may not) be met out of charitable intentions, framing development in terms of rights introduces the element of legal obligation to provide a certain standard of living (Uvin 2004). By the turn of the millennium, rights-based approaches had become the norm in many UN and bilateral development agencies, though the World Bank largely resisted the agenda due to its concern not to engage in politics (Cornwall and Nyamu-Musembi 2004).

Building on the environmental movements of the 1970s, the idea of sustainable development as it is generally understood today was also born in the 1980s. This was fuelled by concerns that both the structuralist approach of the 1950s–1970s and the neoliberal prescriptions of the 1980s privileged economic growth above all else without considering the long-term environmental implications of economic expansion (Redclift 1987). The growing chorus of critiques from environmentalists culminated in the 1987 Brundtland report, which famously introduced a definition of sustainability as 'development that meets the needs of the present, without compromising the ability of future generations to meet their own needs'. Despite some concern about how the term has been appropriated and used in different ways to suit different agendas (Lele 1991), the increasingly apparent scale of human-induced global environmental change has ensured that sustainability is now central to the global development agenda.

By the end of the 1990s it became clear that the neoliberal paradigm, which had transformed development policy over the previous two decades, had not generated the promised economic gains. Indeed, the 1980s and 1990s were dubbed 'The Lost Decades' by development economist William Easterly (2001) due to the widespread economic stagnation experienced in developing countries across the world. Scholars seeking to explain this prolonged crisis in developing countries in the midst of market reforms began to turn their attention to the significance of institutions and governance – the social structures and relationships that underpin the performance of states and markets alike. Rooted in the intellectual trend known as the 'New Institutional Economics', institutional approaches to development tend to define institutions in a particular way: as the 'rules of the game', meaning the laws, regulations and norms that structure and constrain human interaction (North 1990; 1995).

The capacity of governments to provide public goods, effectively regulate markets, formulate sound policies and be responsive to citizens' needs varies markedly across countries, as does the quality of the broader institutional environment (e.g. rule of law). Without a sound institutional foundation and effective state oversight, markets simply do not function as smoothly and efficiently as neoclassical economic theory predicts, nor do they deliver broad-based socioeconomic development. This insight has spawned a large research literature exploring the origins and consequences of differential trajectories of political-institutional change, identifying critical junctures in the past, such as European colonialism, that set countries on different political and economic paths, which ultimately contributed to the wide disparities in socioeconomic welfare we see across countries today (Acemoglu and Robinson 2012). A broad (but not universal) consensus has emerged that 'institutions rule' when seeking to account for cross-country differences in socioeconomic development (Rodrik, Subramanian and Trebbi 2004).

This interest among scholars and practitioners in the importance of governance and institutions was intensified around the turn of the millennium by a heighted awareness of the interconnectedness of human security in an increasingly globalised world. Following the terrorist attacks of September 11th, 2001 in New York City, poverty, inequality and global security threats became linked in the public imagination and attention turned to the problem of 'failed states' where weak governments and entrenched poverty appeared to be ideal breeding grounds and operational bases for terrorist organisations and networks. Aid agencies turned their attention to reaching the 'bottom billion' living in weak or failed states (Collier 2007), while Western militaries embarked upon 'state-building' efforts in the countries they occupied as part of the War on Terror, such as Iraq and Afghanistan. With security at the top of the public agenda, there was a distinct 'securitisation' of development discourse and aid delivery at the dawn of the twenty-first century (Duffield 2001; Beall et al. 2006).

However, the first decade of the new millennium was also a time of unprecedented economic expansion for many developing countries, including large, influential 'emerging market' economies such as Brazil, China, Indonesia and South Africa. By the end of the decade these countries were actively challenging the political and economic hegemony of the traditional 'developed' countries in North America, Europe and Japan by establishing their own 'South–South' development cooperation frameworks and becoming significant aid donors rather than recipients (Quadir 2013). With newfound influence, these emerging powers are actively reshaping the global discourse of development, with decreasing tolerance for the 'us' and 'them' assumptions on which previous conceptions of development cooperation were predicated. Nowhere is this more evident than in the latest round of negotiations concerning international development targets. While the MDGs agreed at the turn of the millennium were explicitly focused on goals for low- and middle-income countries, increasing acceptance that the developed–developing binary is breaking down, coupled with the recognition that key security and environmental threats are global in scope, has led to an entirely new approach to framing development targets. The new Sustainable Development Goals are explicitly intended to be universal – i.e. to apply to *all* countries, not just low- and middle-income ones.

Another important recent trend in development research and policy has been the growing prominence of 'evidence-based' approaches to aid, which has been spurred on by a growing chorus of critics who claim that most foreign aid should be abandoned because it is a waste of money at best and counterproductive at worst (Easterly 2006; Moyo 2009). Concerns were also mounting that aid disbursements tended to be based on political preference rather than evidence about 'what works' (Killick 2004). Demand for evidence of substantive impact grew ever more intense after the global financial crisis of 2008 as government aid agencies and donors became increasingly concerned with achieving 'value for money' to justify continued aid expenditures overseas at a time of socioeconomic crises and fiscal retrenchment at home. Researchers pursuing this agenda have adopted randomised controlled trials, which is standard practice in medical research, to evaluate the impacts of specific development interventions aimed at poverty reduction and improving health and education outcomes. This field of study is exemplified by the work of the Poverty Action Lab based at the Massachusetts Institute of Technology and development economists such as Abhijit Banerjee and Esther Duflo, whose work has not only improved our understanding of which policies and interventions work where, but also helped to debunk some myths about the lives of people living in poverty (Banerjee and Duflo 2011). Despite this, the 'evidence agenda' has not been uncontroversial, not least due to concerns about the politics of producing evidence and ongoing debates about what constitutes valid evidence (Denzin 2009).

Finally, by the end of the first decade of the twenty-first century the world population had reached a critical tipping point: for the first time in history more people were living in urban areas than rural ones. Moreover, global population projections indicate that all of the world's population growth in the coming decades will be absorbed by towns and cities, mostly in the low- and middle-income countries of Africa, Asia and Latin America. Consequently, there is now growing realisation that development is increasingly and intimately linked with urban change (World Bank 2013). As we will explain in the next section, this hasn't always been the case.

It is important to recognise that this periodisation is very crude and doesn't necessarily reflect discrete episodes in the history of development discourse and practice. Indeed, each of these movements continues to influence the way scholars, practitioners and members of the general public think about development. Issues of participation, poverty reduction and empowerment, particularly with regard to gender, have been 'mainstreamed' into projects and programmes run by United Nations organisations, while 'sustainability' has arguably risen to the very top of the global development agenda. Although paradigms such as modernisation theory and dependency theory have generally been cast aside, there are even echoes of these in recent scholarship. Nevertheless, this cursory review of the history of development highlights some of the broad contours of the debate which will crop up throughout this book. In particular, the tension between states and markets in processes of socioeconomic change is a recurring theme, as is the importance of institutions and governance in shaping development trajectories. However, before exploring such themes in an urban context, it is useful to examine the shifting role of cities and urbanisation in development discourse and practice in the post-war era.

Cities as problems, solutions and contexts

Historians and social scientists have generally portrayed cities as incubators and manifestations of social and technological progress (see Childe 1950; Sjoberg 1960; Mumford 1961; Davis 1965). Indeed, in a paper prepared for a United Nations conference in the early 1960s, leading American political scientist Lucian Pye stated that 'urban life is the dynamic basis for most of the activities and processes we associate with modernity and economic progress' (Pye 1969: 401). In the 1950s and 1960s, this generally positive view of the role of cities in cultivating progress translated into active investment in urban and regional development in less economically advanced countries to support industrialisation and nation-building efforts by national governments and international agencies (Friedmann 1967; United Nations, Department of Economic and Social Affairs 1968).

During this period, the process of urbanisation was understood to be a natural consequence of industrialisation, and was recognised as such in formal economic models of development. The work of Arthur Lewis was particularly influential in shaping the way that economists and policymakers conceptualised the process. Lewis (1954) proposed a dual-sector model of economic development in which low-productivity labour in the 'traditional' (rural) agricultural sector is transferred to the 'modern' (urban) industrial sector as an economy grows. The model assumes that there is a large surplus of cheap, unskilled labour in rural areas that is drawn into towns and cities in response to higher wages. This migration is expected to drive down urban wages, thereby allowing for the comfortable profit margins necessary to finance further capitalist expansion in the urban sector.

The Lewis model was complemented by the work of economist Albert Hirschman. In the *Strategy of Economic Development* (1958), Hirschman argued that the expansion of the modern sector was driven by key economic sectors that necessarily involved spatial concentration and inequality. Uneven development and polarisation were, in his view, inevitable features of countries in the early stages of economic development but would eventually be overcome by the 'trickle down' of benefits to surrounding areas. Regional planning theorists and professionals at the time thought that spatial inequalities could be mitigated through targeted interventions. For example, the North American planner John Friedmann (a pioneer in the field of regional planning) devised policies to encourage the growth of medium-sized cities in peripheral regions and the establishment of urban 'growth poles' (Friedmann and Alonso 1975). It was generally believed that urban centres could be used to drive regional development in poorer countries, and that over time cities and urban systems would naturally evolve in such a way as to encourage the integration of national economies.

However, a series of UN-sponsored surveys in Africa, Asia and Latin America in the 1950s painted a somewhat less optimistic picture. These revealed historically unprecedented rates of urban population expansion, burgeoning shanty towns and rising under- and un-employment, particularly among youth (United Nations, Department of Economic and Social Affairs 1968: 3–4). In contrast to optimistic regional planners, some scholars such as Gunnar Myrdal (1957) argued that emerging patterns of spatial inequality would become entrenched. Fears were expressed about the potential for social and political unrest that could materialise if rapid urban population expansion continued to outpace economic development, a situation dubbed 'over-urbanisation'. At a UN conference in 1956 it was claimed that in over-urbanised countries 'urban misery and rural poverty exist side by side with the result that the city can hardly be called "dynamic," as social historians of developed countries generally described the process of urbanization' (cited in Sovani 1964: 113). These concerns about the relationship between cities and development were echoed by dependency theorists, though with an emphasis on urban privilege rather than urban deprivation. They were critical of the notion that development inevitability entails inequality, unbalanced growth and the 'natural' evolution of urban systems over time. They argued that cities in poorer countries were essentially parasitic 'islands of privilege' and outposts of capitalist penetration inhabited by a *comprador* class systematically exploiting the rural masses (Schatzberg 1979; Southall 1979).

By the end of the 1970s a growing number of scholars and policymakers concluded that urbanisation was a problem in many less economically developed countries. Cities were no longer seen as drivers of socioeconomic development, but rather obstacles to progress. This growing negativity about urbanisation and cities crystallised in the form of the 'urban bias thesis' associated with the works of Michael Lipton and Robert Bates. In the dramatic opening paragraph of his book *Why Poor People Stay Poor: Urban Bias in World Development* (1977), Lipton depicted socioeconomic development as a zero-sum game between rural and urban areas:

> The most important class conflict in the poor countries of the world today is not between labour and capital. Nor is it between foreign and national interests. It is between the rural classes and the urban classes. The rural sector contains most of the poverty, and most of the low-cost sources of potential advance; but the urban sector contains most of the articulateness, organization and power. So the urban classes have been able to 'win' most of the rounds of the struggle with the countryside; but in so doing they have made the development process needlessly slow and unfair.

Lipton went on to argue that governments in less economically developed countries distorted prices to favour urban consumers over farmers and invested disproportionately in urban infrastructure and services at the expense of the rural masses. Robert Bates (1981, 1983) extended Lipton's thesis in the 1980s in his analysis of agricultural systems in sub-Saharan Africa by further arguing that urban dwellers were not only economically advantaged under prevailing policy frameworks, but also inherently politically more powerful. Fear of urban food riots and of upsetting proximate and organised urban groups meant that governments acted not as vehicles for maximising the social welfare of all their citizens, but rather as agents that 'accommodate the demands of organised private interests' (Bates 1988: 121). Both Lipton and Bates were of the view that rural areas received too little investment and argued in favour of a major shift in development policy away from urban areas and towards rural areas.

This view was further supported by the work of scholars seeking to explain 'over-urbanisation' and identify appropriate policy responses. In a very influential article, development economists Harris and Todaro (1970) argued that limiting rural–urban migration would lead to productivity and welfare gains overall, and that this could be achieved by either investing more in rural development or putting in place policies to discourage or prevent people from moving into urban areas. According to Michael Todaro,

> It is vitally important that imbalances between economic opportunities in rural and urban sectors be minimised. Permitting urban wage rates to rise at a greater pace than average rural incomes will stimulate further rural–urban migration in spite of rising levels of urban unemployment. This heavy influx of people into urban areas not only gives rise to socio-economic problems in the cities but may also eventually create problems of labour shortages in rural areas, especially during the busy seasons. These social costs may exceed the private benefits of migration.
>
> (Todaro 2000: 310)

Strategies to restrict human mobility in the service of 'development' predate such formal theorising. Some of the more dramatic examples include Indonesia's transmigration programme, which dates back to the colonial era and ultimately resulted in the forced resettlement of millions of people from areas of high to low density in the twentieth century, and China's *Hukou* system of household registration introduced in 1958, which required individuals to obtain government permission to move from their places of birth (Ren 2013). However, the use of legal restrictions on migration, forced relocation and rural investment to discourage out-migration became increasingly mainstream with the weight of academic legitimacy. The World Bank even helped to finance the Indonesian programme in the 1980s. Indeed, as Table 1.4 shows, there was a significant increase in the number of countries that adopted such policies between the mid-1970s and today.

The emergence of an anti-urbanisation bias in development theory and practice from the 1970s can also be seen in other arenas of development policy. For example, recent systematic reviews of

Table 1.4 *Number of countries with policies to reduce rural–urban migration, 1976–2013*

	1976	1986	1996	2005	2013
Africa	18	19	22	38	45
Asia	3	8	16	30	37
Europe	12	8	8	23	27
LAC	7	11	2	16	26
Oceania	0	1	0	8	12
North America	0	0	0	0	1
Total	40	47	48	115	148

Source: United Nations, Department of Economic and Social Affairs, Population Division (2013).

Poverty Reduction Strategy Papers, which countries seeking aid must produce to be eligible for funding from many multilateral agencies, reveal that nearly all have a strong emphasis on rural poverty alleviation and agricultural development while neglecting urban poverty altogether or demonstrating a generally poor understanding of urban poverty and development issues (Mitlin and Satterthwaite 2013). Further evidence of declining international support for urban development can be seen in World Bank shelter lending trends for sub-Saharan Africa between 1972 and 2005 (see Table 1.5). Despite historically unprecedented rates of urban population expansion in the region during this period, shelter lending was cut significantly. In sum, while cities and the process of urbanisation were generally associated with positive change in the 1950s and 1960s, there was a significant anti-urban turn in development discourse and policy beginning in the 1970s and running through the 1990s.

Since the turn of the millennium, attitudes have shifted again in both academic and policy circles. This may in part be a consequence of sober reflections on the urban bias thesis and failures of development policy in the 1980s and 1990s. In particular, it is problematic to frame development around a dualistic framework of 'urban' and 'rural' when human settlements exist on a nuanced continuum; there are clear and positive interlinkages between rural and urban economies; there is little evidence of coherent rural and urban classes with uniform and diametrically opposed interests; and there is little evidence of a systematic bias in public expenditure, particularly since the structural adjustment era (see Becker, Hamer and Morrison 1998; Jones and Corbridge 2010).

A second and perhaps more important stimulus for growing interest in urban poverty and development in recent years has been growing recognition of the seemingly inexorable march of world urbanisation and rapid urban population expansion in less economically advanced countries, despite decades of policy designed to stem the tide. As we will see in Chapter 2, these trends are part of a global demographic transition associated with falling mortality rates. As Figure 1.1 illustrates, the global urban population exceeded the global

Table 1.5 *Trends in World Bank shelter lending in sub-Saharan Africa, 1972–2005*

	1972–1981	*1982–1991*	*1992–2005*
Total shelter lending	$498 million	$409 million	$81 million
Equivalent per capita	$5.20	$2.74	$0.32

Source: Fox (2014).

Figure 1.1 *Global trends in rural and urban population, 1950–2050*

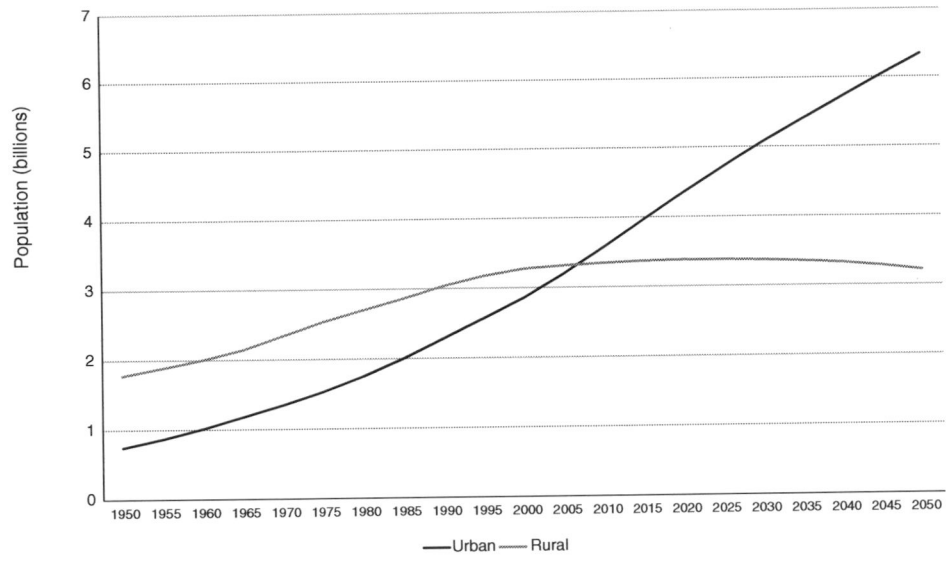

Source: United Nations (2014).

rural population around 2008 and is expected to continue growing apace as the rural population levels off.

As a result, while rural poverty and development remain crucial challenges in many parts of the world, towns and cities are increasingly becoming critical locations where key development challenges such as poverty, inequality, violence and climate change need most urgently to be addressed. Rather than resist the demographic momentum of world urbanisation, many now accept that we are heading towards a predominantly urban future and we need to do our best to cultivate peace, prosperity and sustainability in our towns and cities.

Conclusion

The following chapters are organised around key themes that stand at the intersection between urban studies and development studies. We begin in Chapter 2 with a sweeping overview of 6000 years of urban history, highlighting some of the critical junctures in the process of world urbanisation as well as important themes that emerge from this macro-historical perspective. In Chapter 3 we turn our attention to the dynamic relationships between urbanism and economic development, which remains a central preoccupation in development research and policy, before exploring the multifaceted nature of urban poverty and the challenges of promoting sustainable livelihoods in Chapter 4. Chapter 5 offers an in-depth exploration of the essential material foundations of urban development and planning: land, housing, infrastructure and urban services. We introduce a wholly new Chapter 6 in this edition, which surveys the complex issues associated with urbanisation, urban development and environmental change. Chapter 7 examines the challenges of crime, conflict and violence in cities, followed by a new chapter on the political and institutional forces that shape the towns and cities in which we live. We conclude with a short chapter that reflects on the challenges of shaping city futures and ensuring that the first urban century is characterised by significant progress in our collective efforts to achieve a prosperous, equitable and sustainable future for humanity.

Finally, a disclaimer. This is not a handbook for policymakers and practitioners; nor is it a comprehensive, in-depth treatment of all the issues covered. Instead, we have chosen to concentrate on breadth, bringing together the wide-ranging fields of urban studies and development studies in order to highlight the need for this vast constellation of interconnected issues and debates to be in conversation with each other. Our hope is that the synthesis we provide will serve to sharpen the understanding and engagement of those who have a scholarly or professional interest in cities and development.

Summary

- The world is now more urban than rural, but processes of urbanisation have been uneven and inequality and urban informality are on the rise.
- Cities are human settlements characterised by their size, density and heterogeneity, and by the social, political, economic and cultural effects of these qualities (urbanism). In practice, the definition of 'urban' varies considerably across countries.
- The word urbanisation refers to the proportion of a population living in urban (as opposed to rural) areas, while urban growth is the absolute increase in the number of people in urban areas, and urban expansion is the increase in built-up area of a settlement.

- Development is a contested concept that can be variously understood as a historical process, a project, a social objective and a discourse.
- Development theory has its roots in the Enlightenment, but the term became popular only after 1945. Since then a number of overlapping, and sometimes competing, understandings of development have emerged.
- Key trends in development thinking can be characterised as modernisation theory and structuralism/developmentalism in the 1950s and 1960s; dependency theory and a focus on poverty, inequality, participation and empowerment in the 1970s; neoliberal structural adjustment, sustainability and rights-based approaches during the 1980s and 1990s; and increased emphasis on governance, institutions and state-building in the 2000s.
- More recently, there also been a growing focus on the need for evidence-based aid and 'South–South cooperation', as well as a renewed recognition of the role of cities in development.
- As cities in developing countries grew in the mid-twentieth century, discourses about urbanisation shifted from an initial optimism towards the idea that cities were parasitic obstacles to development, particularly after the 'urban bias' thesis of Lipton and Bates.
- The urban bias thesis placed the rural at the heart of development discourse, and policies to stem further rural–urban migration became commonplace. After this point, cities were relatively neglected in development discourse until very recently.

Discussion questions

1. How has development theory and practice changed over time since 1945? How have different schools of thought influenced each other?
2. Why has the idea of 'development' been critiqued?
3. According to Sen, what is the relationship between development and freedom?
4. Discuss the ways in which attitudes towards cities have changed during the twentieth century, and relate this to the dominant development thinking of the time.
5. What would an urban-centric idea of development consist of?

Further reading

Brett, E. A. (2009) *Reconstructing Development Theory*. New York: Palgrave Macmillan.

Mumford, L. (1937) 'What is a city?', Architectural Record, reprinted in Richard Le Gates and Fredric Stout, eds, *The City Reader*. London: Routledge.

Robinson, J. (2006) *Ordinary Cities*. London: Routledge.

Sen, A. (1999) *Development as Freedom*. New York: Anchor Book (Random House).

United Nations, Department of Economic and Social Affairs, Population Division (2014) *World Urbanization Prospects: The 2014 Revision*. Available online.

Wirth, L. (1938) 'Urbanism as a way of life', *The American Journal of Sociology*, 44(1), 1–24.

Websites

Cities Alliance, a global coalition of cities and their development partners committed to scaling up successful approaches to poverty reduction: www.citiesalliance.org

Eldis, a development gateway site aiming to share the best in development policy, practice and research: www.eldis.org

ODI, a UK development think tank: www.odi.org

This Big City, an online publication with discussion about sustainable cities: www.thisbigcity.net

United Nations Development Programme: www.undp.org

United National Human Settlements Programme: www.unhabitat.org

World Bank urban development site: www.worldbank.org/en/topic/urbandevelopment

2 The global urban transition in historical perspective

- The Urban Revolution and the origins of urbanism
- From the birth of the city to the rise of nation-states
- The demographic transition and the Industrial Revolution
- Colonial urbanism
- Into the first urban century

Introduction

Development as a historical process long predates the rise of development theory and practice. Progressive changes in the social and political institutions that bind communities together, as well as steady improvements in the material conditions of human societies, began some 10,000 years ago with the rise of agriculture. Shortly thereafter, cities arose, and with them the foundations for civilisations and eventually nation-states. It was not, however, until the nineteenth century that the global urban transition gathered pace. Demographic and economic forces created the necessary stimulus that ultimately led to urban growth and urbanisation in Europe and other rapidly industrialising nations. Many of these nations used their expanding economic and military power to colonise territories in Africa, Asia and Latin America, a process which had important effects on trajectories of urban development in these areas. By briefly exploring processes of urban change over the *longue dureé* we can better understand the patterns of urban development that we find today.

We begin this chapter with an examination of the origins of cities some six thousand years ago and the social implications of early urbanism. Next we survey early urban centres across the world and highlight their functions as political–administrative centres and commercial hubs. In Medieval Europe, we find a dynamic tension between urban-based merchant classes and feudal princes that culminated in the formation of nation-states, which came to dominate all other forms of political organisation thereafter. We then turn our attention to the Industrial Revolution, the demographic transition and the process of urbanisation in Europe, where the global urban transition began. This is followed by a discussion of how European colonialism and imperialism radically restructured human settlements across the world as cities served as instruments of imperialism as well as spaces of resistance and confrontation. We close the chapter with a summary of contemporary urban transitions (particularly in Asia and Africa), highlighting the changing shape of urbanism in low- and middle-income countries. It is a necessarily non-comprehensive survey of global urban history – indeed it barely scratches the surface of the wealth of literature available on the subject. It does, however, draw attention to themes directly relevant to our exploration of the link between cities and development.

The Urban Revolution and the origins of urbanism

For millions of years human beings lived in small bands of hunter-gatherers. Dependence upon nature's produce limited the size of these bands and compelled them to move periodically when local supplies of food were exhausted. About 10,000 years ago the situation changed dramatically. What archaeologists refer to as the Neolithic Revolution (circa 8500 BCE) marked a shift away from human dependence on hunting and gathering and towards livelihood strategies characterised by domesticated agriculture and animal husbandry. This momentous event was a necessary precondition for the birth of cities, civilisations, and ultimately nation-states.

The standard 'origin of cities' narrative suggests that the Neolithic Revolution predated the rise of cities by approximately 4000 years. In this time, farmers made gradual improvements in cultivation, eventually resulting in surplus agricultural production. In particular the cultivation of hard grains that could be produced *en masse* and stored for significant periods of time reduced the risk of starvation and made it possible to support larger populations in any given area. Over time, population growth increased the density of agricultural villages, and rising productivity made it possible not only to hedge against the risk of famine, but also to support a class of individuals who were not engaged in agricultural production. The result was an Urban Revolution, to use V. Gordon Childe's famous term (1950).

Archaeologists place the birth of cities around 6,000 years ago in the Sumer region of Mesopotamia (present day Iraq) on the plains that lie between the Tigris and Euphrates rivers, where fertile soils and access to waterways for irrigation and transport facilitated surplus agricultural production (see Figure 2.1). However, cities also arose independently at later dates in Africa (1000 BCE), India (1000–400 BCE), China (700–400 BCE), as well as Mexico and Peru (100 BCE). In each case, there is evidence of domesticated agriculture predating the rise of cities, usually by 2,000 to 3,000 years (Bairoch 1988).

In all early cities, archaeologists find evidence of a socioeconomic order characterised by a division of labour and the development of hitherto unknown degrees of social hierarchy. Indeed, 'for archaeologists and historians the most meaningful difference between a village and a city has nothing to do with size; it is instead a measure of social and economic differentiation within the communities' (Reader 2004: 16). The shift from subsistence agriculture to surplus agricultural production made it possible for a significant number of people to specialise in the production of crafts, the arts of war, the cultivation of philosophy and activities related to trade. As a result, cities became sites of social and technological innovation, giving rise to improvements in irrigation, transportation and metallurgy. A rising demand for inputs into new productive activities stimulated long-distance trade. It was also in cities that writing first emerged, which facilitated the development of mathematical sciences and made bureaucracy and public administration possible. The combined effect of these innovations was a radical transformation in the organisation of society, ultimately resulting in the rise of the first civilisations and empires. As Bairoch notes, 'the city appears to gather together all of the factors conducive to sociotechnical advance' (Bairoch 1988: 96). The standard narrative, then, is that agriculture made cities possible, which in turn gave rise to socioeconomic differentiation, socio-technical innovation, and (eventually) the first civilisations.

Edward Soja (2000), and Jane Jacobs before him (1969), have challenged this narrative by suggesting (based on archaeological evidence) that cities were, in fact, both the impetus for, and the incubators of, the innovations that made the production of an agricultural surplus possible. Excavation sites at Jericho (in present day Palestine) and Çatal Hüyük (in present day Turkey) have revealed

Figure 2.1 *The origins of urbanism in the Middle East*

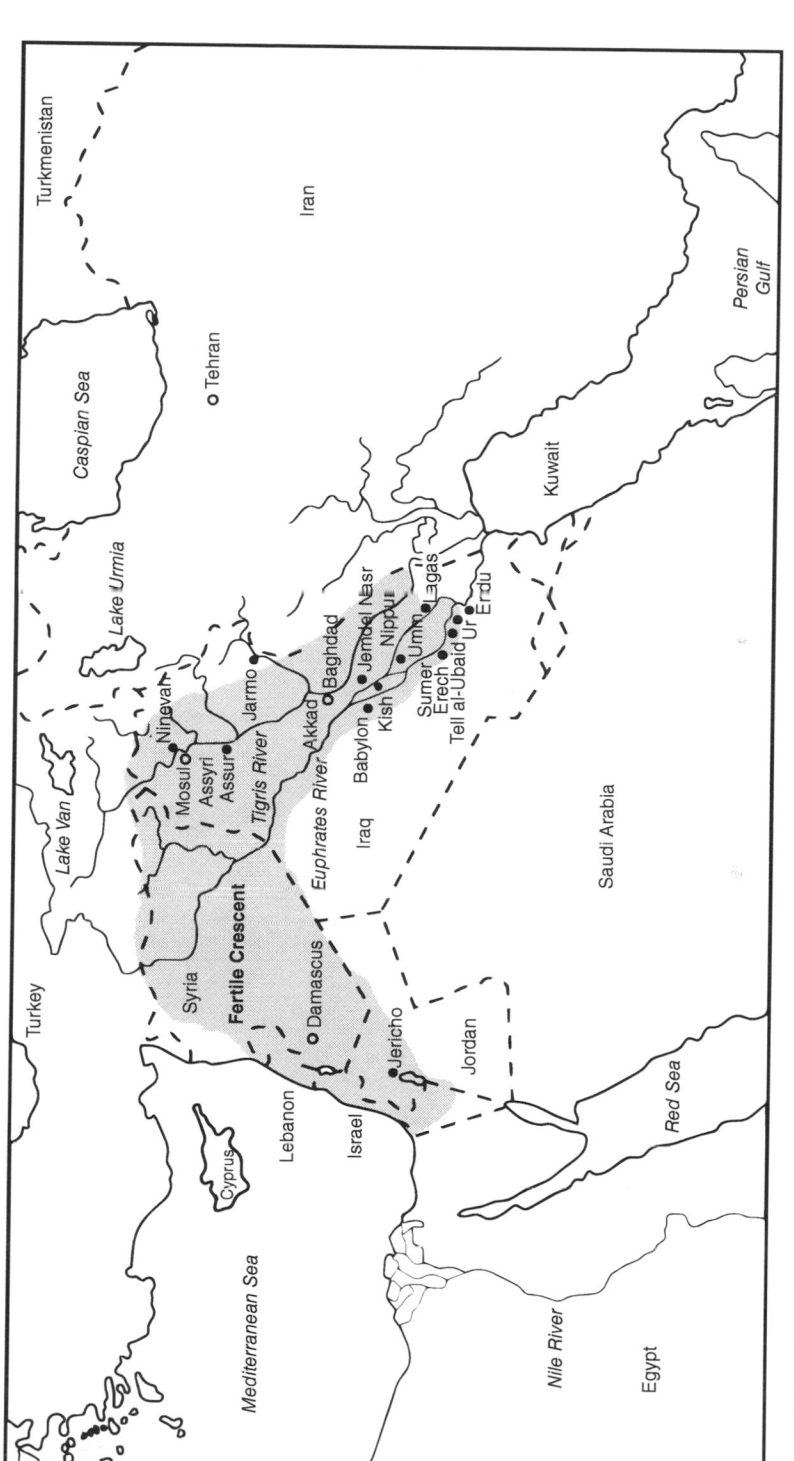

Source: Pacione (2005).

evidence of large, dense, permanent settlements established approximately 10,000 years ago – about the same time that the Neolithic Revolution took place. The inhabitants of Jericho are thought to be the first in the world to live sedentary lives, although they were hunter-gatherers. Çatal Hüyük, a larger and denser settlement, also displays evidence of a hunter-gatherer lifestyle, but there is some evidence of early agriculture in and around the city. Soja argues that

> Jericho and Çatal Hüyük represent a revolutionary leap in the social and spatial scale of human societies and culture…The stimulating interdependencies and cultural conventions created by socio-spatial agglomeration – moving closer together – were the key organising features or motor forces driving virtually everything that followed.
>
> (Soja 2000: 46)

So which came first, farming or cities? The answer (given the available evidence) depends upon one's definitions of 'city' and 'urbanism'. Childe, Mumford and Bairoch dismiss Jericho and Çatal Hüyük as pre-urban towns on the grounds that there is little evidence of socioeconomic differentiation in these settlements given their reliance on hunting and gathering, and that they were relatively small in comparison to the early Sumerian cities. Nevertheless, Soja challenges us to consider the possibility that these settlements were indeed cities on the grounds that they exhibit the 'socio-spatial agglomeration' that characterises urbanism. There is no disagreement on the facts, but rather which aspect of urbanism these scholars choose to privilege.

In making his case, Soja places emphasis on what he calls *synekism*, a respelling of the word 'synoecism' (pronounced 'sin-ee-sism'), which is used in the archaeological and historical literature on ancient cities. Synoecism is derived from the ancient Greek *synoikismos*, which literally means the 'conditions arising from dwelling together in one house' (ibid.: 12). Similarly, synekism refers to 'the economic and ecological interdependencies and the creative – as well as occasionally destructive – synergisms that arise from the purposeful clustering and collective cohabitation of people in space, in a "home" habitat' (ibid.: 12). The power of synekism – also referred to by urban scholars as density, propinquity and proximity – can be conceptualised as a kind of socio-spatial force or stimulus unique to urban agglomerations.

By overlooking the possible role of synekism in Jericho and Çatal Hüyük, the standard narrative sees the rise of cities as a by-product of the Neolithic Revolution as opposed to a contributing factor in this critical juncture in human history. While we may never know the true sequence of events, this controversy highlights two essential aspects of urbanism – synekism and socioeconomic differentiation – that are driving forces behind socio-technical change.

From the birth of the city to the rise of nation-states

The cities that emerged in ancient Mesopotamia were not merely cities, but rather city-states – sovereign urban agglomerations in control of their immediate hinterlands. Their socio-political order was characterised by divine kingship, and the demands of acquiring, defending and distributing an agricultural surplus inspired the evolution of what we recognise today as basic state functions, such as taxation, military conscription, policing and bureaucratic administration. Religion and ritual played an important role in justifying this new order, and temples served as sites of worship, public administration and (importantly) granaries (Childe 1950).

Ur was the first and paradigmatic Sumerian city-state, with an urban population of about 24,000 people (circa 2800 BC) governing and extracting a surplus from some 500,000 farmers in the areas around the city (Bairoch 1988: 26). The city was built on a *tell*, or raised mound, and was surrounded by an oval wall oriented on a North–South axis, located close to the outlets of the Tigris and Euphrates rivers. Its urban centre was dominated by a ziggurat some five stories tall, which contained administrative offices, storage facilities and, at the uppermost level, a shrine. It was a physical manifestation of the centralisation of economic, political, social and spiritual power that would be replicated on increasingly larger scales across the network of city-states that proliferated in Sumer. The Tower of Babel, the ziggurat dominating the skyline of Babylon as it ascended in the ancient world, was some 270 feet (27 stories) tall.

Although the Mesopotamian city-states provide the earliest evidence of the development of social, economic and political structures that would later become the organisational foundations of civilisations, empires and nation-states, urban history is not linear. City-states and urban-based empires with very different characteristics emerged subsequently and independently across the world in the centuries that followed.

Figure 2.2 Other early urban centres

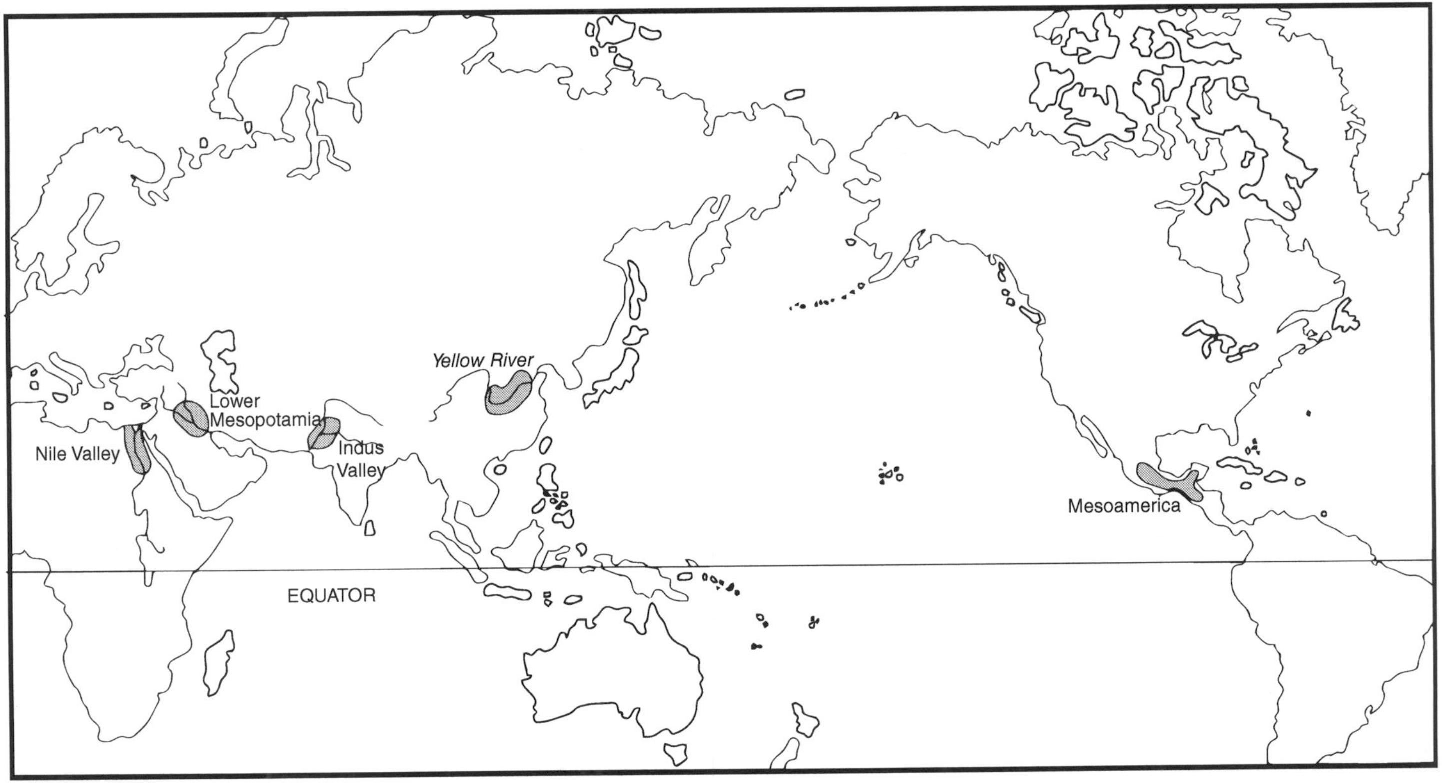

Mesoamerica

Yellow River

Lower
Mesopotamia

Indus
Valley

Nile Valley

EQUATOR

Source: Pacione (2005).

The rich urban history of India began east of Mesopotamia in the Indus Valley (present day Pakistan) where the cities of Harappa and Mohenjo-Daro are thought to have been built in the second millennium BC, were home to up to 40,000 people, had functioning sewers and appear to have been built on a regular plan (Bairoch 1988: 39–40). They were ruled by a single 'priest-king', and appear to have traded with the Sumerian city-states, but their construction does not appear to have been influenced by the birth of cities in Mesopotamia (Pacione 2005: 44). Delhi, the modern capital of India, has been the site of seven cities over the past 2,000 years.

Cities also emerged independently in China as early as 1600 BC – some evidence suggests even earlier – in the Wei River Valley during the Shang dynasty. Similarly to Mesopotamian cities, urban centres in China served important spiritual, economic and political–administrative functions. At its peak, Chang'an (capital of the Sui and Tang dynasties) may have been home to a million inhabitants. It was formally planned and carefully regulated, its morphology and regimented daily life reflecting the rigid political hierarchy of the time. The following imperial age saw a dramatic expansion of Chinese territory and population growth, rendering centralized governance of the daily lives of urban residents difficult. Local elites organized into a range of civic associations and increasingly provided the framework for local governance. As the vast Chinese empire was consolidated through tumultuous episodes of expansion and retraction, cities served critical functions as cultural, intellectual, economic and political hubs (Friedmann 2005).

Not long after the rise of Mesopotamian cities, city-states also appeared in Phoenicia along the eastern shores of the Mediterranean (present day Lebanon, Syria and Israel). These cities were among the first truly commercial or merchant cities in the world, trading extensively throughout the Mediterranean. They served as hubs of inter-regional commerce making extensive use of maritime trade as far West as the Atlantic Ocean. There is evidence of a significant shipbuilding industry, as well as the production and export of glassware and dyed cloth. Unlike their Sumerian cousins, they were not ruled by kings, but by councils of elders drawn largely from the merchant class (Parker 2004; Bairoch 1988: 30). Although they never reached the size of the Sumerian city-states, their function as centres of commerce – indeed their reliance on commerce – is a recurrent theme in urban history and one that later played an important role in the evolution of nation-states.

North across the Mediterranean, the Greek *polis* was a particularly influential urban form, serving as a centre of cultural innovation that generated a canon of intellectual, architectural and artistic works that continue to inform contemporary social, political, scientific and aesthetic endeavours (Hall 1998). Rome, which began as an Etruscan colony, became an independent city-state in the sixth century BCE and grew into one of history's greatest empires, reaching an unprecedented size (some say one million inhabitants by AD 100), compelling its rulers to 'devise complex systems of international food supplies, to grapple successfully with long-distance delivery of water and with complex systems of waste disposal, even to formulate rules of urban traffic management' (Hall 1998: 621). The Romans were arguably the first masters of urban planning, recording their strict imperial order in the geometry of new cities with grid-iron planning. These satellite cities were the urban nodes of a networked empire, connected to the core by an impressive road network. Still today the ancient infrastructure of the empire can be found in cities across Europe, Northern Africa and Western Asia.

In the Americas, great cities were established in Mesoamerica and the Andean mountain range long before Spanish colonisers arrived. The Aztec capital of Tenochtitlán, with its sparkling pyramids, famously inspired awe in Fernando Cortés and his men as they entered the Valley of Mexico for the first time; the Mayan city of Tikal covered more than 123 square kilometres and was home to some 45,000 people in circa AD 550; and the Incan city of Cuzco, which reached a peak population of perhaps 300,000 inhabitants, was known as the 'city of bureaucrats', who managed an extensive road network and 170 administrative satellites (Butterworth and Chance 1981).

While the urban history of Africa has received less attention from archaeologists, there is evidence of cities in the kingdoms of Ghana and Cush before the turn of the millennium, as well as large urban settlements such as Aksum (circa AD 100–600) in the Horn of Africa and Great Zimbabwe (circa AD 1000–1500) in Southern Africa serving as important centres of regional governance and inter-regional commerce (Anderson and Rathbone 2000). One of the more enigmatic discoveries has been the remains of Jenne-Jeno in present day Mali, which was established in the third century BC and had perhaps 20,000 residents by AD 800. There is evidence from the site of the economic specialisation typical in all cities, but unlike so many other ancient cities, there is no evidence of political centralisation or social stratification (McIntosh and McIntosh 1981; Freund 2007).

Everywhere that cities 'crystallized', to use Mumford's term (1961), they served as nuclei around which new socio-political orders revolved, as centres of trade and incubators of new technologies. Their political influence often extended far beyond their immediate hinterlands, and their socio-technical innovations were diffused through trade. For thousands of years, cities reigned supreme as 'proto-states', sometimes competing with one another, sometimes forming strategic federations. But in medieval Europe, a very particular kind of collaboration emerged between urban-based merchants and territorial monarchs that culminated in the rise of nation-states.

Charles Tilly (1994: 6) has argued that 'the variable distribution of cities and systems of cities by region and era significantly and independently constrained the multiple paths of state formation' in Europe. He observed that different kinds of states emerged in regions with few cities as opposed to densely urban ones, that organisationally advanced urban centres played a significant role in national politics, and that urban merchants and financiers played an integral part in the financing and provisioning of new states through their control of capital and markets. According to Tilly, the rise of the territorially defined and centrally governed nation-states that we are accustomed to today was essentially the product of a strategic collaboration between urban-based capital and rural-based coercion. Urban-based merchants and financiers were specialists in acquiring, managing and deploying capital, as they relied heavily on trade to accumulate wealth. By contrast, feudal landlords and petty despots were specialists in the use of coercion, or armed force, and amassed their wealth through taxing peasants.

Before 1500 or so, monarchs maintained armies drawn from their own subjects who owed them personal service. But as the frequency and intensity of territorial conflicts increased in the period 1500–1700, it became necessary to employ mercenaries, which in turn required capital. Monarchs turned to wealthy urbanites to finance their wars, but this support carried a cost. The urban classes demanded certain securities in return, which led to a bargaining process. In Tilly's own words:

> In Europe before 1800 or so, most important changes in state structure stemmed from rulers' efforts to acquire the requisites of war, from resistance to those efforts, and from bargains that ended – or at least mitigated – that resistance. Courts, treasuries, representative assemblies, central administrations, fiscal

> structures, and much more formed and reformed in response to the creation of
> military force, the pursuit of war, and the payment of its costs.
>
> (1994: 10)

In effect, the institutions developed over time in cities to accumulate and manage financial resources, and the institutions developed by monarchs to ensure coercive dominance over their populations and territories came together to create a powerful political unit. In the nineteenth century, rulers

> continued to bargain with capitalists and other classes for revenues, manpower,
> and the necessities of war. Bargaining, in its turn, created numerous new claims
> on the state: pensions, payments to the poor, public education, city planning, and
> much more. In the process, states changed from magnified war machines into
> multipurpose organisations.
>
> (ibid.: 9)

The result was the political form of the nation-state, which remains today the pillar of the international political–economic system, despite the forces of globalisation. The gradual consolidation and solidification of nation-states in Europe and across the globe continued well into the twentieth century. At the same time, beginning in the latter half of the eighteenth century and continuing to this day, a critical tripartite transition began: a demographic transition, an economic transition and an urban transition.

The demographic transition and the Industrial Revolution

Despite the fact that cities had existed for thousands of years, truly urbanised societies did not emerge until the twentieth century. Cities had grown in both number and size over the previous millennia, but they remained relatively small by modern standards and were home to a very small minority of the world's population. Then, beginning in the mid-nineteenth century, a second urban revolution began, with the global urban population growing rapidly in both absolute and relative terms (see Figure 2.3). In 1800 the world population was roughly one billion, only 3–5 per cent of which lived in towns and cities; by 1900 the world population had grown to 1.6 billion, with about 14 per cent residing in urban areas. By the turn of the second millennium world population had reach six billion with nearly 50 per cent living in urban settlements. In 1800 there was just one city in the world with

Figure 2.3 *Global population change, 1000 CE to 2025 CE*

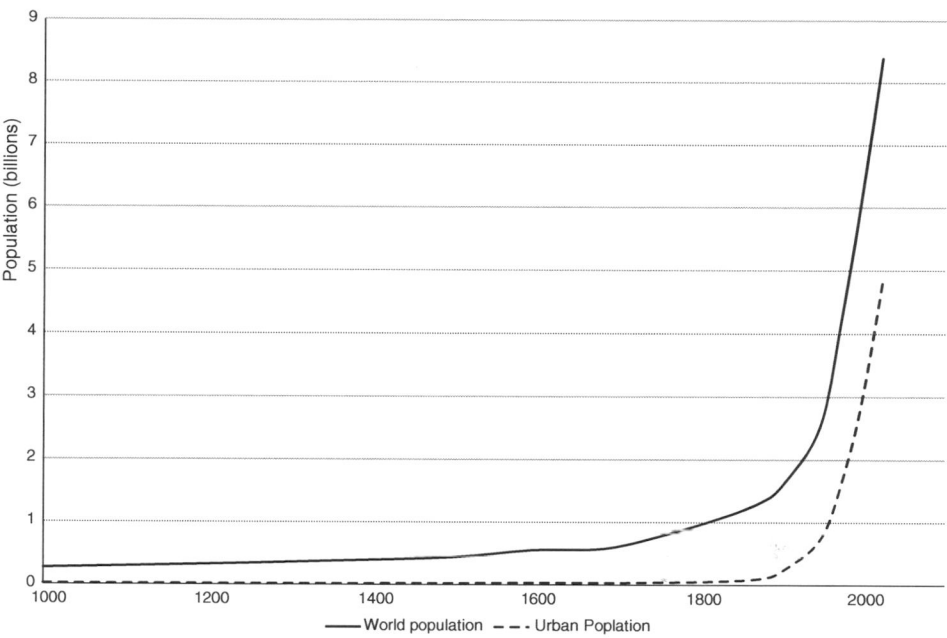

Source: Adapted from Fox (2012).

one million inhabitants or more; by 2000 there were over 360 (United Nations 1980; United Nations, Department of Economic and Social Affairs, Population Division 2014). To understand the driving forces behind this second urban revolution it is useful to consider the key constraints on urban growth and urbanisation in the pre-industrial era: disease and food security.

Before the nineteenth century, cities everywhere were demographic 'sinks'. High population densities coupled with poor sanitation and hygiene created an ideal environment for infectious and parasitic diseases to flourish. High disease-related mortality rates effectively placed a 'ceiling' on urban growth, and cities depended on regular in-migration from rural areas to reproduce themselves (Dyson 2001; Fox 2012). This disease constraint was augmented by limitations on the availability of surplus food supplies to support non-agricultural populations. As human settlements grow and expand they require an increasing amount of food to sustain their populations, which can only be acquired by either producing more food locally (i.e. increasingly agricultural productivity) or importing food from other regions. This largely explains the geography of early urban settlements, which emerged in areas naturally conducive to surplus food production (e.g. fertile river valleys) or locations with good 'natural' transport infrastructure (i.e. on coasts and along major rivers) (Childe 1950, Davis 1965, Bairoch 1988). But even in such favourable locations poor agricultural productivity and high transportation costs placed a limit on just how much food any one city could secure to support its population (Fox 2012). For the vast majority of people in the world, farming was a necessary way of life.

The second urban revolution began in Northern Europe, where a flurry of technological and social innovations in the eighteenth and nineteenth centuries dramatically improved food security and disease control, and stimulated fundamental shifts in economic activity. Innovations such as nitrogen fertilisers, crop rotation and mechanisation boosted agricultural productivity; the harnessing of inanimate sources of energy to fuel railroads, steamships and eventually automobiles led to a dramatic reduction in transportation costs; and improvements in hygiene, medical knowledge, the discovery of vaccines, maternal education, urban planning practices and organised public health interventions facilitated substantial and sustained declines in mortality in both urban and rural areas (Fox 2012).

Declining mortality rates, in turn, triggered what is known as 'the demographic transition'. The demographic transition is now recognised as a global process, although its onset has varied significantly across the globe since it began in Northern Europe in the eighteenth century. In brief, a demographic transition occurs when a population experiences a shift from a 'Malthusian' cycle of reproduction in which many babies are born but many die, to one in which fewer babies are born and many more survive to become parents themselves. In other words, a demographic transition occurs when mortality rates and fertility rates both decline substantially thereby creating an increasingly efficient cycle of reproduction. Importantly, a key feature of demographic transitions everywhere is a delay between the onset of mortality decline and the onset of fertility decline, which creates a period of rapid population growth.

In Britain, the demographic transition and its associated population boom coincided with the onset of the Industrial Revolution: a period of historically unprecedented economic change characterised rapid technological innovation, the rise of mass manufacturing, and the emergence of new industries requiring immense amounts of energy and labour (Bairoch 1988: 269). In the early days of the Industrial Revolution it was small towns and rural areas that contributed the most of both. This can largely be explained by the technology of the time: machines were driven by water-power and the iron and steel industries depended heavily on coal. This resulted in a restructuring of Britain's urban hierarchy as firms chose to locate close to these resources (and the cheaper labour available in these areas), thereby stimulating employment and construction outside of the major urban centres of the day (ibid.: 331). Later, the invention of the steam engine released industry from a reliance on water-power, permitting the construction of the first great factories with purpose-built housing for workers close by. The development of railroad networks, which reduced the costs of transporting food, fuel and raw materials (not to mention people), further facilitated urban growth and expansion (ibid.: 276–278). And rapid improvements in agricultural productivity reduced labour demand in rural areas, spurring workers to migrate into towns and cities in search of new employment opportunities. The cumulative effect of these changes was profound: the United Kingdom alone was responsible for 35 per cent of all urban population growth in the 'developed world' (i.e. Europe, North America and Japan) between 1800 and 1850, and between 1800 and 1900 the urban population of the UK grew from 3.1 million to 27.8 million (Bairoch 1988: 290).

As the Industrial Revolution spread across Europe, North America and Japan in the late nineteenth and early twentieth centuries, the growth and expansion of towns and cities contributed to further economic and demographic change. New ideas travel fast in cities, bolstered by better access to formal and informal education, improved social mobility, and access to mass media, all of which facilitate the production, diffusion and cross-fertilisation of knowledge and technologies. Furthermore, the enhanced division of labour characteristic of urban economies opens up new possibilities for productive innovations by specialists in particular trades or industries. Indeed, historical evidence indicates that innovation was directly correlated with levels of urbanisation and with city size once the Industrial Revolution was underway in Europe and North America (Bairoch 1988: 323–327).

The dramatic social, cultural and economic changes associated with the Industrial Revolution and increased urban habitation also contributed to a decline in fertility rates, as women took on new roles in the labour market and the costs of having children rose along with the benefits of investing more in their health and education. This decline began in cities and spread to rural areas and to less economically advanced nations, thereby progressing affected populations into the latter stages of the demographic transition (Livi-Bacci 2001: 94; Galor 2005). Figure 2.4 illustrates the dynamic interactions between these demographic, industrial and urban transitions.

The unprecedented pace of demographic and economic change in Europe in the nineteenth century brought myriad problems. Towns and cities were simply unprepared for such a massive shift in the balance of human settlements. In the 33 years between 1847 and 1880, for example, London grew from two million to five million inhabitants, and the human consequences were severe (Bairoch 1988: 285). Friedrich Engels (1845) described the conditions of London's slums in the middle of the nineteenth century thus:

> The streets are generally unpaved, rough, dirty, filled with vegetable and animal refuse, without sewers or gutters, but supplied with foul, stagnant pools instead. Moreover, ventilation is impeded by the bad, confused method of building of the whole quarter, and since many human beings here live crowded into a small space, the atmosphere that prevails in these working-men's quarters may readily be imagined.

Figure 2.4 *Industrialisation, urbanisation and the demographic transition*

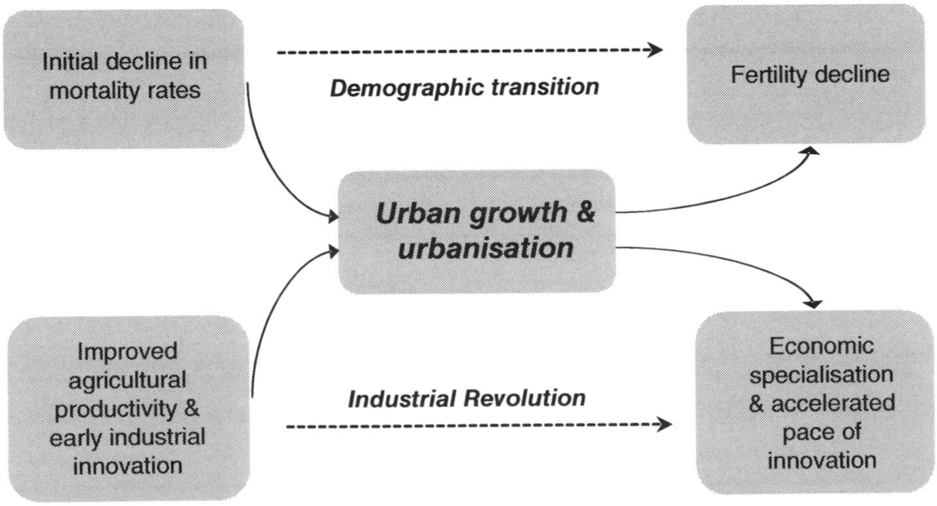

Such conditions were hardly unique to English cities; across Europe and North America similar squalor could be found. In some cases, wretched living conditions sparked political resistance that resulted in significant changes in public policy. For example, around the turn of the twentieth century, Glasgow was overflowing with labourers working in the shipbuilding and naval industries. Organised labour groups fought for over a decade for legislation to improve housing conditions, but were repeatedly thwarted by powerful landlords. However, a grassroots initiative supported by local housewives led to the Rent Strike of May 1915. Strikers refused to pay increased rents, protected one another from forced eviction (by force if necessary) and took to the streets to support Labour reform efforts. By November there were 20,000 city residents participating in the rent strike, forcing the Municipal Corporation to engage the national government. On 25 November of the same year, a Rents and Mortgage Interest Restriction Bill was presented in Parliament, and the following years saw the development of national legislation to improve urban housing conditions throughout the country (Castells 1983: 29–30). Theoretically speaking, the Glasgow Rent Strike can be understood as a case of synekism at work. The intensification of economic specialisation associated with the expansion of the industrial transformation drew increasing numbers of labourers into cities such as Glasgow, and the conditions in which they found themselves collectively sparked the formation of organisations and networks that ultimately transformed the political–economic institutions governing their housing conditions.

The power of the city to foment political reform has been observed by many scholars. Figures 2.5a and 2.5b from Dyson (2001) illustrate the positive relationship over time between urbanisation and democratisation across the world. While not claiming that urbanisation is *the* decisive factor in bringing about democracy, Dyson notes that 'Urbanisation focuses attention on the distribution of political power in society, so helping to bring about the rise of modern democracy' (2001: 17). Similarly, Borja and Castells (1997: 251) argue that cities are 'privileged places for democratic innovation', providing 'a chance to build a democracy of proximity, of participation by all in the management of public affairs' (246). The urban milieu is clearly conducive to the organisation of political action (see Chapter 8), and may very well be partially responsible for the spread of democracy worldwide. At the same time, cities can and have been instruments of imperial control and colonial exploitation.

Figure 2.5a *Urbanisation and democratisation, Europe and Asia*

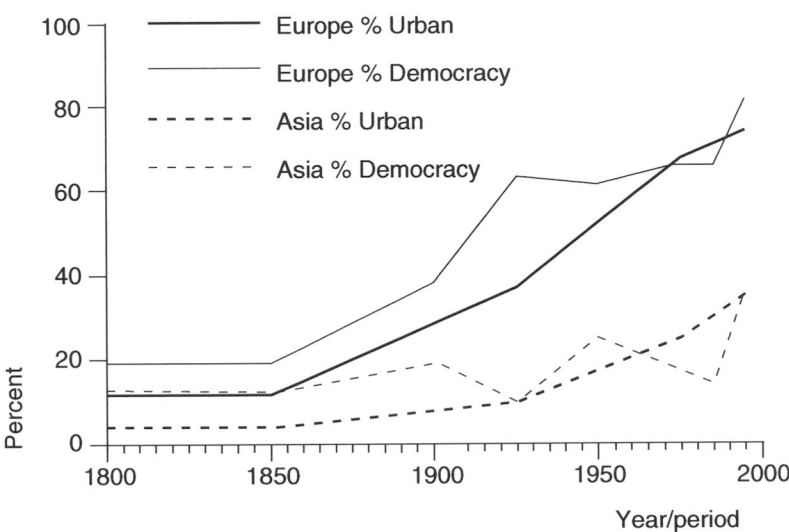

Source: Dyson (2001).

Figure 2.5b *Urbanisation and democratisation, Latin America and Africa*

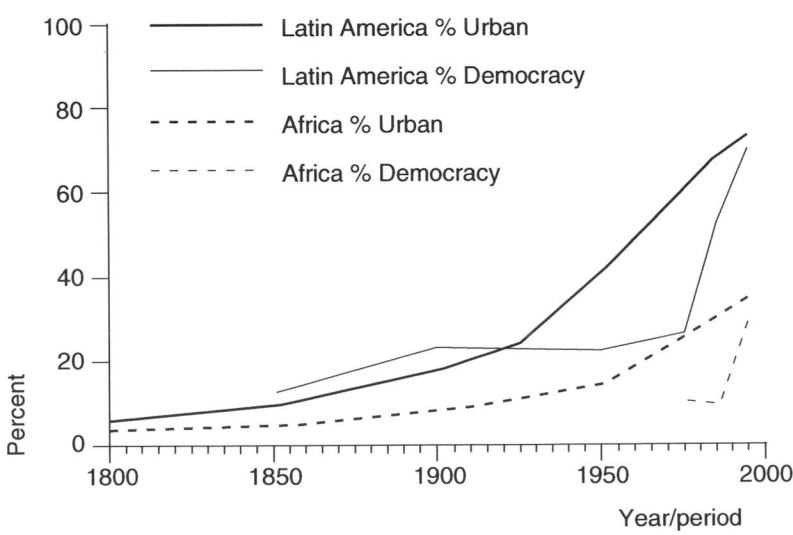

Source: Dyson (2001).

Colonial urbanism

Beginning in the sixteenth century European powers had begun to expand their spheres of influence, establishing settlements and imperial outposts in the Americas, Asia, Africa and across the Pacific. Over the next few hundred years, mercantilist enterprise gave way to imperial conquest and colonialism. Colonialism – defined as the political control of a people and territory by a foreign state (Bernstein 2000) – was characterised by political domination, social control, racism and exploitative economic relations, all of which have left their mark on the development trajectories of former colonies. Throughout the colonial era 'the city and its institutions were a major instrument of colonisation' (King 1990: 29), serving as 'the major links between core and peripheral economies…articulating the flow of capital, people, commodities, and culture that flowed between them…they were "global pivots of change"' (ibid.: 7). Colonial relations of domination were literally inscribed in the form of the built environment: in the architecture, institutions and infrastructure of the city. Today, many cities in low- and middle-income countries continue to grapple with the legacy of colonial urbanism.

In the sixteenth century, European powers began to support the exploits of individual adventurers seeking profits in the natural wealth of foreign lands, such as gold and silver, spices, silks and sugar. In general, early European settlements in foreign territories were small; companies essentially established bases for linking into existing local trade networks or tapping into mineral resources (such as silver mines in South America). But as profits grew, so too did the need to establish more permanent arrangements with warehouse facilities and military garrisons for protection.

Around this time, conquistadors received instructions from the King of Spain (first Ferdinand and then Philip II) concerning the appropriate locations and forms of cities to be built in Central and South America. In an interesting case of continuity in urban form, the Spanish kings based their urban planning instructions on Roman traditions, apparently drawing on the works of Vetruvius – a famous Roman architect and engineer who lived during the first century BC. The location and orientation of cities were dictated, as well as the geometry of plazas, street plans, and the placement of cathedrals, town halls and other administrative buildings (Stanislawski 1947). After Hernán Cortés destroyed Tenochtitlán and most of its inhabitants, he rebuilt it in accordance to the wishes of the Spanish

king and in line with Vetruvius' orderly vision (see Plate 2.1). This
was an urban plan that was repeated across Central and South
America and can be observed today in the older cities of Latin
America.

While most Latin American countries achieved independence in the
first two decades of the nineteenth century, their political and
economic trajectories over the subsequent two centuries were heavily
influenced by the colonial episode. The governments that took over at
independence generally comprised the elite descendants of European
colonisers who had amassed great fortunes in mining and plantation
agriculture. The indigenous populations that had managed to survive
the initial colonial onslaught of military conquest and European
diseases were relegated to subsistence agriculture, or forced to work
in mines or on large farms. Powerful landlords continued to dominate
politics, perpetuating the gross inequalities established in the colonial
era. Even today, legacies of economic inequality and racial
discrimination remain.

In other regions, the Dutch, French, British and Portuguese began
establishing small commercial and residential enclaves, adapting
architectural forms and planning norms from their homelands to local
climates and available materials. While in some cases initial colonial
encounters led to the destruction of existing populations and urban
centres (as in Spanish-America), the early phases of colonialism
generally left existing cities and urban hierarchies intact (Drakakis-
Smith 2000). In the late nineteenth and early twentieth centuries,
however, interests in the colonies began to change, and so too did the
role of cities. War in Europe reduced the availability of venture
capital to be spent on colonial enterprises, and many of the companies
that spearheaded mercantilist ventures were forced to hand over
control of their overseas possessions to state governments. Not long
after, the Industrial Revolution kicked into gear in parts of Europe,
increasing the demand for food and raw materials to supply
bourgeoning urban populations and new industries. Colonies, in turn,
played an important new role in the economic expansion of Europe,
providing both raw inputs for production, as well as markets for
manufactured goods (ibid.).

Over the course of the late nineteenth century, European powers
significantly expanded their presence in what had previously been
small colonies through the active acquisition of territory and
investments in infrastructure. This industrial phase of colonialism has

Plate 2.1 ***Plan of Mexico City in 1794 by Manuel Ignacio de Jesus del Aguila***

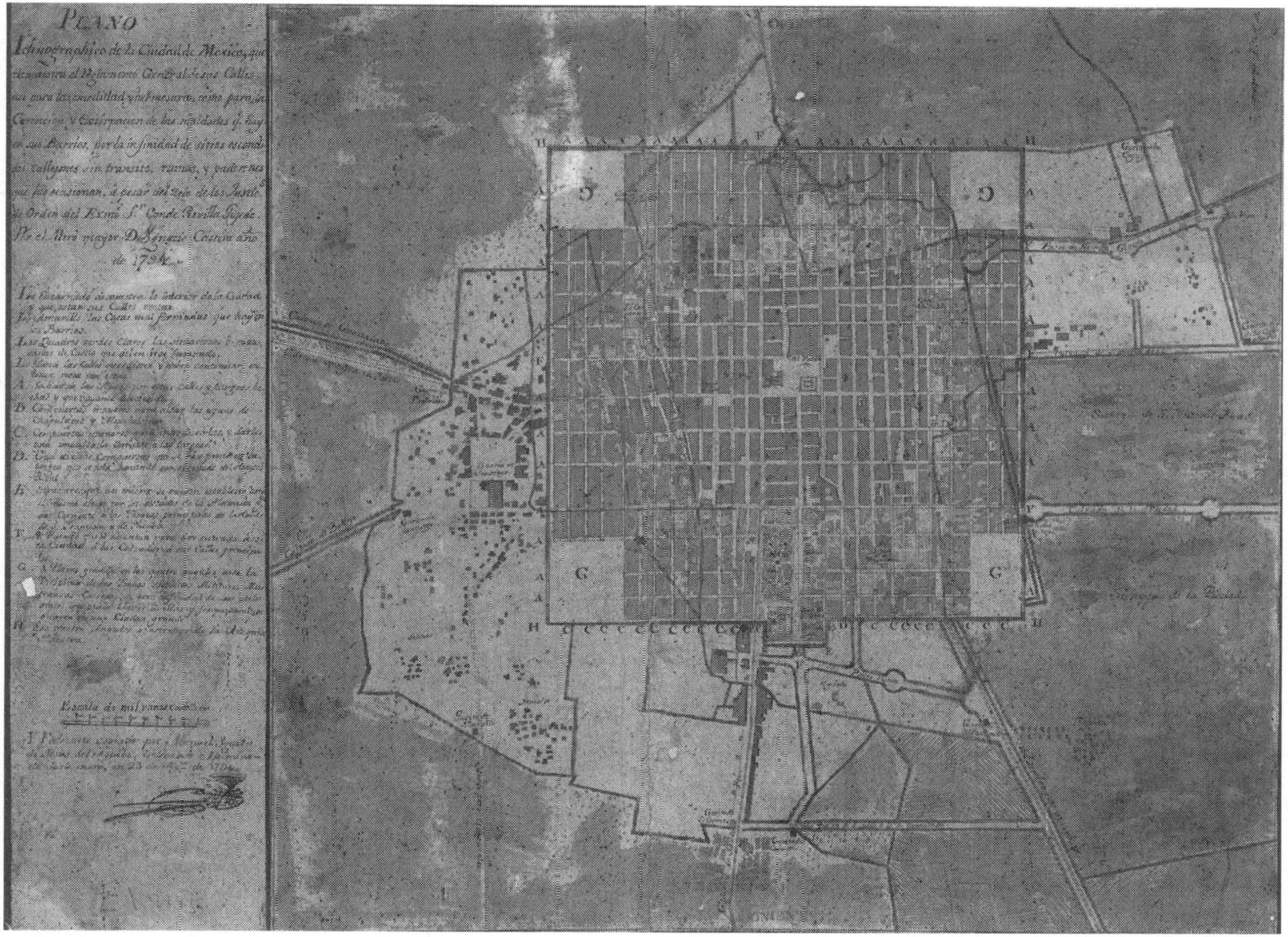

Source: Library of Congress, Geography and Map Division.

received much attention from dependency theorists, who see the history of colonialism as one of integrating Latin America, Africa and Asia into a capitalist world system in which peripheral countries produce primary commodities for core countries while absorbing the excess productive capacity of core countries by purchasing manufactured goods. For example, by 1931 two-thirds of British exports went to its colonies, and just less than half of all imports came from its colonies (Christopher 1988 cited in King 1990: 5). Some have even argued that a particular type of city was itself among the exports foisted onto the colonies; that as 'the first modern industrial power, Britain was the chief exporter of municipalities... Sporting pastimes apart, and the English language, urbanism was the most lasting of the British imperial legacies' (Morris and Winchester 1983: 196). In this context of increasing economic interdependence, colonial cities served as political, administrative and military nodes of empire, as well as new spaces of consumption and accumulation. Urban spaces were transformed through the collision of cultures, the form of political domination and the technologies imported from industrialising powers (King 1976).

Within cities, racial and ethnic segregation became a standard feature of colonial urban morphologies. New planning concepts imported from Europe were used to create 'sanitary' districts in colonial urban settlements, and reinforce notions of European superiority. Across many British colonies, the central districts of European-designed cities were usually surrounded by a green belt, a *cordon sanitaire*, within which no colonial subject was allowed to live. In certain Nigerian towns, the green zone was to be at least 440 yards (402.3 metres) wide since it was then believed that this was further than a malarial mosquito could fly (Hardoy and Satterthwaite 1989: 20). In Lusaka, Zambia, African inhabitants were made to live in village-like compounds within the city (see Box 2.1). In Delhi, India, urban planning driven by sanitary concerns in the late nineteenth century was accompanied by 'nuisance laws' to regulate the activities of urban residents. One could be found guilty of public nuisance for any

negligent act likely to spread infection of disease dangerous to life...[any] malignant act likely to spread infection of disease dangerous to life...fouling water of public spring or reservoir...making atmosphere noxious to health...and negligent conduct with respect to poisonous substance.

(from the penal code of 1862, cited in Sharan 2006: 4906)

Box 2.1 Eric Dutton's vision for colonial Lusaka

Lusaka was chosen as the site of a new administrative capital for Northern Rhodesia (now Zambia) because of its location halfway between the settler-dominated city of Livingstone in the south and the economic lifeblood of the Copperbelt in the north. At the time, Lusaka was just a railway watering station on the line linking the two, and as such was built almost entirely from scratch. Eric Dutton, the assistant chief secretary at the time, was hugely influential in terms of how the colonial capital was constructed.

Lusaka was in many ways designed as an archetypal 'garden city', with rows of trees and hedges strategically planted to divide compounds, hide African areas from the traveller's eye and generally – in Dutton's own words – conceal 'a multitude of sins'. The African compounds were based on elements of the traditional African village, carefully placed within the larger plan of Lusaka. The servants' compounds consisted of several units of four small huts looking in on themselves, emphasising a clear notion of inside and outside and reinforcing the idea that African residents belonged in one clearly defined section of their 'village', and should concern themselves only with that part of it. There were no central gathering places, and strategically located hedges were used to spatially segregate units and prevent the emergence of a broad-based – and potentially political – urban consciousness. These compounds were also built on low floodable land looking up at the Ridgeway, the location of the Government House, which was located so as to be visible to the African population as frequently and widely as possible.

Other than these 'showpiece' compounds, little regard was paid to Africans in terms of the way Lusaka was designed, and most Africans ended up in the expanding unplanned areas of 'Old Lusaka'. In effect, these settlements undermined Dutton's planning, and as Myers highlights, it is the way in which the urban majority, excluded from planning processes, reframed colonial cities that ultimately had the more significant impact on their evolution. Indeed, 'Even as the colonial regime seemed to try and make Lusaka readable like a book…the poetry of that book was being rewritten from underneath' (Myers 2003: 56).

Source: Myers (2003).

Street traders, bathers and fishermen could all be guilty of nuisance for various reasons. The physical organisation of city-space and the regulation of activity in that space were complimentary strategies of control and segregation in Delhi.

Almost without exception, the economies of colonial possessions were not encouraged to diversify and grow internally. Indeed, they were actively discouraged from doing so. Colonial powers ruthlessly exploited the natural resources and peoples under their control, extracting what profit they could without concern for the human or environmental consequences. This extractive economic system was reflected in the restructuring of urban hierarchies. Colonial capitals and ports grew disproportionately, with their political and economic power derived from direct links with the colonial metropole (Home 1997). Pre-colonial urban settlements were often destroyed or co-opted. In sub-Saharan Africa, pre-colonial urban centres were generally found inland as most trade was conducted overland as opposed to by sea. The building of colonial ports on the Western, Southern and Eastern coasts of Africa inverted the traditional urban hierarchies, drawing power and economic activity towards the colonial cities on the coasts, or along critical transport lines (e.g. Nairobi).

This shift in urban hierarchies was facilitated by the pattern of infrastructure development in this industrial phase. Linked infrastructure, such as railways and telegraph lines, were extended more deeply into colonial territories, but these networks 'were designed mainly to evacuate exports. There were few lateral or intercolonial links, and little attempt was made to use railways and roads as a stimulus to internal exchange' (Hopkins 1973, quoted in Graham and Marvin 2001: 84). In effect, infrastructure was designed to funnel all goods to a port. As a consequence of this approach, many of the largest cities in the world today – including Mumbai, Kolkata, Hong Kong and Lagos – are port cities that were 'creatures of British colonialism' (Home 1997: 62).

The lack of attention to basic water and sanitation infrastructure within such cities led to acute public health crises, including outbreaks of the bubonic plague which killed thousands every week in cities such as Bombay at the turn of the twentieth century (Home 1997). Meanwhile, colonial cities were subject to dramatic changes in fortune depending on developments in global commodity markets. For example, in the late Victorian period Calcutta (now Kolkata) was the 'second city of the British Empire', widely renowned as the 'city of palaces'. Just a couple of decades later it was rapidly deteriorating and dubbed 'the city of dreadful night' by Rudyard Kipling, largely due to the global decline in demand for jute, the material on which its economy had been based (Chakravorty 2000).

In the latter phase of colonialism, there was a significant boost in urban infrastructural development, reflecting the modernist ideals and moral discomfort with imperialism emanating from Europe and the demands associated with attempts to improve local productivity through land reforms and mechanisation. Urban migration accelerated and cities began to swell with indigenous populations

(Drakakis-Smith 2000). In response, many colonial governors felt the need for more extensive urban planning. The 'Garden City' movement in England influenced British colonists in particular, who used urban planning to re-enforce segregation and reflect, in the built environment, their perception of their own superiority (Graham and Marvin 2001: 82).

Cities where settler colonialism was a feature, such as Harare, Nairobi and Lusaka, share a similar morphology in the strict separation of administrative, commercial, industrial and residential space, a separation typical of the modernist concepts being promoted in Europe at the time (O'Conner 1983: 199). Yet colonial cities were also sites for experimentation with forms of social and spatial organisation that governments in the metropole could or would not try at home (Home 1997). More often than not, social segregation was a central aspect of this. In Nairobi, for example, residential areas were segregated by race, with white Europeans, Asians and Africans living in separate quarters of the city. Some colonial governments – such as in Northern Rhodesia, or present day Zambia – increased investment in indigenous quarters in an attempt to improve the stability and productivity of labour to support industrialisation and 'civilise' native populations (Mabogunje 1990: 127; Heisler 1971). These investments had consequences for colonial rulers. The urban milieu was home to an increasingly educated and disgruntled urban African population who found new ways of organizing resistance:

> Associational life…was getting richer and more varied, built around occupational and residential affinities, around connections to common regions of origin, around churches or mosques or indigenous religious institutions, around mutual aid needs of various sorts, around the new forms of music and artistic creation. Most important was the mix of urban-born and migrant youth, a category marked, as Rémy Bazenguissa-Ganga puts it, by its 'availability' – a vibrant, volatile force that could be channelled in different directions.
>
> (Cooper 2002: 34–35)

Resistance movements emerged and spread across the remaining colonial empires in Africa, Asia and the Middle East, leading to a long period of decolonisation in the aftermath of World War II. And although independence struggles were not a strictly urban phenomenon, the urban milieu provided an essential space for the organisation of resistance, for the acquisition of essential resources and as information and communication hubs.

Despite dramatic differences in the nature of colonial experience and timing of independence movements across the world, most former colonies were left with infrastructure designed to funnel goods abroad instead of encourage domestic circulation, with national boundaries that did not reflect pre-colonial political geographies, with unbalanced urban hierarchies, with cities designed to segregate, and with a variety of institutions – such as property regimes and regulatory frameworks – unfit for inclusive development.

Into the first urban century

The end of World War II, and the wave of decolonisation that it spawned, coincided with an acceleration of international economic integration and the commencement of the Cold War. At first, post-colonial governments, taken with modernisation theory and intent upon establishing a sense of national identity and ownership, embarked upon large-scale infrastructure and economic development projects to redress underinvestment during the colonial era. Airports, multi-lane highways, power stations, public housing and monuments to the leaders of the independence struggle were built (Graham and Marvin 2001). But these extensive investments failed to generate the benefits imagined by their designers. A post-war boom in the global economy turned into a global economic downturn in the early 1970s, undoing what little progress had been made in previous decades, and sending many low-income countries into heavy debt. The political legacies of colonialism began to reveal themselves, thwarting early nationalist ambitions by exacerbating conflicts within and between nations. In Africa (and arguably elsewhere), colonial governments had behaved as 'gatekeeper states' standing 'astride the intersection of the colonial territory and the outside world. Their main source of revenue was duties on goods that entered and left the ports' (Cooper 2002: 5). Post-colonial governments, inheriting economies oriented toward extraction,

> realized early on that their own interests were served by the same strategy of gatekeeping that had served the colonial state before World War II: limited channels for achievement that officials controlled were less risky than broad ones which could become nuclei for opposition. But the post-colonial gatekeeper state, lacking the external capacity of its predecessor, was a vulnerable state, not a strong one.
>
> (ibid.)

Corruption, coups, conflict and outright war became regular features of the post-colonial era. At the same time, many former colonies found themselves caught up in the power struggle of the Cold War, with the United States and the Soviet Union (among others) intervening in the political and military affairs of nations struggling to consolidate political order and stimulate economic development. In many cases, these Cold War enemies propped up dictators and financed proxy wars in Africa, Asia and Latin America with devastating consequences. The brief spells of hope that had attended the early days of independence quickly faded.

Through all of this, cities continued to serve as nodes in the international political economy and today many cities in low- and middle-income countries remain more economically and politically bound to cities in Europe, North America or East Asia than to their own hinterlands. The legacy of extractive economic relations has been difficult to overcome, with many countries continuing to rely on exports of primary commodities (such as agricultural products and mineral resources) that perpetuate economic vulnerability and dependency. Although the political bonds of colonialism have been broken, many scholars argue that contemporary economic relations between industrialised and less industrialised nations are a form of 'neo-colonialism' exercised through the organisation of the global economy; through trans-national corporations and multilateral agencies such as the World Bank and IMF, which exert enormous influence in low- and middle-income countries. Thus while a city such as Lusaka may not appear on any ranking of 'global cities', it is deeply implicated in the regional and global economy both through Zambia's natural resource exports and the major presence in the city of international donors and NGOs (Robinson 2002; 2006).

Within cities, colonially imposed 'laws, norms and codes governing housing, building and planning...remain today, largely unchanged... Not surprisingly, they are very poorly suited to [independent nations] where rapid urban growth and increasingly urban societies have become the norm' (Hardoy and Satterthwaite 1989: 20), as we discuss in further detail in Chapter 5. While well-serviced urban quarters once served to segregate Europeans from indigenous populations, they now serve to insulate domestic political and business elites, who have maintained a high standard of living by capitalising on their role as gatekeepers while allowing the formerly 'indigenous' quarters of their cities to deteriorate (Mabogunje 1990: 141). In this way, the colonial experience and its patterns of domination and oppression live on in many cities (Porter 2010). As Ananya Roy has argued, 'Empire is not simply an unfortunate backdrop to planning, one that can be simply denied allegiance. Rather, empire is planning's "present history"', and continues to haunt urban development trajectories (Roy 2006: 8).

Against this background, towns and cities in Africa, Asia and Latin America experienced explosive growth in the latter half of the twentieth century in the face of historically unprecedented rates of population growth. As Figure 2.6 shows, the demographic transition in Europe and North America drove a significant increase in population growth between 1750 and 1850, after which population growth has steadily slowed (apart from a brief post-war baby boom). In contrast, the demographic transition began later and has been far more intense in Africa, Asia and Latin America. Livi-Bacci (2001) explains this stark difference in the timing and intensity of the demographic transition across regions thus:

> Slow mortality decline...was the result of an accumulation of knowledge, especially medical knowledge, which helped bring infectious diseases under control...beginning in the mid-twentieth century, that knowledge slowly accumulated by the rich countries was rapidly transferred to the poor ones and mortality dropped dramatically.
>
> (129–130)

The rapid decline in mortality in less economically advanced countries that began the late colonial period was not immediately accompanied by a decline in fertility rates, which explains the obvious global population boom apparent in the graph. Figure 2.7 illustrates how inter-regional variation in the timing of this population boom translated into the differential timing of urban transitions across major sub-regions. The share of Europe's population living in urban settlements began steadily rising in the nineteenth century, followed by countries in Latin America, which achieved independence from European powers and began industrialising much earlier than most African and Asian countries (albeit with inconsistent success). It was not until the middle of the twentieth century that the majority of countries in Africa and Asia began to urbanise, and when they did the pace was considerably faster than that experienced in Europe (especially in East Asia).

Figure 2.6   Regional trends in population growth rates, 1750–2050

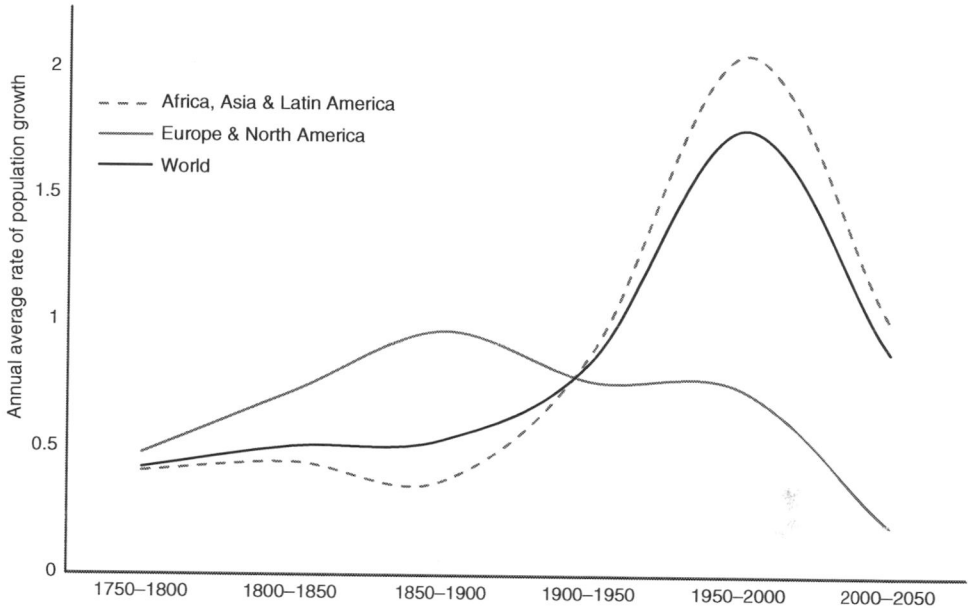

Source: Data for 1750–1900 from United Nations Department of Economic and Social Affairs (1999); Data for 1950–2005 from United Nations Department of Economic and Social Affairs, Population Division (2014).

Figure 2.7   Levels of urbanisation in major world regions, 1850–2050

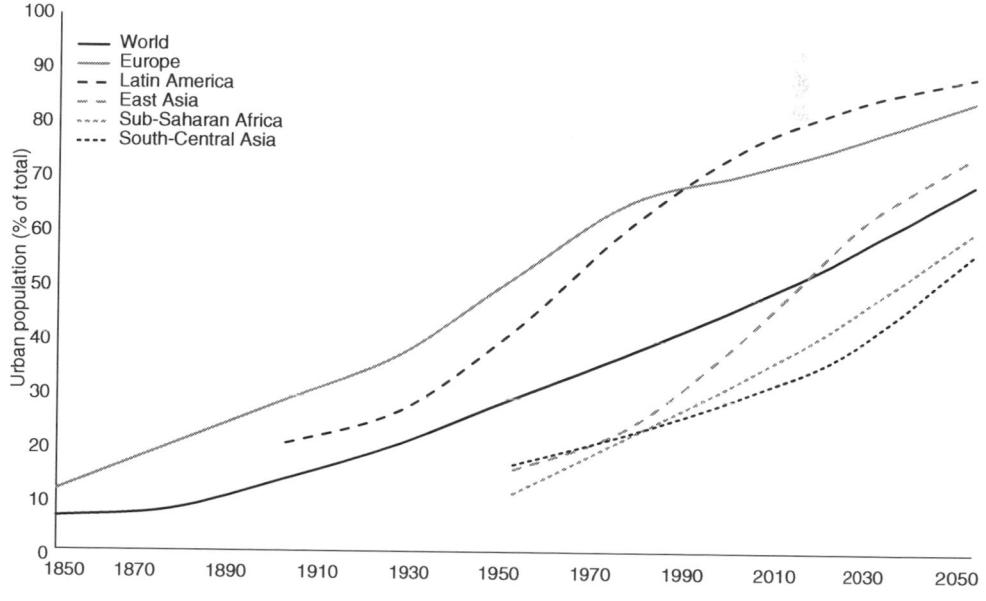

Source: Adapted from Fox (2012).

As levels of urbanisation rise in these regions fertility rates will almost certainly fall, eventually completing the global demographic transition. Indeed, in many East Asian countries fertility rates have already fallen to replacement level – i.e. the number of births each year is roughly equal to the number of deaths. However, in the meantime populations continue to expand at breakneck pace in many less economically advanced countries, which in turn is driving very rapid urban population growth. It is important to appreciate the scale of these changes, as it is ultimately the absolute number of people being added to cities that presents the greatest urban development challenges—not the relative number of people (i.e. rate of urbanisation). As Figure 2.8 shows, the urban population of Asia grew by over one billion people between 1950 and 2000; projections indicate that towns and cities in the region will absorb nearly two billion more people between 2000 and 2050. Similarly, urban populations in Africa and Latin America grew by roughly 250 and 325 million people respectively between 1950 and 2000; between 2000 and 2050 Latin America will likely add a further 275 million to its towns and cities while Africa's urban areas will add over one billion.

Figure 2.8 *Urban population growth by major world region, 1950–2050*

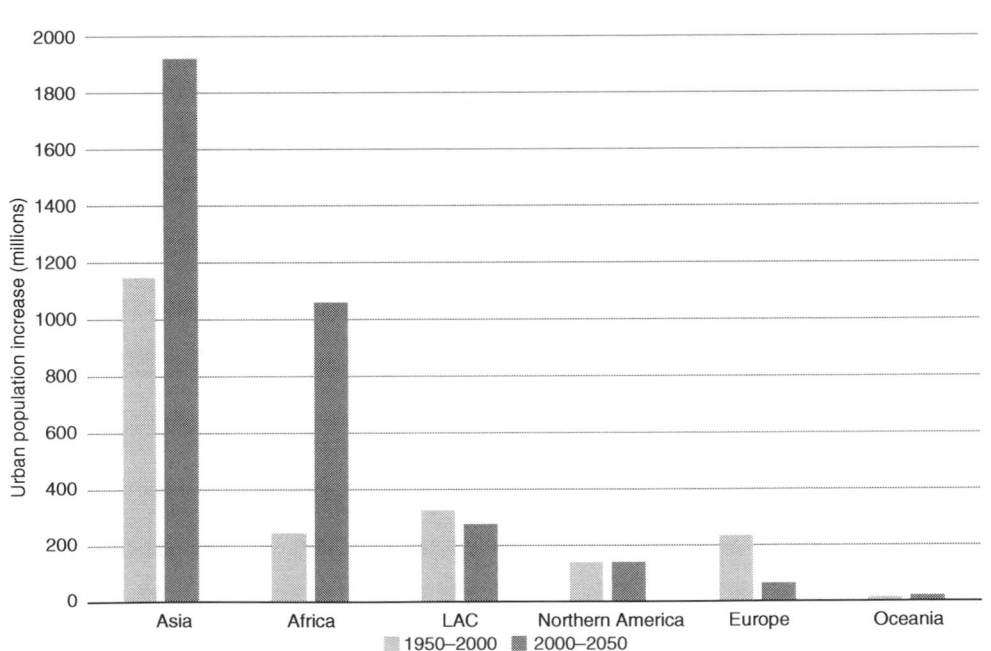

Source: United Nations, Department of Economic and Social Affairs, Population Division (2014).

The scale of this spatial-demographic change is historically unprecedented, and in contrast to the European, North American and Japanese experiences, many countries in these regions are suffering the pangs of urban growth and expansion without the potential benefits of simultaneous and extensive industrialisation. Even in those countries which have had economic growth to match urban growth, such as China, the challenges of providing secure livelihoods and adequate housing while minimising environmental impact and social dislocation are acute, as we discuss in Chapters 5 and 6. But the geography of these challenges is not necessarily in line with popular images of dystopian megacities. As Figure 2.9 shows, by 2025 the majority of the world's urban population (56 per cent) will live in urban settlements of one million or less. A further 22 per cent will live in large cities of one to five million. In other words, it is the small and medium-sized cities in Africa and Asia that will be absorbing the majority of the world's population growth in the years ahead.

Figure 2.9 *Distribution of the global urban population by city-size class, 1950–2025*

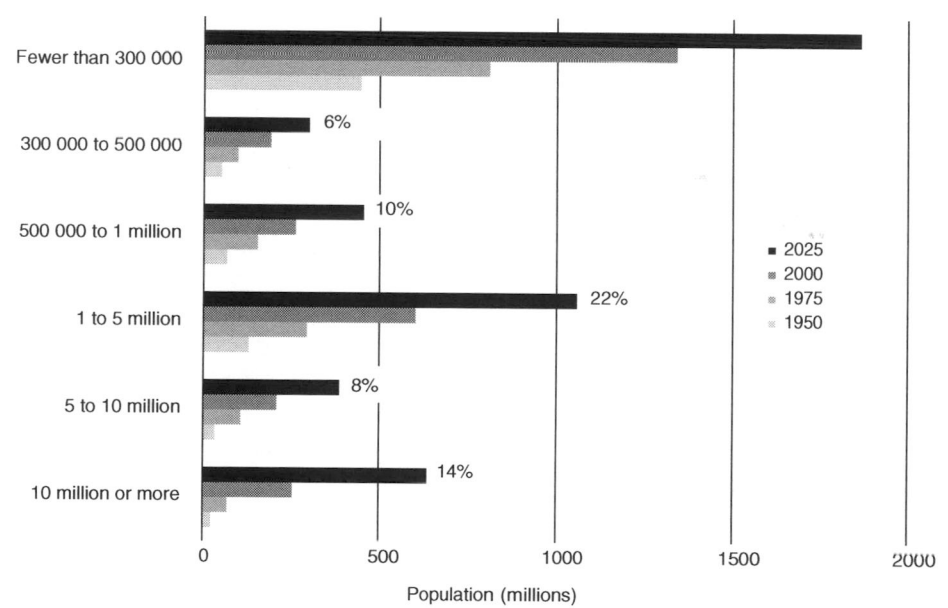

Source: United Nations, Department of Economic and Social Affairs, Population Division (2014).

Conclusion

Throughout history cities have been innovative spaces. Urban habitation is defined both by specialisation of social, economic and political activity, and by the nearness of diverse people. The result is change – for better or worse. There is no doubt that early urban societies were highly unequal, and that living conditions were bad. But in the long run, urban innovations such as metallurgy, coined money, writing and bureaucratic administration contributed to the rise of increasingly complex and wealthy societies. From their very origins, cities have been focal points for the accumulation and exchange of information, knowledge, capital and goods. One could make the case that modern society represents the sum total of urban innovations and exchanges across the millennia.

As instruments of empire and colonisation, cities served as command and control centres for the global ambitions of expansive powers from Babylon to Rome to London. In some cases, the residues of segregation and exploitation associated with these episodes in history remain visible in the layout of urban areas, the nature of urban hierarchies, and in the shape of national infrastructure. But exploitation and repression are difficult to maintain in cities unchallenged: the urban-based independence movement in Zambia and the Glasgow Rent Strike both demonstrate the urban potential for political mobilisation and transformation. Indeed, the global urban transition set in motion by the demographic transition and Industrial Revolution in Europe have generally been associated with increasing wealth, welfare and democratisation worldwide – albeit unevenly distributed. As the twenty-first century progresses, the demographic and economic forces that drive urban growth and shape urban spaces will continue to be conditioned by global forces set in motion in the eighteenth and nineteenth centuries. To fully understand and appreciate the complex dynamics shaping urban spaces, societies and economies today, an appreciation of the historical continuities and discontinuities that link cities across the globe is imperative.

Summary

- The earliest cities emerged around six thousand years ago in Mesopotamia. There is some disagreement as to whether urban centres were a response to the innovations of agriculture or whether they predate and provided the impetus for these innovations.

- These early cities and those that subsequently emerged in other regions were essentially 'proto-states', sites of political and economic innovation and centres of regional domination and influence.
- In medieval Europe the relationship between urban-based capital and rural-based coercion culminated in the rise of the nation-state, as monarchs struck bargains with wealthy urbanites to finance wars and expansion.
- The second urban revolution in Europe was set in motion by both the Industrial Revolution and demographic transition, but subsequently acted as a catalyst for both.
- Social and technological innovation made the sustained expansion of urban populations possible by a) reducing the disease burden that had historically rendered cities 'demographic sinks' and b) providing sufficient food to support an increasingly non-agricultural (i.e. urban) population.
- Cities in many developing countries evolved somewhat differently due to their role in colonial processes of extraction and domination.
- Colonial urbanism led to the growth of ports and capitals, as well as increasing use of urban planning, segregation and the inequitable development of urban infrastructures. The legacy of these practices endures to the present day.
- Urban transition is occurring at unprecedented rates in developing countries but without the benefits of the economic advances that accompanied it in Europe. The majority of growth is in small and medium-sized cities, and is linked with informality and inequality.

Discussion questions

1. Discuss the relationship between cities and the emergence of agriculture.
2. What role did cities play in the evolution of states in Medieval Europe?
3. In the nineteenth and twentieth centuries in Europe, what was the relationship between urbanisation and a) the demographic transition; b) industrial development? Why are these relationships

different in many low- and middle-income countries currently urbanising?

4. What are the most significant legacies of colonial urbanisation for development in contemporary low- and middle-income countries?

5. What can be learnt from the historical experience of urbanisation to inform the future development of cities in low- and middle-income countries in the first urban century?

Further reading

Bairoch, Paul (1988) *Cities and Economic Development: From the Dawn of History to the Present*. Chicago: University of Chicago Press.

Childe, V. Gordon (1950) 'The Urban Revolution', *Town Planning Review*, 21(1), 3–17.

Dyson, Tim (2010) *Population and Development*. London: Zed Books.

Hall, Peter (1998) *Cities in Civilization*. London: Weidenfeld and Nicolson

King, Anthony D. (1990) *Urbanism, Colonialism and the World-Economy*. London: Routledge.

Livi-Bacci, Massimo (2001) *A Concise History of World Population (3rd edition)*. Oxford: Blackwell Publishers.

Mumford, Lewis (1961) *The City in History*. London: Martin Secker and Warburg Ltd.

Tilly, Charles (1994) 'Entanglements of European cities and states', in *Cities and the Rise of States in Europe: A.D. 1000–1800*. Colorado: Westview Press, pp. 1–27.

Websites

H-Urban: www2.h-net.msu.edu/~urban

International Planning History Society: www.planninghistory.org

The European Association for Urban History: www.eauh.eu

Urban History Association: The City in History: urbanhistorians.wordpress.com

3   Urbanism and economic development

- Economic growth and development
- Industrialisation, urbanisation and urban concentration
- Cities as engines of growth
- Rural–urban linkages and regional development
- Cities and regions in a global economy

Introduction

Understanding the process of economic development remains a central preoccupation in the field of development studies and is a core concern for policymakers. Economic growth and diversification improve living standards and enhance freedom by giving people greater choice over what they do, where they go and what they consume. Although there are many scholars and professionals who feel that the global development agenda often places too much emphasis on economic growth, few would argue with the assertion that 'long-term sustainable and inclusive growth is the driving force for poverty reduction in developing countries' (Lin 2011, 2).

We begin this chapter with a discussion of the foundations of economic growth and development, identifying key mechanisms such as specialisation, investment and innovation, which were key ingredients in European industrialisation. We then explore the relationships between industrialisation, urbanisation and urban concentration, highlighting some of the problematic assumptions that became embedded in development theories largely informed by European history. We then examine the ways in which cities, by their very nature, can facilitate economic growth by bringing workers, firms and consumers together. The clustering together of people in space eases flows of information, people, goods and new ideas, thereby enhancing productivity and stimulating innovation, but also generating costs that can undermine such benefits. We then turn to the question of how towns and cities can contribute to broader regional and national development by serving as important nodes in domestic economies. The final section of the chapter examines the role that large agglomerations – i.e. metropolitan areas and city-regions – play in regional and global economic systems, producing, consuming and trading the world's capital, goods, people and ideas. While these 'world cities' can play a pivotal role in national development, they are also characterised by intense inequalities that raise concerns about the potential disconnect between global economic forces and local material realities.

Economic growth and development

The terms 'economic growth' and 'economic development' are often used interchangeably, and they are closely linked. But there are some distinct conceptual differences. The term 'economic development' is used broadly to denote sustained improvement in living standards for a country's population (Arndt 1987, Todaro 2000). This obviously includes reliable access to basic necessities such as food, shelter, clothing, bodily health and physical security, but it also encompasses freedom from oppression, education, reliable employment and opportunities for cultural engagement and expression (Sen 1999, Todaro 2000). For early development scholars in the 1950s and 1960s, the path to achieving these goals was synonymous with industrialisation and the structural transformations in production associated with this process (Myrdal 1957; Hirschman 1958; Rostow 1960). Drawing on the histories of the industrialised countries of the post-World War II era, these scholars conceptualised economic development as a transition from an economy dominated by subsistence agriculture towards one engaged in manufacturing, heavy industries and services such as retail and finance. As Table 3.1 shows, in economically advanced regions agricultural employment represents less than 5 per cent of total employment; by contrast in sub-Saharan Africa – the poorest major world region – agriculture still accounts for over 60 per cent of employment.

Table 3.1 *Sectoral distribution of employment in major world regions, 2002 and 2012*

	2002			2012		
	Agriculture	*Industry*	*Services*	*Agriculture*	*Industry*	*Services*
Developed Economies & EU	5.0	26.3	68.7	3.6	22.5	73.9
East Asia	47.3	22.6	30.1	31.0	30.9	38.1
South East Asia & Pacific	48.2	17.3	34.5	39.2	19.8	41.1
South Asia	56.9	17.0	26.1	48.5	22.2	29.3
Latin America & Caribbean	20.6	21.2	58.2	15.7	21.1	63.2
Sub-Saharan Africa	65.0	8.3	26.7	61.1	8.9	30.0

Source: ILO (2014).

As the field of development economics evolved in the latter half of the twentieth century, attention shifted in mainstream circles away from understanding and promoting structural change in production and towards a near myopic concern with economic growth. Economic growth refers specifically to an increase in the total value of goods and services produced in an economy (e.g. a country or region). In the field of development studies, it is generally understood that the goal is growth relative to population, usually expressed and measured as income per person, or Gross Domestic Product (GDP) per capita. An economy can grow simply by adding more people, but this doesn't mean the population as a whole is any better off. It is only when there are more goods and services available for each person that material conditions can be said to be improving. Efforts to explain why some countries have grown more rapidly and more consistently over time than others have yielded a variety of theories but no definitive explanation. Here we provide a cursory summary of some of the key insights that have emerged from the quest to explain why some countries have managed to get rich while others remain comparatively poor.

Adam Smith provided the first pivotal insight back in the eighteenth century when he noted the economic benefits of specialisation and exchange. Imagine if you were to try to grow your own food, build your own house, make your own automobile and assemble your own mobile phone. It would take a lifetime to learn all of the necessary skills and acquire all of the necessary materials to produce these things. It is thanks to specialisation in production processes that it is possible to have access to such a diverse range of very useful goods. The more general insight that emerges from this line of reasoning is that productivity growth – i.e. making more of a good or service with fewer inputs – is key to long-run, sustained economic growth and diversification in production.

What drives productivity growth and diversification? For the first generation of development economists advising governments and multilateral agencies (e.g. the UN, IMF and World Bank) the answer was investment. Expanding the quantity and range of output requires investing in direct inputs such as raw materials, tools, buildings, or machines, as well as public goods such as infrastructure and education, which facilitate more efficient production processes. This simple theory laid the foundations in the post-war era for the establishment of 'development assistance' (or foreign aid), which revolves around financial transfers and technical assistance where investment resources are scarce. In theory, poor countries remained poor because they didn't have enough surplus capital to invest in economic growth. By making up for this capital deficit through aid, it was believed that the global income gap could be narrowed relatively quickly (Todaro 2000, Lin 2011).

While variation in investment did help to explain some of the differences in income per capita across countries, the model proved insufficient in itself. A key weakness in this simple model of growth is that there is a limit to how much investment can support productivity growth over the long run. Imagine a carpenter who invests in upgrading from a hand saw to an electric one. This will dramatically improve her productivity by allowing her to cut a lot more wood in much less time. However, investing in a second electric saw won't have the same effect: she can only use one saw at a time, so the second one will mostly be left idle. This is what is known as 'diminishing returns to investment' in the field of economics. This observation is particularly relevant to more advanced economies that are using the most up-to-date technology. In such contexts it was argued that further productivity gains can only be achieved through innovation (or technological progress) (Todaro 2010, Lin 2011). Innovation in the form of new ideas and technologies can improve productivity and generate new kinds of goods and services, allowing an economy to continue expanding when all of the potential economic gains to old ideas and old technologies have been realised. For low- and middle-income countries, the adoption of technologies developed in more advanced economies can substitute temporarily for domestic innovation as these countries 'catch up' with the global technological frontier. Importantly, innovation is not considered random, but rather a product of factors within government control, such as education (i.e. human capital) and investment in basic research (ibid.).

Investment and innovation are therefore central to theories of economic growth. For less economically advanced countries, investments in upgrading knowledge, technology and physical infrastructure are essential; as countries 'catch up' to the technological frontier, innovation becomes increasingly important for sustaining growth. However, while these insights are conceptually and practically valuable, they don't provide a wholly satisfactory explanation for the dramatic divergence of economic performance across countries over long periods of time. Why have some countries serially failed to invest and innovate, while others have done so consistently over decades or even centuries?

The most recent generation of development economists have sought to answer this question by exploring the institutional foundations of economic growth and development. As noted in Chapter 1, institutions can be thought of as 'the rules of the game' that shape human interactions. These can be formal, such as laws and regulations, and informal, such as social and cultural norms (North 1990). While early development economists effectively took the institutional context of a country for granted, basing their assumptions on the way things worked in Europe and North America, more recent scholarship has focused on identifying the nature, causes and consequences of institutional variation across countries (e.g. Rodrik 2003, Acemoglu and Robinson 2012). Institutions matter because they provide a predictable framework for exchange between individuals and organisations and fundamentally shape incentives for investment and innovation.

For example, if a farmer owns her land and has confidence that it won't be arbitrarily confiscated, she may choose to invest in improving irrigation and soil quality, or perhaps planting crops that are lucrative but require several years to mature (e.g. many varieties of fruit tree). If she does not have confidence in her right to remain on her land she will likely forgo such investment. This is an example of the importance of property rights – one of the key institutions economists have focused on. A parallel example can be provided for incentives to innovate. In the pharmaceutical industry, patents (or 'intellectual property rights') give firms exclusive rights to sell medicines that they develop, essentially guaranteeing firms a healthy profit for a fixed period of time. This institution – i.e. the patent

– therefore provides firms with strong incentives to invest in developing new medicines.

The significance of 'good' institutions extends far beyond economic ones such as property rights. Political institutions, for example, shape how decisions are made about taxation, the expenditure of public funds for health and education, business regulation, civil rights, etc. If political institutions do not enable populations to effectively hold their governments to account, this increases the likelihood of the government behaving in a predatory and self-seeking manner. Judicial institutions are also crucial: if people do not feel that they will be treated fairly in legal disputes, they are less likely to take risks, including in the form of investment. In other words, institutions are not only critical for achieving sustained growth, but also for translating growth into the more holistic goal of economic development.

While the intellectual quest to understand the origins and mechanics of economic growth has usefully identified the importance of investment, innovation and institutions, it also drew attention away from two of the primary concerns of early development economists: the need to promote structural changes in production and output, and the spatial-demographic implications of such changes. However, there has been a resurgence of interest in both issues in recent years, with important implications for urban development research and policy (see Box 3.1).

Box 3.1 Beyond growth: rediscovering structural change

In the 1950s and 1960s economists conceptualised economic development as a process of structural change involving industrial upgrading – i.e. the accumulation of physical infrastructure and technologies required for large-scale manufacturing and industry. As discussed in Chapter 1, governments were seen as having an important role to play in stimulating structural transformation through strategic investments and infant industry policies.

This 'structuralist' view fell out of favour in the 1970s and was replaced by the market-centred neoliberal paradigm of economic development theory. Mainstream development economists abandoned structuralist interpretations of economic development and instead came to focus almost exclusively on understanding and promoting economic growth. The policy prescriptions that emerged from this new line of investigation in the 1980s and 1990s proved highly controversial and largely unsuccessful in generating sustained growth in many of the poorest countries in the world. At the same time, governments in

East and Southeast Asia (most famously China) eschewed the advice of many mainstream economists and played an active role in stimulating technological upgrading and economic diversification in their economies. This has led to a recent resurgence of interest in the role governments can play in facilitating structural transformation, epitomised by the publication in 2011 of a paper on this topic by the Chief Economist of the World Bank at the time, Justin Yifu Lin.

It is important to remain mindful of the difference between growth and some other key goals of economic development, such as diversification and accumulation of assets and infrastructure which support production and exchange. Growth can be very one-dimensional. For example, a country that produces oil may benefit from rising prices in international markets. If the volume of oil being produced remains constant but the price of oil rises, the value of the country's exports will rise, the economy will grow in monetary terms, and average per capita income will rise in statistical tables. However, unless the profits from such a windfall are re-invested in infrastructure, social development (e.g. health and education), or in promoting diversification in production, this growth may not result in any substantive economic progress. Diversification matters because it renders an economy less vulnerable to unforeseen shocks, such as a dramatic fall in prices of major exports (e.g. coffee or oil) or the effects of a natural disaster on output in one sector of the economy. More diverse economies are less vulnerable to such shocks, and are more likely to deliver consistent growth over the long run.

Growth matters because it can translate into real improvements in people's daily lives. But a myopic emphasis on understanding and promoting growth has tended to obscure the importance of targeted government interventions to ensure a sound platform for achieving sustained growth, as well as translating this growth into genuine welfare improvements for a broad cross-section of a society.

Sources: Lin (2011), Mobarak (2005), Herzer and Nowak-Lehmann (2006), Aditya and Acharyya (2013).

Industrialisation, urbanisation and urban concentration

Historians and economists have traditionally understood the process of industrialisation as the primary driving force behind urban growth and urbanisation. The economic histories of European, Japanese and North American economies clearly indicate that the growth of employment opportunities in towns and cities, coupled with the intensification and mechanisation of agricultural production, drew large numbers of peasants into the urban labour force, lured by the higher wages on offer (Lewis 1954). The result was a more productive, diversified and prosperous economy with a largely urbanised population.

Post-war development scholars such as Myrdal and Hirschman also believed that industrialisation and urbanisation would be accompanied by a spatial evolution in the organisation of economic activity within a country. In the early stages of development, when transportation and communications infrastructure is poor, skilled labour is scarce, and financial institutions are limited in scale and scope, it makes sense for firms to cluster together in a single city or small number of large urban centres to maximise efficiency. The extent to which people and economic resources are concentrated in one or two large cities is referred to as urban concentration; when a single city dominates a national urban network it is known as a primate city. Urban concentration or the emergence of a primate city at the earlier stages of development was to be expected; at later stages of development it was anticipated that firms would distribute themselves more evenly across a nation's urban centres as the costs of operating in large and congested cities came to outweigh the benefits (Williamson 1965).

These early models of economic development established a set of assumptions about how the process should unfold – and how it should affect urban geography – which continue to be influential today in both research and policy circles. However, in reality the relationships between economic development, urbanisation and urban concentration have proven somewhat more complicated in developing countries throughout the post-World War II era.

The theory that urbanisation is driven by industrialisation is only half correct. There is abundant evidence that growth in manufacturing and industrial sectors encourages people to move into cities (Fay and Opal 2000; Henderson 2003) and that richer countries are generally more urbanised than poorer ones, as illustrated in Figure 3.1. Yet as early as the 1950s concerns were raised about 'over-urbanisation' in some developing countries where urban population growth seemed to be outpacing industrial development (Sovani 1964). This seeming mismatch in the pace of economic and demographic change became even more apparent in the 1980s and 1990s when many countries (particularly in sub-Saharan Africa) experienced rapid urban growth and persistent urbanisation despite economic crises, stagnation and in some cases decline. This phenomenon has been dubbed 'urbanisation without growth' (Fay and Opal 2000) and challenges the assumption that economic development is the primary stimulus for urbanisation. Indeed, a closer inspection of Figure 3.1 reveals that countries at similar levels of income can have very different levels of urbanisation.

Figure 3.1 *Correlation between national GDP per capita and level of urbanisation*

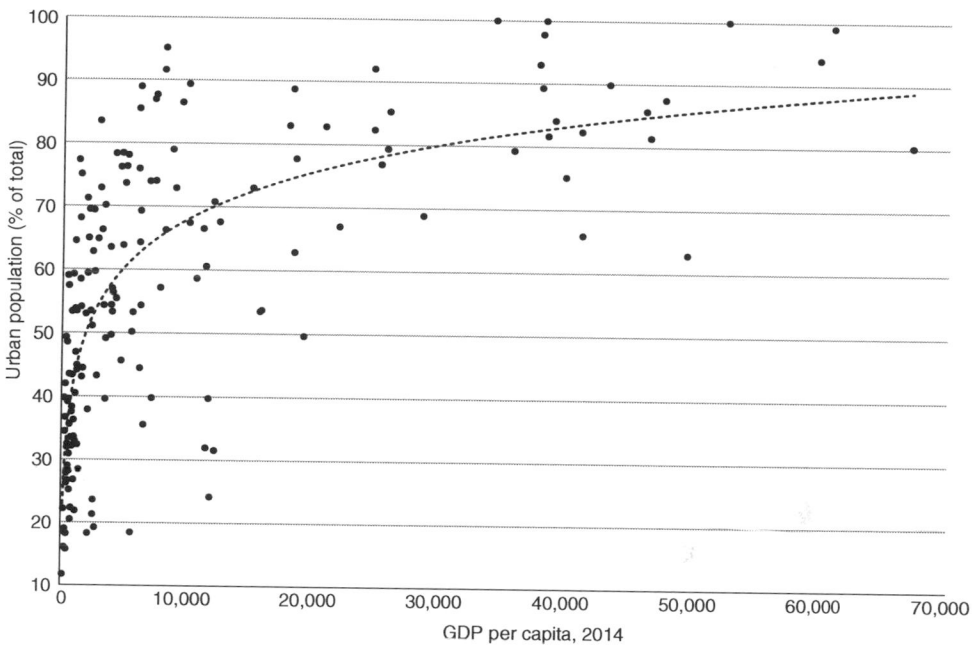

Source: World Bank (2015).

In practice, people move to cities for a wide variety of economic and non-economic reasons, and high rates of population growth in developing countries in the latter half of the twentieth century meant there were a lot more potential urban migrants. Research in the 1960s and 1970s by sociologists and anthropologists found many other 'pull' factors beyond potential job opportunities that influence peoples' decision to move to a town or city, such as a desire to take advantage of the 'thick' market for spouses in urban areas, acquire the social prestige associated with urban life or simply pursue adventure in the bright lights of the big city (see Byerlee 1974, Mazumdar 1987, Jamal and Weeks 1988, Becker and Morrison 1995). Conversely there are many potential 'push' factors that may drive people from the villages where they were born, such as the desire of youth to escape the control of elders, of women to escape gender discrimination, and of families to escape the ravages of armed conflict or ecological degradation due to population pressure or climate change (Fay and Opal 2000, Barrios, Bertinelli and Strobl 2006, Fox 2012). Given these myriad potential push and pull factors, it is not surprising that towns and cities continue to grow even in the absence of industrial development in urban areas, especially in a context of rapid population growth. Once these dynamics are recognised, the concepts of 'over-urbanisation' and 'urbanisation without growth' are revealed as subjective judgements predicated on false assumptions or overly simplistic models of the nature of urban transitions.

As with over-urbanisation, concerns about 'excessive' urban concentration and urban primacy emerged in the 1960s and have had an important impact on development policy for decades. Despite theoretical models suggesting that urban concentration could benefit countries at an early stage of industrialisation, the unprecedented pace of urban population growth in development countries, and subsequent 'overcrowding', called into question the economic benefits of large cities. Growing pessimism was compounded by concerns about spatial equity, leading many planners and policymakers to conclude that settlement patterns should 'be adjusted so as ultimately to provide a regional structure that meets the demand for equality of status for people in different parts of the country' (Hägerstrand 1978: 556). In extreme cases governments built entirely new cities in territorially central locations to encourage more regionally balanced economic growth. Examples include countries as diverse as Brazil, Tanzania and Malawi.

Empirically speaking, less economically advanced countries in Africa, Asia and Latin America have generally exhibited higher degrees of urban concentration and urban primacy than wealthy countries in Europe, East Asia and North America (Linsky 1965, Sassen 2012). Evidence from relatively recent research also indicates that low-income countries do indeed benefit economically from urban concentration and primacy, but that more economically advanced economies do not (Henderson 2003, Brülhart and Sbergami 2009). However, this is another case where the prevailing models and assumptions have proven incomplete and contributed to misinformed policy. Empirical studies have also demonstrated that natural geography, economic history and political institutions play a significant role in shaping human settlement patterns across countries and over time (Linsky 1965, Smith 1995, Davis and Henderson 2003; Ades and Glaeser 1995). It therefore makes little sense to compare urban systems in countries as diverse as China, Bangladesh, Kenya and Mexico. One only need compare Britain and Germany to see that countries with very different urban geographies can thrive: London is a paradigmatic primate city while population and economic activity are spread far more evenly across the regions of Germany.

The implicit suggestion embedded in phrases such as 'over-urbanisation' and 'excessive urban concentration' is that there is a 'normal' path to prosperity that all countries should follow, one that roughly conforms to a very stylised version of the historical experiences of today's wealthier countries. The result has often been the adoption of policies designed to 'correct' apparent deviations from the normal or ideal path (e.g. by trying to limit rural–urban migration or restructure settlement systems) with little discernible positive impact. The facts simply show that urban transitions have unfolded differentially across countries due to variations in their individual demographic, technological, economic and political histories.

Cities as engines of growth

So far the discussion about economic growth and development has been framed as a national challenge. Yet it is not clear that this is the appropriate way to approach the conceptualisation and investigation of economic life. Figure 3.2 presents a classic illustration of cross-country variation in levels of economic development, as measured by per capita income in 2012. Figure 3.3 provides a very different perspective by focusing on a single country: India. The first panel on the left shows that there is substantial variation in per capita income levels across states within India, while the right hand provides an even more geographically nuanced view by presenting estimates of GDP per kilometre squared (instead of per person). This reveals the dramatically uneven distribution of economic activity *within* Indian states, with high concentrations in major cities and towns that taper off into the surrounding countryside.

It is precisely this kind of dramatic sub-national variation in wealth, with cities serving as focal points of economic activity, which inspired Jane Jacobs to question our analytic attachment to nation-states as the most important unit of economic analysis in development economics (1984). She argued vigorously that cities, not nations, are the salient unit of analysis for understanding economic processes:

> Nations are political and military entities, and so are blocs of nations. But it doesn't necessarily follow from this that they are also the basic, salient entities of economic life or that they are particularly useful for probing the mysteries of economic structure, the reasons for rise and decline of wealth.

(ibid.: 31)

Figure 3.2 *GDP per capita by country in 2012*

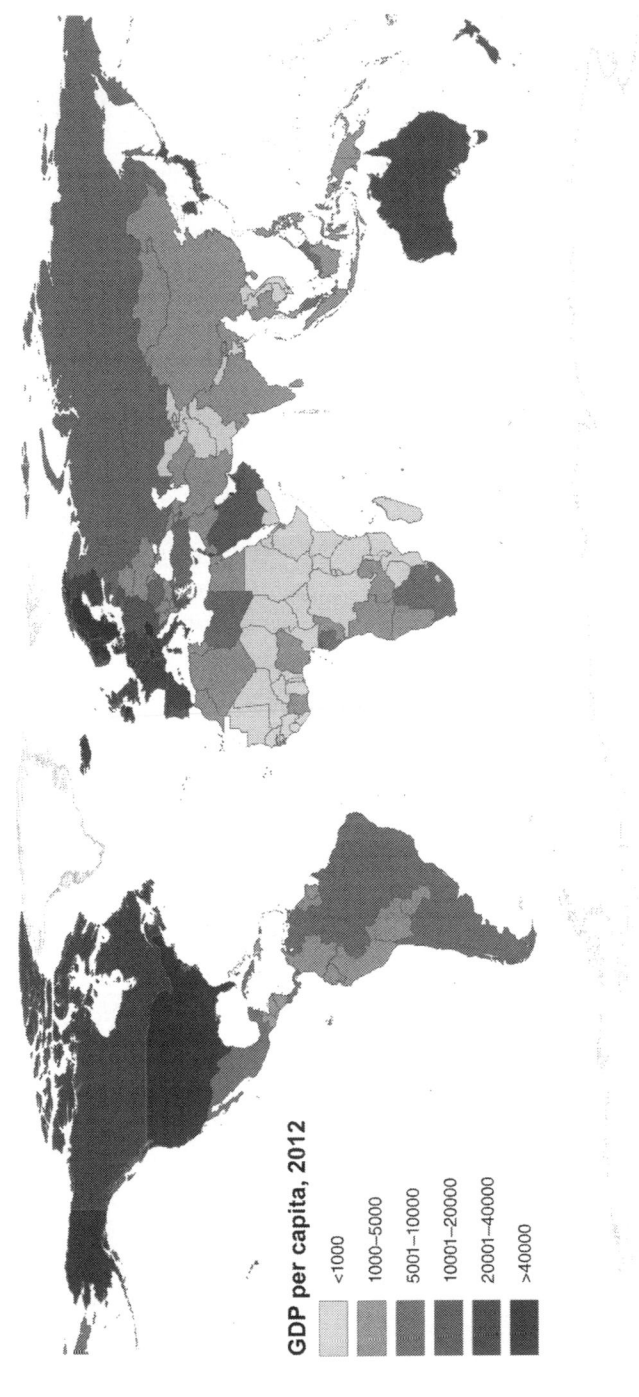

GDP per capita, 2012

	<1000
	1000–5000
	5001–10000
	10001–20000
	20001–40000
	>40000

Note: GDP per capita data are in constant 2005 US$ and taken from the World Bank World Development Indicators online database, accessed July 2015.

Figure 3.3 *The distribution of income in India by state and km² in 2012*

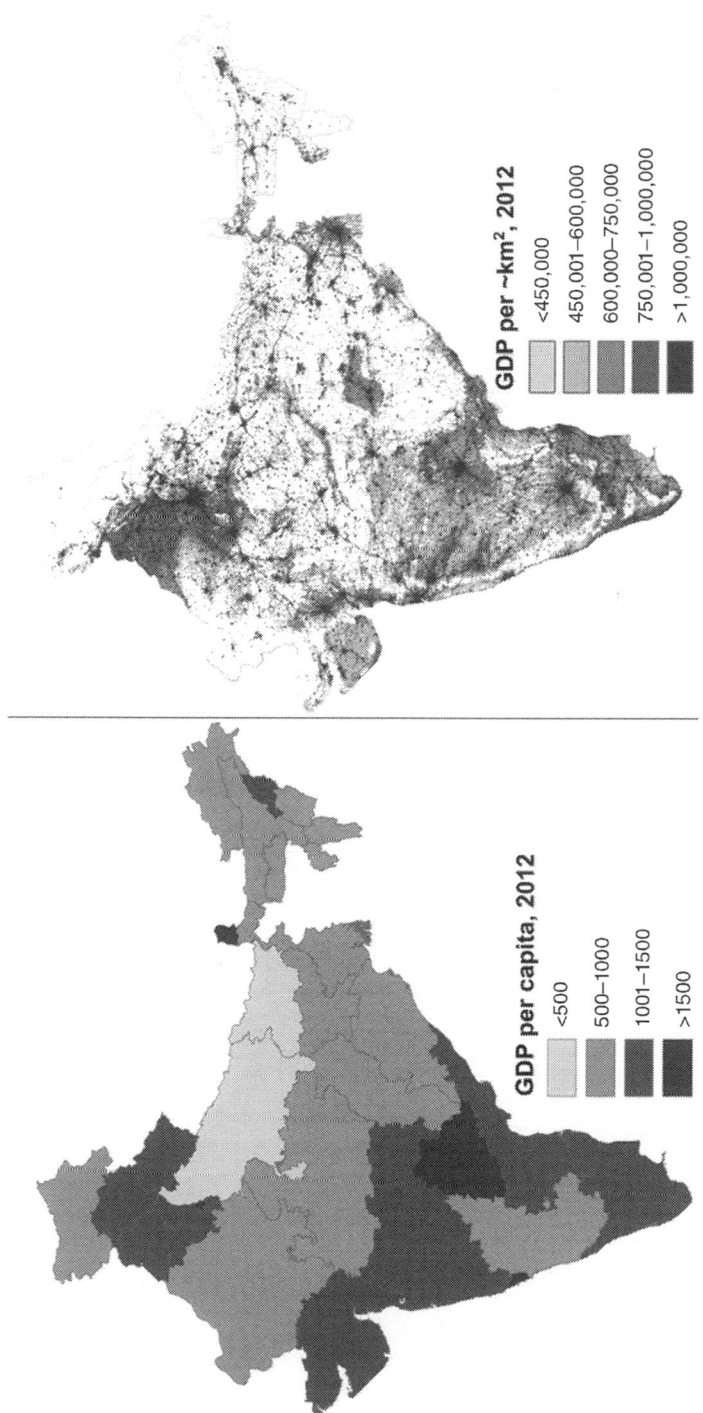

GDP per capita, 2012

<500
500–1000
1001–1500
>1500

GDP per ~km², 2012

<450,000
450,001–600,000
600,000–750,000
750,001–1,000,000
>1,000,000

Notes: State level estimates of GDP per capita are from the Planning Commission, Government of India. Estimates of GDP per square kilometre were calculated from luminosity measurements derived from satellite data collected by the National Oceanic and Atmospheric Administration's National Geophysical Data Centre (NOAA NGDC). See Henderson et al. (2012); Sutton et al. (2007) and Sutton and Costanza (2002) for more about this technique.

Jacobs went on to note that 'most nations are composed of collections or grab bags of very different economies, rich regions and poor ones within the same nation' (ibid.: 32). In order to understand *inter*-national and *intra*-national disparities of economic development, Jacobs suggests that we examine the economies of cities, which 'are unique in their abilities to shape and reshape the economies of other settlements, including those far removed from them geographically' (ibid.: 32). Indeed, cities everywhere generate a disproportionate share of economic output relative to their populations, as Table 3.2 illustrates. For example Nairobi, Kenya was home to about 8 per cent of the country's population in 2007 but generated nearly 20 per cent of GDP; similarly Mumbai hosts less than 2 per cent of India's population but accounts for over 6 per cent of national GDP.

The study of urban economic life centres around the effects of agglomeration: the clustering together of people and firms in space. Agglomeration facilitates the circulation of goods, people, knowledge and ideas. By facilitating these flows, agglomeration generates what economists refer to as *external economies of scale*, which enhance efficiency and stimulate innovation. External economies of scale refer to improvements in the productivity of a firm due to factors external to that firm. An individual firm can improve its own productivity through investment, perhaps by buying new technology to improve the efficiency of a production process. But productivity gains can also

Table 3.2 *Population and GDP output for a selection of cities in 2007–2008*

City	Country	City share of national population in 2007 (%)	City share of national GDP in 2008 (%)
Buenos Aires	Argentina	32.4	63.3
Casablanca	Morocco	10.2	24.1
Istanbul	Turkey	13.4	17.7
Lagos	Nigeria	6.4	11.1
Bangkok	Thailand	10.5	22.9
Mumbai	India	1.6	6.2
Nairobi	Kenya	8.0	19.7
Santiago	Chile	34.4	49.6
São Paulo	Brazil	9.8	19.6

Source: Sassen (2012).

Box 3.2 Key definitions in urban economics

Economies of scale refer to the productivity gains realised when making more of a good (i.e. producing it on a larger scale) results in lower unit costs.

Externalities are external benefits or costs that accrue to others as a result of decisions made by an individual or firm. In other words, the individual doesn't capture the full benefit or incur the full cost of that decision. These can be positive (e.g. localisation economies) or negative (e.g. air pollution emitted from private automobile use).

External economies of scale refer to the productivity gains realised by a firm due to factors external to a particular producer or firm. The clustering of economic activity in space generates external economies of scale.

Localisation economies refer to the productivity gains that a firm in a particular sector receives by locating in a place with other, similar firms. By clustering together in space these firms can share suppliers, support a large, specialised labour force and benefit from knowledge spillovers and the rapid diffusion of new innovations in their industry.

Urbanisation economies refer to the productivity gains that a firm receives by locating in a large and economically diverse agglomeration. Scale facilitates specialisation while diversity can stimulate innovation through the cross-fertilisation of ideas, thereby providing a platform for sustained growth. Also known as *Jacobs economies* in recognition of Jane Jacobs' valuable insights on this topic.

---

be a result of factors that an individual firm has no immediate control over. Individual firms do not control the size, density and diversity of a city, and yet these factors improve individual firm performance by generating external economies of scale. Economists generally distinguish between two kinds of external economies that are a direct result of urbanism: *localisation economies* and *urbanisation economies*. These, in turn, facilitate sharing, matching and learning (Duranton and Puga 2004, Puga 2010).

Localisation economies arise from the spatial clustering of firms engaged in the production of a similar good or service. The concept of localisation economies grew out of economist Alfred Marshall's observation that a) proximity allows firms to share inputs; b) a large pool of labour improves matching between the needs of firms and the skills of workers; and c) in densely populated areas people share information and ideas, facilitating learning, innovation and diffusion (1920: 267–277). All of these factors – a direct consequence of the close physical proximity of economic agents – can improve the performance of a firm or industry.

The most straightforward advantage of agglomeration for firms is the sharing of basic infrastructure. Ready access to reliable water supplies, electricity, roads and telecommunications infrastructure significantly enhances productivity for firms in virtually every sector in an economy. This kind of infrastructure is expensive to build and maintain, and is also 'lumpy' in the sense that it is more efficiently provided in areas of high population density (Arnott and Gersovitz 1986). For example, a single kilometre of road or water piping will naturally be able to serve more people in a densely populated area than a sparsely populated one.

The sharing benefits of agglomeration extend far beyond physical infrastructure. Consider a laptop computer manufacturer, which requires hundreds of different kinds of parts to make a single unit, such as microprocessors, circuit boards, cooling fans, displays, keyboards, etc. While there are many companies that sell laptops, none produces and assembles all of the component parts. The complex nature of the good requires specialisation in the supply chain to produce laptops *en masse* at affordable prices. As a result, different computer companies share many of the same suppliers. While it is theoretically possible for computer companies and their suppliers to be spread across many towns and cities, particularly in today's global economy, there are in fact just a handful of cities that dominate the global computer manufacturing industry, such as Silicon Valley in the USA, Taipei in Taiwan, Beijing in China and Tokyo, Japan. By clustering in a few places, big computer firms are able to share suppliers, which in turn have an incentive to locate near as many big firms as possible to maximise their potential customer base.

The colocation of multiple firms in a single industry also improves matching between firms and workers. If a microchip manufacturer were to locate in a small town or a sparsely populated area they would have a very limited pool of labour to draw upon and might struggle to find people with the specific expertise needed in the production process. But by locating in a densely populated urban area, a firm can draw from a larger talent pool and therefore has a greater chance of finding someone with the right kind of skills. Equally, skilled workers looking for a job have an incentive to migrate to places where there are several potential employers, while local job-seekers have an incentive to gain skills needed to work in the firms nearby. A cluster of firms in a particular industry can therefore encourage the emergence of a critical mass of skilled labour and improve matching between workers and firms.

The third key benefit of agglomeration is the way in which proximity facilitates learning. As Alfred Marshall observed, in cities

> The mysteries of the trade become no mysteries; but are as it were in the air, and children learn many of them unconsciously. Good work is rightly appreciated, inventions and improvements in machinery, in processes and the general organization of the business have their merits promptly discussed: if one man starts a new idea it is taken up by others and combined with suggestions of their own; and thus it becomes the source of further new ideas.
>
> (1920: 271)

In other words, physical proximity lubricates the flow of information and ideas, thereby enhancing learning that can have industry-wide effects on productivity. When a firm discovers a more efficient way of doing something, or acquires information that might improve planning, this information will spread relatively quickly in a place where workers and managers are moving between firms, socialising in the evening, and developing personal relationships.

The benefits of sharing, matching and learning go a long way toward explaining why producers in a single industry choose to cluster together in space, but there are further benefits (i.e. external economies) associated with locating in large settlements with a diversity of firms and workers that are not industry specific. These are known as urbanisation economies or 'Jacobs economies', in recognition of Jane Jacobs' insights in the field of urban economics.

Large cities can encourage deeper specialisation due to simple scale effects. For example, a small town may have a doctor, but its small population won't justify the presence of specialists such as oncologists and neurologists. There would simply be too little business to keep such specialists employed. But in a big city with a lot of people there is a much greater possibility that such specialists would be gainfully employed. The size of a local market can therefore have a direct effect on the degree of local specialisation, and hence overall productivity in an economy. Indeed, physical proximity to potential consumers is itself a strong incentive for firms to locate in large urban centres, regardless of their particular industry.

Jacobs was also a great believer in the way in which urban diversity serves to stimulate innovation. She argued that 'economic life develops by grace of innovating' (1984: 39) and that 'development is a process of continually improvising in a context that makes injecting improvisations into every day life feasible. Cities...create that context. Nothing else does' (ibid.: 154). In particular, Jacobs (like Marshall) argued that cities encourage the re-combination of old ideas into new ones, thereby stimulating innovation and economic growth.

There is robust empirical evidence that localisation and urbanisation economies exist and have a substantial effect on productivity (Puga 2010). Bigger cities are more productive than smaller ones, and diversity is good for growth at both the industry and city level (see Glaeser et al. 1992; Henderson et al. 1995; Rosenthal and Strange 2004; Duranton 2014). Duranton and Puga (2001: 1455) have further demonstrated theoretically and empirically that large 'diversified cities act as a "nursery" for firms' by providing an ideal environment for the development of new products, while smaller, less diversified cities provide benefits to established firms who seek to avoid the congestion costs associated with large centres.

Globalisation has, of course, significantly reduced the costs of transporting goods, people, information, knowledge and money. Mobile phones, email and the internet have dramatically reduced the costs of and time involved in communication, allowing customers to order goods direct from manufacturers, and making it possible for workers to work from home. Does this portend the death of the city? According to Storper and Venables (2004):

> Face-to-face contact remains central to coordination of the economy, despite the remarkable reductions in transport costs and the astonishing rise in the complexity and variety of information – verbal, visual, and symbolic – which can be communicated near instantly.
>
> (ibid.: 352)

They go on to explain how face-to-face contact is:

> a highly efficient technology of communication; a means of overcoming coordination and incentive problems in uncertain environments; a key element of the socialisation that in turn allows people to be candidates for membership of 'in-groups' and to stay in such groups; and a direct source of psychological motivation.
>
> (ibid.: 365)

If face-to-face contact remains essential to the economies of highly urbanised and industrialised nations (as Storper and Venables suggest), it is even more important in less urbanised and less advanced economies in low- and middle-income countries where communication and transportation costs remain high due to limited infrastructure, and insecure institutional environments make long distance transactions more risky.

Face-to-face contact also plays an important role in building relationships between economic actors and organisations within cities and regions. These relationships can play an important part in realising the potential benefits of agglomeration and play a critical role in supporting *clusters*. Porter (2000: 15) defines clusters as geographic concentrations of interconnected companies, specialised suppliers, service providers, firms in related industries, and associated institutions (e.g. universities, standards agencies, trade associations) in a particular field, which compete but also cooperate.

Porter argues that economic policy should focus on clusters as opposed to sectors, industries or companies because 'a good deal of competitive advantage lies *outside* companies and even outside their industries, residing instead in the locations at which there business units are based' (ibid.: 16). He observes that 'connections across firms and industries are fundamental to competition, to productivity and (especially) to the direction and pace of new business formation', and that governments (at local and national levels) have an important role to play in cultivating the links and spillovers that cut across firms and industries (ibid.: 18). In particular, he advocates investment in public goods that benefit all firms in a cluster (e.g. infrastructure and education), maintenance of a business-friendly institutional environment and efforts to mobilise and provide support to firms within an identified cluster. Bangalore, India, provides an excellent example of cluster emergence in the IT sector. Effective cooperation and coordination between urban planners, local government authorities, national government policy, private firms and educational institutes produced a dense web of mutually re-enforcing relationships and incentives that supported the industry's growth (see Box 3.3).

There are, however, many critics of the cluster approach to urban economic policy, which has become popular in developing and developed countries alike. Nathan and Overman (2013) point out that cluster theory essentially emphasises all the good things that agglomeration brings without paying suitable attention to the downsides, and that there is in fact little evidence of success with cluster policies. Efforts to cultivate clusters without addressing the fundamental constraints to growth in urban areas are unlikely to prove effective in the long run (ibid.).

Box 3.3 The birth of an IT cluster in Bangalore

Bangalore provides a particularly successful example of cluster development in the IT sector. The city, which is the capital of Karnataka state, was the site of many large-scale investments in manufacturing and industry in the early twentieth century. The infrastructure developed to support these industries, and the presence of skilled engineers in the region, made Bangalore an appealing location for early investments in the growing IT sector beginning in the 1970s. The development of a globally competitive cluster was encouraged by a variety of national- and local-level policies and initiatives, including financial incentives for firms, targeted education and training to cultivate an appropriately skilled labour force, provision of land and basic infrastructure (including buildings) for potential investors, investments in basic research and development, and promotion of linkages between firms in the sector. Karnataka state was the first in India to develop a comprehensive IT policy, worked cooperatively with city planners and the private sector to develop several industrial parks equipped with the kind of high-tech infrastructure demanded by IT firms, and actively markets Bangalore abroad as an attractive site for foreign investors. Strong links have developed between firms and various educational institutions in the region, who provide appropriate skills training as well as specialized research, some of which has been funded by the local branches of transnational IT firms such as Texas Instruments. As a result of these proactive policies, effective coordination and cooperative linkages between firms and educational institutes, Bangalore has become a globally competitive city in the IT sector and a major contributor to the national economy of India. It has, in other words, become a paradigmatic cluster.

Can Bangalore's success be replicated elsewhere through cluster policy? The confluence of historical, economic and political forces that gave rise to Banglore's IT cluster cannot be copied and pasted to another city with a different history. Despite widespread interest and adoption of cluster policies, and several well-known success stories, a broad survey of the research on hundreds of such initiatives shows little evidence of positive and sustained impact on local economic development. Clusters are clearly good for a city economy, but they appear to be very difficult to cultivate directly through public policy.

Sources: Basant and Chandra (2007), van Dijk (2003), Patibandla and Peterson (2002), Nathan and Overman (2013).

Economic clustering also needs to be considered in the context of the physical and human agglomeration that accompanies it, which can have negative as well as positive consequences for development. Agglomeration can produce myriad negative externalities that undermine productivity and welfare, including congestion, pollution, disease and crime. By inhibiting the flow of people and goods, traffic congestion undermines the immediate benefits of physical proximity and contributes to atmospheric pollution in the form of particulate matter that can have negative effects on the health of local residents.

In densely populated cities with poor water and sanitation infrastructure, infectious and parasitic diseases can thrive, further undermining the health (and hence productivity) of workers. Crime rates are also generally higher in urban than in rural areas, and higher in larger cities than smaller ones. A higher density of potential victims and potential customers of stolen or illicit goods, and a lower probability of arrest (due to a higher number of potential suspects), attract the criminally minded (Glaeser and Sacerdote 1999).

We will address all of these issues in detail in subsequent chapters. The important point, for now, is that cities can be engines of growth, but this potential cannot be taken for granted. Overall, the benefits of agglomeration appear to outweigh the costs for most people – otherwise cities wouldn't exist at all. But getting the most from cities requires active government efforts to maximise the benefits of agglomeration and minimise the costs.

Rural–urban linkages and regional development

The significance of urban economies extends far beyond city boundaries. The tendency to conceptualise development challenges as either rural or urban (e.g. through the lens of the urban bias thesis discussed in Chapter 1) is highly problematic, especially when we consider the flows of goods, people and resources between rural and urban areas (Tacoli 1998; Lynch 2005). Human settlements are better understood as existing on a continuum, from isolated households in far-flung rural areas all the way up to mega-cities, which collectively form an interconnected and interdependent network.

Cities ultimately depend upon surplus food production in rural areas, but cities are not simply parasites extracting surplus from rural agriculturalists. Historically, urban settlements have helped boost agricultural productivity through the production and distribution of farm equipment and fertilisers (Bairoch 1988), as well as through their role as regional 'service centres' providing access to goods and services (such as banking, medical, repair and information services) that support rural enterprises. Networks of towns and cities 'form an essential marketing network through which agricultural commodities are collected, exchanged, redistributed' (Rondinelli 1994: 373). Producers sell their goods in small market towns where distributors collect and bundle them into larger units, which are sold in larger towns and cities. There they may be consumed, processed or re-bundled into yet larger units for sale on international markets. In other words, networks of towns and cities 'are essential to the whole chain of exchange on which commercial agriculture depends' (ibid.).

Rural and urban dwellers also rely on each other as markets. The sale of agricultural surplus to urban populations provides extra income to farmers who would otherwise rely on subsistence production. This income may, in turn, be used to purchase inputs, but also consumer goods (such as clothing, furniture, cookware, radios, televisions, etc.) that are produced in urban areas and travel through the very same urban networks to reach rural consumers. The terms of trade between rural and urban areas have been distorted in the past through government intervention, as argued by Lipton and Bates (see Chapter 1) but this does not imply that the economic relations between urban and rural areas *must* necessarily be of an exploitative nature.

There is also a tendency to use the terms 'rural' and 'agricultural' synonymously. In fact, many rural households have diverse livelihood strategies that include producing, trading and marketing a range of non-agricultural goods. Similarly, urban residents – particularly those living in peri-urban areas with access to larger plots of land – often engage in urban agriculture as a component of their livelihood strategies. Moreover, households may straddle the rural–urban divide, taking advantage of livelihood and entrepreneurial opportunities available in each. In some cases, this may entail the permanent migration of a rural family member to a city, who then sends home remittances from his/her waged employment. In other cases, it may involve seasonal migration, with individuals moving between urban and rural areas at different times of the year to take advantage of waged employment while still participating in agricultural production during labour-intensive seasons, such as harvest time (Tacoli 1998; Potts 2010). In yet other cases members of families or larger groupings engage in exchange and trading activities between rural and urban areas.

The potential for dynamic interactions between rural and urban areas to stimulate local/regional development was recognised by development scholars and policymakers in the 1950s to the 1970s, when urban and regional planning was seen as a way to encourage the integration of domestic economies and facilitate national development more broadly (Friedmann 1967). Investment in 'propulsive industries' and infrastructure in select urban centres was expected not only to benefit urban populations, but to drive regional development more generally by serving as 'growth poles' for wider city-regions (Parr 1999a). Regional planners thought that encouraging decentralised industrialisation in peripheral regions could reduce spatial inequalities. However, growth pole strategies, popular in the 1950s and 1960s, were replaced by the rural development paradigm in the 1970s and 1980s.

Dissatisfaction with industrial decentralisation and growth pole strategies stemmed from inconsistent results in practice. In theory, towns and cities can be regional drivers of change by facilitating agricultural development while providing alternative income-generating opportunities for rural dwellers. However, in practice, growth poles were often set up in depressed areas with low levels of infrastructure and a poor fiscal base. At worst, small towns and cities have been deemed 'vanguards of exploitation' (Southall 1988), serving as gateways for the penetration of opportunistic global capital into the countryside. More moderate perspectives suggest that the relative success of such initiatives in cultivating regional development largely depends on the local political and economic dynamics of the regions affected. Land ownership structures, national government policies, global economic trends and local social relations may all confound well-intentioned efforts to stimulate local or regional development (Tacoli 1998; Hinderink and Titus 2002; Faguet 2004, 2003). Parr (1999b) attributes the general failure of industrial decentralisation to the uncritical way in which the idea of growth poles was adopted and put into practice. Often planning horizons were too short or plans were overambitious, poorly designed and demanded unrealistic capital outlays. Another handicap was that decisions on the location of growth poles were made on political rather than economic grounds and in many cases the administrative capacity required to see a growth pole strategy through was simply lacking (ibid.).

Despite the mixed record of industrial decentralisation and growth pole strategies, there have recently been calls for a revaluation of urban and regional development strategies. Satterthwaite (2006a: 35) has suggested that we need to 'forget the rural–urban divide and see all settlements as being on a continuum'. Friedmann (2007) argues that we must recognise the 'organic relation' between cities and countryside and makes a plea for policies and interventions that promote the endogenous development of city regions through the cultivation of their unique assets. Scott and Storper (2003: 580) claim that reigning paradigms in development studies have generally overlooked the 'geographical foundations of economic growth' and undervalued the role of cities and regions as 'springboards of the development process'. There is, however, surprisingly little empirical evidence of the effects of urban economic development on territorial (including rural) development and poverty reduction more broadly. Two recent studies indicate that there are diverse impacts depending on the size and distribution of towns and cities across a territory. Both, however, highlight the need to better understand and support the role of 'the missing middle' – small and medium-sized settlements – in fostering inclusive regional economic development and poverty reduction (Berdegué et al. 2015; Christiansen and Todo 2013).

Cities and regions in a global economy

Globalisation – or the increasing social, political and economic integration of the world's population – arguably began thousands of years ago when inter-regional trade between settlements in Asia, Europe and the Middle East began. But revolutions in transportation and communication technologies over the past two hundred years, as well as the emergence of new international institutions governing trade and finance in the post-World War II era, have led to a dramatic acceleration of the process of global integration. Today, financial markets allow capital to range freely with little regard for national boundaries; advancements in transportation and communications technologies permit an ever-expanding range of goods and services to be traded in unprecedented volumes; transnational and multinational corporations (TNCs and MNCs) have restructured production processes, spreading operations across the globe to exploit the comparative advantage of different countries and regions; and workers – from the unskilled to the highly skilled – are increasingly mobile. Urban scholars have taken a particularly keen interest in the processes and effects of globalisation, exploring both the ways in which cities serve as agents of globalisation, as well as the impact that global integration has on individual cities.

As we observed in Chapter 2, cities have functioned as nodes in inter-regional and international trade networks since their first appearance some 6000 years ago. Indeed, as renowned social and economic historian Fernand Braudel observed, 'A world-economy always has an urban center of gravity, a city, as the logistic heart of its activity' (1984: 27). This idea was further developed by a range of scholars working within the 'world cities' paradigm (see Hall 1966; Friedmann and Wolff 1982; Friedmann 1986; Knox and Taylor 1995; Taylor 2004a). Eschewing the idea that contemporary globalisation is simply a variation on a much longer playing theme of capitalist development, these scholars have argued that the current era of globalisation represents a 'historically unprecedented' moment in which capitalist institutions have managed to 'free themselves from national constraints' leading to a radical reorganisation of the international space economy (Friedmann and Wolff 1982: 310). In this new global landscape, cities are believed to serve as 'basing points' for the accumulation and deployment of international capital in the hands of TNCs and MNCs, while simultaneously experiencing social and spatial transformations linked to their specific role in the new international division of labour (Friedmann 1986).

This body of research has largely been focused on the ways in which certain major urban centres (primarily in advanced economies) have come to serve as command and control centres in the world economy, concentrating the corporate headquarters of global firms in the financial, transport, business service and communications sectors. Much of this research has been preoccupied with developing league tables and mapping the international hierarchy of world cities based on the concentration of global service firms in cities, the interconnectedness (in terms of transport and communications) of cities, and of course the concentration of financial capital in cities. In these formulations, cities are often ranked according to their apparent role as primary/secondary, core/peripheral or alpha/beta/gamma world cities.

At the top of the world city hierarchy are 'global cities' – a term coined by sociologist Saskia Sassen in her study of New York, London and Tokyo (Sassen 1991). According to Sassen (2012: 8), there are about 40 global cities in the world today, which serve as

> (1) command points in the organization of the world economy; (2) key locations and marketplaces for the leading industries of the current period – finance and specialized service firms; (3) major sites of production, including the production of innovations, for these industries.

For Sassen, the phenomenon of global cities is not just about where major international service sector firms cluster, but also represents a re-scaling of strategic territory and relative weakening of national states. Many of these cities are more intimately connected to the global economy than their regional or national economies (Sassen 2002). The high concentration of wealth and power in these world/ global cities has been accompanied by stark socioeconomic polarisation within them. Indeed, Friedmann and Wolff (1982: 322) claimed that 'the primary fact about world city formation is the polarization of its social class divisions'. Dense clusters of firms in the financial, business service and IT sectors have emerged while manufacturing industries have declined in many of these cities, leading to a decline in demand for semi-skilled labour. This shift has been accompanied by a rising demand for low-wage jobs to service the new global elite and their 'expensive restaurants, luxury housing, luxury hotels, gourmet shops, boutiques, French hand laundries' (Sassen 1991: 9). The result has been a dramatic stratification of incomes in these 'strategic nodes' in the global economy. Social polarisation in cities has always existed, but the degree of stratification that has emerged in the past three decades is historically unprecedented 'engendering massive distortions in the operations of various markets, from housing to labour' (Sassen 2012: 11).

The pervasive fascination with world/global cities has inspired the development of multiple indices that rank individual cities according to their perceived power and connectivity, such as the Global Cities Index, the Global Economic Power Index and the Global City Competitiveness Index. Urban politicians across the world have adopted the global city discourse and actively seek to attract the kind of investment to make their way into these league tables, despite concerns about the possible polarisation effect. Far less attention has been paid to the 'large number of cities around the world which do not register on intellectual maps that chart the rise and fall of global and world cities' (Robinson 2002: 531). Cities in less economically advanced nations have generally been regarded as peripheral to the governance and functioning of the global economy, yet they have also undergone significant transformation in the contemporary global era and do play a role in its functioning.

As Jennifer Robinson has argued, it is one thing to acknowledge that globalisation is heightening the power of certain cities and the links between them, but 'quite another to suggest that poor cities and countries are irrelevant to the global economy' (2006: 101). She points out that a city like Lusaka in Zambia, for example, plays a very significant role in the regional and even global economy, albeit one that sits below the radar of those concerned with the clustering of major multinational firms and the idea of global 'command and

Plate 3.1 *Efforts to promote foreign investment in Kigali, Rwanda*

control'. As the capital of a major producer of raw materials Lusaka facilitates the distribution of copper and agricultural goods around the world, as well as being a site for extensive trade in second-hand clothing and other goods. Such cities are also connected globally through the large amounts of international aid that flows there and the presence of a high concentration of international agencies such as the UN, World Bank and IMF. Moreover, there are significant 'subterranean' economic networks that penetrate many cities in the developing world, often involving international criminal organisations as well as illicit negotiations between transnational firms and those who produce and export goods illegally. Viewed through this lens, there are many other important international economic flows that fall 'off the map' of global city discourse (Robinson 2002).

In addition to this focus on what Robinson terms 'ordinary cities' there have recently been some attempts to balance the world cities/global cities paradigm through an examination of globalising cities in low- and middle-income countries, such as Mexico City, São Paulo, Cairo, Johannesburg, Mumbai, Delhi, Shanghai and Jakarta (see Segbers 2007; Gugler 2004; Dupont 2011; Bunnell and Miller 2011). These 'second-tier' world cities are playing increasingly important roles in the global economy as centres of production, consumption, regional financial networks and specialised service provision, and some have even risen to the top of the global city league tables in recent years. For example, Shanghai has become a critical strategic gateway into the rapidly expanding Chinese economy, serving as a regional financial and service centre. At the same time, it is exhibiting many of the symptoms of socioeconomic polarisation observed in the world/global cities of more economically advanced nations (see Box 3.4).

Shanghai, however, is not merely an isolated pretender to world city status, but rather is part of an emerging 'global city-region' that encompasses many settlements in China and increasingly in neighbouring countries. The concept of a 'global city-region' can be thought of as a corollary to the world or global city concept, but on a broader geographical scale. Global city-regions are 'dense nodes of human labor and communal life' (Scott 2001: 1) that manifest as very large metropolitan centres, as well as regional networks of cities (Scott et al 2001: 11). Generally speaking, research on global city-regions is somewhat more inclusive than that related to world/global cities. Using a standard threshold in this literature – i.e. a population of one million inhabitants or more as a rough approximation of

Box 3.4 Shanghai: a resurgent world city

In 1843, China was forced to open itself up to trade with the West with the Treaty of Nanjing. Shanghai, a small coastal fishing village with a particularly appealing natural port, was dramatically transformed over the following years, eventually coming to be known as the 'Paris of the East.' As an emerging centre of industry and international trade, Shanghai's population mushroomed: by 1900 the city population had reached one million people, and by 1936 it was ranked the seventh largest city in the world with a population over three million. Over this period Shanghai produced half of China's industrial output, accounted for half of the nation's trade and become the nation's leading financial centre, host to some 90 per cent of China's banks.

In 1949 China took a dramatic turn and closed itself to trade with the West. Shanghai was again transformed. From an international centre of production and trade, the city's economic prowess was directed towards serving the interests and needs of a highly centralised socialist state. As one of China's most productive regions, over 85 per cent of Shanghai's revenue was appropriated by the central state over the subsequent three decades, contributing approximately one-sixth of the central state's revenue over that period.

Beginning in the late 1970s, China began slowly reopening its economy to the world, and again, Shanghai was transformed. Having established a strong industrial base, the city was well placed to take advantage of new opportunities in a globalising economy, becoming a major centre for export manufacturing. In 1990, the central state announced plans to develop a new economic hub in the Shanghai city region, accompanied by a series of reforms handing greater to power to Shanghai's local government. Over the subsequent decade, Shanghai began its re-emergence as a world city. Changes in national laws concerning foreign investment, massive investments in local infrastructure, a wide range of locally promulgated policies to provide incentives for investment, and reforms in land management practices have ushered in a flood of foreign investments. More than half of the world's top 500 transnational corporations and 57 of the largest industrial enterprises in the world have set up shop in Shanghai, contributing to an annual regional growth rate of over 20 per cent – more than twice the national average. In 2007, Shanghai was once again ranked the seventh largest city in the world with a population over 15 million, and by 2015 the population of greater Shanghai had reached nearly 25 million, making it the third largest metropolitan region in the world.

The rapid transformation of Shanghai's economy – and skyline – over the course of the 1990s was accompanied by a marked shift in the socioeconomic profile of the city. Rising prosperity stimulated migration to the city, attracting over two million new migrant workers over the course of the decade – a trend which continues to this day. At the same time, the city's manufacturing sector contracted (shedding almost one million jobs) while the business services, finance and real estate sectors expanded. With rising demand for high-skilled labour and extensive in-migration of low-skilled labour, the gap between the rich and poor has risen dramatically. A rapid expansion of high-end luxury apartments and gated communities has been accompanied by the demolition of informal settlements and working-class neighbourhoods and the forced relocation of their residents. Shanghai's experience does lend support to the general hypothesis that world city formation is inevitably accompanied by socioeconomic stratification.

Sources: Li and Wu (2006), Wu (2000), He (2010), Li and Song (2009).

metropolitan status – the number of such agglomerations has grown from 197 in 1980 to 501 in 2015 – the overwhelming majority of them in low- and middle-income countries (United Nations, Department of Economic and Social Affairs, Population Division 2014). These city-regions are attracting increasing attention not only because of their importance to the functioning of the global economy, but also because they are increasingly recognised as representing 'distinctive political actors on the world stage' (Scott et al. 2001: 11). Even Africa, as the world's least economically developed and urbanised continent, now has a large number of agglomerations considered to be 'city-regions' (Beall et al. 2015). Since the rise of nation-states in Europe several hundred years ago, states have generally been seen as the most important actors influencing the territorial organisation and regulation of production and exchange. In the contemporary era, some see cities and city-regions as reasserting their historic role in directing the world economy (Segbers 2007: 13).

While some city-regions contain within them the kind of global cities that have attracted so much attention, others do not, but they nevertheless play important roles in the global economy. For example, the El Paso-Juárez city-region, with a combined metropolitan population of approximately 2.57 million people, consists of two cities sitting just across from each other on the Rio Grande – the river that defines the US–Mexico border. This pair of cities, along with many others scattered along the frontier, play a vital role in managing a large volume of cross-border production and exchange activities. They also provide an interesting illustration of just how important institutions and governance can be in shaping the destiny of cities (see Box 3.5).

Box 3.5 The economy of the El Paso-Juárez city-region

Linked by a handful of bridges that straddle the Rio Grande, the economies of El Paso, Texas and Ciudad Juárez, Mexico are deeply intertwined. This border city-pair is one of many scattered along the 2,000 mile frontier that separates the USA from Mexico. Beginning with a Border Industrialization Program launched in 1965, and expanding in the 1980s as the government of Mexico began a process of liberalising trade and reducing restrictions on foreign investment, cross-border trade has stimulated the growth of dynamic border city-regions. In particular, manufacturing plants known as *maquiladoras* sprung up all along the Mexican side of the border, serving as export assembly plants for a wide range of goods destined for US markets including textiles, automobiles and electronics.

The development of the *maquiladora* industry in Mexico has made significant contributions to the economies of both Mexico and the south-western United States. Employment in *maquiladoras* represents around 50 per cent of all manufacturing employment in Mexico, and the industry produces over US $196 billion worth of exports annually – an increase of more than 50 per cent since 2009. On the US side, the industry has generated demand for finance, accounting and legal services, as well as a host of other services such as hotels, car rentals and restaurants. It has also increased demand for US manufacturers that are used as inputs in the *maquiladoras*.

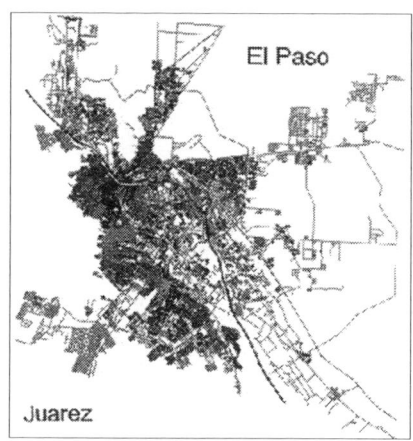

Behind the San Diego-Tijuana city-region, El Paso-Juárez is the second largest metropolitan region in the Western Hemisphere, managing an annual cross-border trade worth over US $88 billion – equal to nearly 16 per cent of all US–Mexico border trade. On average, everyday more than 2,200 trucks, 28,500 cars and 17,000 pedestrians cross the border into El Paso. The growth of the *maquiladora* industry has created over 200,000 jobs in Juárez, and hundreds of thousands of jobs in related services and industries. In turn, apart from a spike following the global economic crisis, El Paso has seen a steady decline in unemployment rates, despite a steep decline in manufacturing employment due to the growth of other sectors related to border trade.

However, between 2007 and 2012 Juárez experienced declining economic growth and productivity and a surge of violence. In 2010 it became known as 'the most dangerous city in the world' with extremely high rates of homicide and femicide. In turn these problems – and the reputation that came with them – had further negative impacts on the economy of the city. In stark contrast, El Paso became the 'safest city in the USA' during this period, illustrating the profound importance of institutions and governance in shaping the welfare of urban populations. El Paso-Juárez may be geographically and economically intertwined, but when viewed through the lens of human development the city is clearly divided into two very different worlds separated by an invisible political line in the desert.

Sources: Hanson (2001), UTEP (2012), Heid, Larch and Riaño (2013), Payan (2014), Rex (2014), Niño et al. (2015). *Image source:* City of El Paso Planning Department.

A notable feature of the El Paso-Juárez regional economy is the high incidence of 'informality' – a phenomenon highlighted in Sassen's analyses of global cities and a subject of much research and debate among scholars of urbanism in low- and middle-income countries. Research conducted in the mid-1990s suggests that somewhere between 30–40 per cent of the residents of the El Paso-Juárez city-region rely on 'informal' economic activities to get by (Staudt 2001). As we will see in Chapter 4, informality is a pervasive phenomenon in low and middle-income countries, and one that seems to have grown in response to the combined forces of globalisation and the era of neo-liberal reform.

Conclusion

There is no question that economic growth and development can stimulate urbanisation, but history has shown that this is no longer the primary driving force behind urban transitions in developing countries. Rapid demographic change can generate rapid urban population expansion even in countries with faltering economies. Yet while urbanisation without economic growth is clearly possible, economic development without urbanisation is not: agglomeration is a natural complement to economic growth.

Cities facilitate sharing, matching and learning between workers and firms. The result is productivity growth, innovation potential and the prospect of sustained economic growth. Moreover, networks of settlements support broader territorial development by acting as service centres and hubs in regional and global markets. Indeed, cities play a critical role in the functioning of global economic processes. Yet the very characteristics that make cities uniquely dynamic economic spaces also create costs such as congestion, pollution, health risks and crime. Getting the best from cities therefore requires active planning, investment and targeted interventions to address these drags on urban productivity and human welfare. As we will see in the coming chapters, these are particularly acute challenges in many developing countries today.

Summary

- Development is not simply about economic growth but also structural societal change and the diversification of economies. This has important implications for cities.
- While urbanisation has traditionally been seen as an indicator for 'economic progress', this is not necessarily still the case. Much of sub-Saharan Africa experienced 'urbanisation without growth' in

the 1980s and 1990s, highlighting the importance of demographic processes in driving urban change in developing countries.

- Cities are increasingly seen as important units of analysis in development economics. Urban areas are potential engines of economic growth due to the benefits of agglomeration and the various kinds of economies of scale they foster.
- Factors such as proximity, high concentrations of skilled labour, the exchange of ideas and diversity of firms in a city contribute to the *localisation* and *urbanisation* economies, and can generate *clusters*, such as the IT sector in Bangalore.
- There are also significant costs with such agglomeration, such as congestion, pollution, health risks and crime. However, for most, the benefits of agglomeration seem to outweigh the costs, otherwise cities would not exist at all.
- Urban and rural economies are intrinsically linked, and small and medium-sized towns can play an important role in supporting rural and regional economic development.
- Many large settlements serve as 'world cities' or 'global cities', managing global flows of capital, goods and information. At the same time they exhibit extreme inequalities.
- There are also 'city-regions' that do not necessarily involve world cities, yet still play an important role in the regional and global economic processes (e.g. El Paso-Juárez).
- Many of these city-regions, especially in low- and middle-income countries, contain high levels of informal economic activities.

Discussion questions

1. What are the key drivers of sustained economic growth?
2. How does economic growth differ from economic development?
3. Is sub-Saharan Africa 'over-urbanised'?
4. How does agglomeration affect economic growth?
5. With reference to 'global cities' and 'ordinary cities', explain why it is important to think about the role of smaller and poorer cities and city-regions in the global economy.

Further reading

Berdegué, J. A., Carriazo, F., Jara, B., Modrego, F. and Soloaga, I. (2015) 'Cities, territories, and inclusive growth: Unraveling urban–rural linkages in Chile, Colombia, and Mexico', *World Development*, 73(September), 56–71.

Duranton, G. and Puga, D. (2004) 'Micro-foundations of urban agglomeration economies', *Handbook of Regional and Urban Economics*, 4, 2063–2117.

Friedmann, John (1986) 'The World City Hypothesis', *Development and Change*, 17(1), 69–84.

Jacobs, Jane (1984) *Cities and the Wealth of Nations*. New York: Vintage Books.

Nathan, M. and Overman, H. (2013) 'Agglomeration, clusters, and industrial policy', *Oxford Review of Economic Policy*, 29(2), 383–404.

Sassen, Saskia (2012) *Cities in a World Economy* (fourth edition). Thousand Oaks, California: Pine Forge Press.

Segbers, Klaus (ed.) (2007) *The Making of Global City Regions: Johannesburg, Mumbai/Bombay, São Paulo, and Shanghai*. Baltimore: Johns Hopkins University Press.

Websites

Global and World Cities Research Network (GaWC): www.lboro.ac.uk/gawc/index.html

International Institute for Environment and Development (IIED) rural–urban linkages page, with links to numerous publications: www.iied.org/rural-urban-linkages

World Bank, World Development Indicators, presenting current development data including national, regional and global estimates: data.worldbank.org/data-catalog/world-development-indicators

4 Urban poverty, livelihoods and informality

- Understanding poverty: definition, measurement and trends
- Urban poverty and vulnerability
- Income, assets and sustainable livelihoods
- The informal economy

Introduction

After more than 50 years of concerted efforts to reduce global poverty, fundamental material and social deprivation remains disturbingly widespread. The most recent World Bank figures indicate that the total number of people living on $2 or less per day – a crude but popular measure of poverty – fell only slightly between 1981 and 2011, from 2.59 billion to 2.2 billion. While the majority of the poor still live in rural areas, there is growing evidence that the geography of poverty is changing with the steady march of world urbanisation. It is widely agreed that economic growth is a necessary condition for lifting people out of poverty, but insufficient on its own: pockets of poverty can be found in every country of the world, rich and poor. Why, and what can be done about it? These questions have motivated decades of research into the nature, causes and extent of poverty in rich and poor countries alike.

This chapter begins with a review of the various ways in which poverty has been conceptualised and measured in the past. Definitions and metrics are important for identifying the location and scope of poverty, but can obscure important differences among those classified as poor. We then examine the nature of urban poverty and vulnerability, which is a multidimensional phenomenon intrinsically linked with the physical and institutional characteristics of the urban built environment, such as access to secure housing and basic services. We then focus specifically on the challenge of building sustainable livelihoods that provide regular income, which is vital in the highly monetised urban context. Finally, we consider the nature, scale and significance of the informal economy, which is pervasive in many of the world's least economically developed countries. We conclude by highlighting the complex ways in which hazards and insecurity in cities can interact to trap the urban in a position of persistent vulnerability.

Understanding poverty: definitions, measurement and trends

Poverty can be defined most simply as a lack of one or more of those things that determine quality of life. But what are 'those things' and how do we define 'quality of life'? Over time, conceptualisations and measurements of poverty have evolved from a concern with subsistence, to a focus on basic needs, to the notion of relative poverty or deprivation. The notion of absolute or subsistence poverty, which dates back to the nineteenth century, defines poverty in relation to the minimal physical requirements of human beings, such as food, clothing and shelter. To be poor, from this perspective, is to lack the basic means to survive. The most common way of measuring poverty understood in this way is to measure the number of people who fall below a specified level of income, usually called the poverty line, to get what is called a poverty headcount measure. There are a variety of ways of calculating poverty lines, but by far the most popular approach is to use an income threshold that represents the minimum required to purchase enough calories to survive (usually set at between 2,000–2,400 per day) or a basket of essential goods (e.g. food, clothing and access to shelter). The analytical strength of this approach to poverty assessment rests in the clarity of definition, allowing ease of measurement through identifiable indicators such as income. In policy terms it provides a clear basis for setting the threshold of, for example, a minimum wage or targeted income transfers such as welfare grants.

Nevertheless, this approach has its shortcomings. For example, if 40 per cent of a country's population falls below the poverty line, we have no idea of where particular individuals fall – are their incomes close to the poverty line or close to zero? Furthermore, if resources are targeted so that they reach the poorest of the poor rather than those closer to the poverty threshold, the result may be an improvement in the lives of the least well-off, but the aggregate number of poor living below the poverty line (i.e. the poverty headcount) may not shift at all. These kinds of indices can also mask differences in the way poverty is experienced. For example, rural livelihoods are often dependent on access to natural assets such as fertile land while urban dwellers are more dependent on waged employment opportunities and often require more cash to satisfy basic needs than their rural counterparts (Mitlin and Satterthwaite 2013). As a result, direct comparison of poverty headcount measures between rural and urban areas can be highly problematic and may significantly underestimate the extent of urban poverty (ibid.). There are also the usual problems associated with gathering accurate data, especially given that people tend to be reticent about revealing their true economic status.

It was in part as a result of the shortcomings of poverty lines and the absolute poverty approach that a focus on basic human needs gained prominence in the 1970s. It was argued that policies directed at addressing minimum requirements for basic consumption were inadequate and that in addition to food, shelter and clothing people needed support in making a living as well as access to decent services, health care and education. The critique also recognised that human beings are more than machines and have more complicated needs than the means of sustenance and reproduction, including for example, social and psychological needs (Hulme and McKay 2005). While compelling, the basic needs approach suffered conceptually from vague definitions and the application of inconsistent criteria for establishing basic human needs and valuing intangible ones such as dignity, respect and relationships of love or friendship. In policy terms the basic needs approach offered income-generating projects and sought to deliver basic services such as safe water and adequate sanitation, primary education and basic health care. While admirable, it did not distinguish between one context and another and did little to identify or address the underlying causes of poverty and inequality.

Neither the concepts of absolute poverty nor basic needs reveal much about income distribution within a context or group, which is important for a deeper understanding of the nature and causes of poverty in specific settings. A corrective was the notion of *relative poverty,* with a person understood to be in relative poverty when his or her income falls below that of others in a particular country. Relative poverty measures were developed with the poverty line being based, for example, on half the medium income or a fraction of average national income. The advantage of a relative approach is that it captures inequality and hence gives us a clearer sense of relative deprivation within a country. Relative deprivation can also be assessed by identifying whether people lack the minimum requirements for both material and social needs in a particular time or place. According to Townsend (1993: 36):

> People are relatively deprived if they cannot obtain, at all or sufficiently, the conditions of life – that is, the diets, amenities, standards and services – which allow them to play the roles, participate in the relationships and follow the customary behaviour which is expected of them by virtue of their membership of society. If they lack or are denied resources to obtain access to these conditions of life and so fulfil membership of society they may be said to be in poverty.

For example, in many Asian societies, social interaction is predicated on being able to offer visitors to the home a cup of tea. Tea has little nutritional or caloric value but not to be able to offer it excludes people from social participation according to customary behaviour. The relative deprivation approach to poverty is echoed in the work of Martha Nussbaum and Amartya Sen (see Chapter 1). Sen defines poverty as *capability deprivation,* by which he means the condition of lacking one or more of the capabilities required for a person to live the kind of life they have reason to value (1999: 87). Within this understanding of poverty, Sen famously argued – drawing on Adam Smith – that highly context-specific factors such as 'the ability to appear in public without shame' should be understood as a fundamental capability, without which people should be considered poor.

The relative deprivation and capability deprivation perspectives offer an important corrective to simplistic conceptualisations of poverty based only on income or consumption measures because they concede that social and political factors are often important in determining the experience of poverty. For example, two people – a man and a woman – may have precisely the same income, but the woman may have less choice about how she earns or spends her income, or may not be allowed to express personal or political opinions if she lives in a society that persistently discriminates against women. In this case, although the man and the woman may appear to have equal access to financial resources, the socio-political context renders the woman poorer than the man from a relative deprivation perspective.

Efforts to measure poverty have evolved roughly in line with these changing conceptualisations of poverty. Generally, however, reliance is still placed on subsistence or basic needs approaches that use indicators such as income, caloric intake or other proxies for consumption, particularly for cross-national comparison. Perhaps the most well-known poverty index is the International Poverty Line – or 'dollar-a-day' measure – produced by the World Bank, which is currently $1.25 per person per day. Those who live on less are said to live in extreme poverty; those living on less than two dollars a day are classified as moderately poor. Many national governments and international agencies still use these measures, although there have been a proliferation of alternatives in response to criticism of these thresholds. One of the first substantial and influential quantitative alternatives was the Human Development Index (HDI), which was developed by UNDP in the early 1990s under the guidance of an international group of scholars drawing on the capability deprivation conceptualisation of poverty. But this index is more properly understood as an alternative index of economic development. In a more recent effort to provide a similarly holistic quantitative indicator of poverty, UNDP introduced the Multidimensional Poverty Index in 2010, which uses indicators of education, health and standard of living rather than a simple income metric (see Box 4.1). These are generally recognised as common characteristics of poverty regardless across contexts. Furthermore, by limiting the dimensions of poverty to three key areas, they are easily researchable, communicable and comparable. Proponents of this approach to measuring poverty champion its ability to account for the multiple reasons that people live in poverty, and the way in which it acknowledges that there may be more than one factor that causes a household to be deprived, and highlight the material conditions of poverty. The household data may be aggregated up to both regional and national levels to provide useful data for geographical comparisons of multidimensional deprivation. The most recent data, covering 91 countries (roughly 75 per cent of the world's total population), indicates that 1.5 billion people live in multidimensional poverty (see Alkire and Santos, 2014; Alkire et al. 2014; Potter et al. 2012; UNDP 2014; UNDP 2015).

Box 4.1 The Multidimensional Poverty Index

Dimension	Indicator	Measure (Deprived if...)
Education	Years of schooling	No household member has completed at least one year of schooling
	Child school attendance	No children are attending school up to the age at which they should finish class 6
Health	Child mortality	2 or more children have died in the household
	Nutrition	Severe undernourishment of any adult or any child in the household
Standard of living	Electricity	The household has no electricity
	Sanitation	There is no sanitation facility
	Water	The household does not have access to drinking water, or safe water is more than a 45 minute (round trip) walk away
	Floor	The household has a dirt, sand, or dung floor
	Cooking fuel	The household cooks with dung or wood
	Asset ownership	The household has no assets (radio, mobile phone, refrigerator, etc.) or car

Source: UNDP (2015).

The reason indices such as these matter is because there is strong congruence between how poverty is conceptualised and measured and how it is addressed. This can be illustrated by comparing the World Bank's *World Development Report (WDR) 1990* and its *WDR 2000*, both of which had poverty as their theme (World Bank 1990, 2000). The *WDR 1990* was driven by a conceptualisation of poverty based on a subsistence approach, where poverty was defined as 'the inability to attain a minimum standard of living' (World Bank 1990: 26). The influence of the basic needs approach was also evident in that the report focused in addition on shortfalls in education and health. The policy recommendations involved a three-pronged approach to poverty reduction: economic growth, social safety nets for the very poorest and human development through improvements in health and education.

If we turn to the *WDR 2000* we can observe a broadening of the World Bank's perspective illustrated by the following extract from the report:

Poverty is pronounced deprivation in well-being. But what precisely is deprivation? The voices of poor people bear eloquent testimony to its meaning. To be poor is to be hungry, to lack shelter and clothing, to be sick and not cared for, to be illiterate and not schooled. But for poor people, living in poverty is much more than this. Poor people are particularly vulnerable to adverse events outside their control. They are often treated badly by the institutions of state and society and excluded from voice and power in those institutions.

(2000: 15)

There are continuities of analysis between the two reports, and growth remains centre stage as *the* main solution to problems of poverty, along the provision of basic services. However, it is acknowledged here that simply providing resources to poor people, whether money or services, is not sufficient and other factors enter the equation. Institutional reform needs to accompany economic growth in recognition of the fact that there are institutional barriers that prevent the poor from accessing public goods, services and decision-making arenas. It is noted too that vulnerability matters. Lastly, inequities are recognised, not only across society as a whole but within deprived groups as well, such as those based on gender, ethnic, racial or religious differences.

This shift towards a relative deprivation or capability deprivation approach not only inspired a more comprehensive set of indicators for measuring poverty, it also changed the way it was analysed. This included drawing on the expertise of a wider range of specialists both from within and outside the discipline of economics. For example, anthropologists and sociologists offered important insights into poverty processes, highlighting the way in which people fall into and escape from poverty through their lifetimes, and how intergenerational dynamics affect socioeconomic mobility. One example of how such approaches were mainstreamed was the *Voices of the Poor* (VoP) project (Narayan et al. 2000a, 2000b). This was a World Bank initiative that informed the analysis adopted in the *WDR 2000* by providing an in-depth exploration of the multidimensional nature of poverty through qualitative analysis. The VoP project sought to understand the nature of poverty by talking to those who actually experience it, by collecting, cataloguing and analysing interviews with 60,000 people across 60 countries. Holistic approaches to poverty research such as this helped to re-shape approaches to poverty reduction and the international development agenda more generally.

This is apparent in the content of the Millennium Development Goals (MDGs). The MDGs were a list of eight international development targets which sought to combat poverty, hunger, disease, illiteracy, environmental degradation and discrimination against women. They were adopted by 189 countries following the September 2000 United Nations Millennium Summit, with tangible targets to be achieved by the end of 2015. While the MDGs served to focus on the minds and efforts of national governments and international aid agencies on tangible goals, success has been mixed, with some areas of strong progress, but also many areas with woeful progress made towards reducing poverty and deprivation (see Box 4.2).

For example, globally the number of people living in extreme poverty halved between 1992 and 2010, with this goal proclaimed to have been met. However, several major regions have not met this target. Much of the progress in reducing extreme poverty can be attributed to China and East Asia, with most other regions lagging behind on key indicators. For example, sub-Saharan Africa and South Asia – where the majority of the extreme poor live – will both miss targets for reducing child mortality, and almost all regions will miss the target for reducing maternal mortality.

While the MDGs have been highly influential, particularly in policy circles, they have also been controversial. Critics argued that the MDGs were generated by a technocratic elite, inattentive to vast differences in contexts across countries and regions, unrealistic in scope, devoid of any articulation of the underlying *causes* of poverty, and wholly inadequate in formulation with regards to the realities of urban poverty (see Fehling et al. 2013; Mitlin and Satterthwaite 2013). Indeed, the only specific mention of urban development challenges was Goal Seven: to improve the lives of 100 million slum dwellers by 2020. Ironically, this goal has been met, but the population living in slums has increased substantially over the same period (see next section).

In September 2015 the MDGs were replaced with a new set of Sustainable Development Goals, due to be achieved by 2030. These were decided upon through a far more extensive and inclusive process of global consultation, focus more directly on causes of poverty, seek to account for the contextual and multidimensional nature of poverty, and include a specific goal to make cities and human settlements inclusive, safe, resilient and sustainable. This last goal is particularly timely given contemporary demographic trends and growing evidence that these are being accompanied by the 'urbanisation of poverty'.

Box 4.2 Millennium Development Goals progress report

Goal and Target	Africa		Asia				Oceania	Latin America & Caribbean	Caucasus & Central Asia
	Northern	Sub-Saharan	Eastern	South-Eastern	Southern	Western			
Goal 1: Eradicate extreme poverty and hunger									
Reduce extreme poverty by half	Y	N	Y	Y	Y	N	N	Y	Y
Productive and decent employment	N	N	Y	Y	Y	Y	N	N	Y
Reduce hunger by half	Y	N	Y	Y	N	NN	N	Y	Y
Goal 2: Achieve universal primary education									
Universal primary schooling	Y	N	N	Y	Y	N	Y	N	N
Goal 3: Promote gender equality and empower women									
Equal girls' enrolment in primary school	Y	N	Y	Y	Y	N	N	Y	Y
Women's share of paid employment	N	N	Y	N	Y	N	N	Y	Y
Women's equal representation in national parliaments	N	N	N	N	N	N	N	N	N
Goal 4: Reduce child mortality									
Reduce mortality of under five-year-olds by two-thirds	Y	N	Y	Y	N	Y	N	Y	Y

Goal and Target	Africa		Asia				Oceania	Latin America & Caribbean	Caucasus & Central Asia
	Northern	Sub-Saharan	Eastern	South-Eastern	Southern	Western			
Goal 5: Improve maternal health									
Reduce maternal mortality by three-quarters	N	N	Y	N	N	N	N	N	Y
Access to reproductive health	N	N	Y	N	N	N	N	N	N
Goal 6: Combat HIV/AIDS, malaria and other diseases									
Halt and begin to reverse the spread of HIV/AIDS	NN	Y	N	N	Y	NN	Y	Y	N
Halt and reverse the spread of TB	N	N	Y	Y	Y	Y	N	Y	Y
Goal 7: Ensure environmental sustainability									
Halve proportion of population without improved drinking water	Y	N	Y	Y	Y	N	N	Y	NN
Halve proportion of population without sanitation	Y	N	Y	Y	N	Y	NN	Y	Y
Improve the lives of 100 million slum dwellers	Y	N	Y	Y	Y	NN	n/a	N	n/a
Goal 8: Develop a global partnership for development									
Internet users	Y	N	Y	Y	N	Y	N	Y	Y

	Goal met, or due to be met by December 2015
	Some progress, but goal not met and not expected to be met by December 2015
	No progress and/or situation deteriorated
	No data

Source: UNDP (2014).

Urban poverty and vulnerability

It is difficult to accurately assess the scale and scope of urban poverty today due both to a paucity of basic, comparable data and difficulties in capturing the contextual complexity of urban vulnerability (Mitlin and Satterthwaite 2013). Nevertheless, recent evidence indicates that the global geography of poverty may be shifting. A World Bank study in 2007 found that between 1993 and 2002 the number of people living on US\$ 1 a day or less fell by 150 million in rural areas but rose by 50 million in urban areas (Ravallion et al. 2007). Overall, the study concludes that world urbanisation is contributing to a reduction in poverty overall, but that 'poverty is clearly becoming more urban' (ibid.: 693). More recent research has found that the proportion of the global population in poverty has fallen considerably since 1990, but declines have been more rapid in rural areas (where the majority of the poor live) than urban areas, and that 'poverty is becoming more urban in more urbanised countries' (World Bank 2013: 87). This provides further evidence of an 'urbanisation of poverty'.

The study also highlights an important and previously overlooked dimension of the distribution of urban poverty: in most regions the majority of the urban poor live in small and medium-sized settlements, not the megacities that make headlines. For example, in Brazil 72 per cent of the population classified as poor live in urban areas, but only 9 per cent live in Rio de Janeiro and São Paulo while 56 per cent live in small and medium-sized town and cities (World Bank 2013). The exception to this general finding is sub-Saharan Africa, where the majority of the urban poor are clustered in large cities (ibid.).

While these represent some of the most nuanced macro-analyses of urban poverty trends to date, they likely underestimate the scale of urban poverty for two key reasons. First, data aggregated at this level, with average indicators of 'rural' and 'urban' poverty based on randomised surveys, gloss over substantial *intra*-urban variation, which can be highly misleading in contexts of intense socioeconomic inequality (Haddad et al. 1999). Indeed, many detailed city-level studies have found that core health indicators of poverty, such as infant mortality, are often as high or higher in the poorest urban districts than in rural areas (Pryer 2003; Mitlin and Satterwaite 2013). Second, the standard approach of using income as the core poverty metric – or more recently, visible assets, as in the MPI – is that these are insufficient to capture the complex vulnerabilities faced by the urban poor (Mitlin and Satterthwaite 2013).

It is interesting to note the substantial difference in the percentage of the global urban population considered poor using the dollar-a-day metric and the percentage of the global urban population living in 'slums', which for measurement purposes are defined as housing units that lack one or more of the following: access to improved water supply; access to improved sanitation; sufficient living area (i.e. no more than three persons per room); and a dwelling made of durable materials (World Bank 2013). Using these indicators, UN-Habitat developed a 'slum population index' to monitor the number and proportion of urban dwellers who live in such conditions.

In many countries, slum conditions are found in the sprawling shantytowns surrounding a built-up urban core, but they can also be found in inner-city tenements that have fallen into disrepair, much like the slums that emerged in Europe and North America in the wake of the Industrial Revolution. Indeed, in places with a relatively long urban history such as Turkey (Ünsal and Kuyucu 2010) and China prior to its drive to renew dilapidated inner-city areas (Junhua 1997), some of the most squalid slums in recent years have been in inner-city buildings rather than squatter settlements on the urban fringe.

Historically, the existence of slums and informal settlements has been explained as a natural and temporary by-product of industrialisation and urbanisation as workers flock to urban areas in pursuit of jobs before sufficient housing is available, or before they have enough money enter the formal housing market (Frankenhoff 1967; Turner 1969). With time, it was assumed that they would work their way up the income ladder and move out of such conditions or upgrade their housing units. Slums and informal settlements were assumed to be cheap housing for poor people working their way into the urban economy. However, more recent research has shown that the scale of the slum phenomenon today reflects a combination of rapid demographic change in contexts of urban poverty and weak institutions that have roots in the colonial past (Fox 2014). In other words, in many places slums are not transitional spaces but permanent homes for their residents.

According to recent estimates, nearly 900 million people – roughly 30 per cent of the urban population in developing countries – live in slum conditions today (United Nations 2015). This is considerably more than the 11.6 per cent estimate of global urban poverty

Plate 4.1 *A riverside slum in Mumbai*

Source: Philipp Rhode.

incidence in 2008 derived from the dollar-a-day approach (World Bank 2013). It is also worth considering the disparity in estimates of trends of absolute versus relative deprivation: while the percentage of the global population living in slums declined between 1990 and 2014, the absolute number of people living in slums has in fact increased due to rapid population growth (see Figure 4.1). This raises an interesting question: which matters more, the actual number of people in poverty, or the proportion of people in poverty?

The slum population index is perhaps no more appropriate or accurate as a means to quantitatively evaluate the scale and scope of urban poverty than income-based indices, but it does usefully reflect the way in which material and environmental context shape well-being. Slum conditions expose individuals, families and whole communities to a range of vulnerabilities that undermine the development of capabilities that permit human flourishing. It is this persistent vulnerability, perhaps more than economic deprivation, which

Figure 4.1 *Slum population trends, 1990–2014*

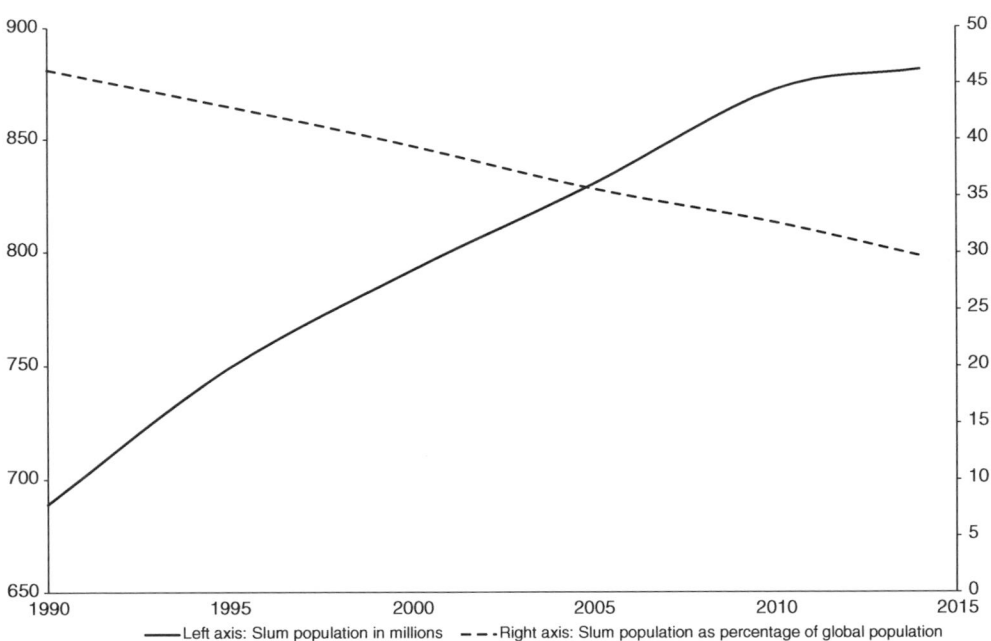

Source: United Nations (2015).

characterises urban poverty. Moreover, vulnerability has its roots as much in social and institutional exclusion as in material and environmental context.

It is important not to draw a distinction between rural and urban poverty too rigidly, given that many drivers of social disadvantage operate at national and even the international level (Wratten 1995). Nevertheless, there are some key differences between life in a village and life in a town or city that bear consideration when conceptualising, measuring and seeking to alleviate urban poverty. These are summarised in Box 4.3. Note the difference between this list and the list of MDGs in Box 4.2. The MDGs emphasised the manifestations of poverty (i.e. low incomes, poor health); whereas the characteristics listed in Box 4.3 emphasise the persistent vulnerabilities experienced by the urban poor.

When the slum population index was originally conceptualised, there was in fact a fifth criterion for classifying a household as a slum household: lack of security of tenure. This was later discarded because it was too difficult to accurately measure (World Bank 2013),

Box 4.3 Key characteristics of urban poverty and vulnerability

- Poor quality housing

- Insecurity of tenure (for both owners and tenants)

- Lack of access to basic infrastructure and affordable services

- Susceptibility to diseases and accidents

- Environmental hazards, including the impacts of natural and man-made disasters

- Social fragmentation

- Exposure to violence and crime; fear of violence

- Increasing exposure to armed conflict and terrorist attacks

- Reliance on a monetised economy

- Reliance on employment in the informal economy

but it is widely believed that there is a very high correspondence between tenure security and slum settlements. Indeed, the terms 'slum' and 'informal settlement' are often used interchangeably and those living in them are frequently identified as squatters – i.e. illegal occupiers of public or private land. These settlements are often tolerated where there is low demand for the occupied land, but residents live under the constant spectre of eviction. Indeed, large-scale evictions have been documented in cases as diverse as India (Bhan 2009), Rwanda (Goodfellow 2014a), Vietnam (Harms 2012) and Zimbabwe (Potts 2006). In some cases these have been state-driven in the name of the 'public interest', while in others the impetus has been real estate speculation or development interests motivated by increases in the value of previously undesirable land due to urban growth and expansion (Durand-Lasserve and Royston 2002; Lombard 2012). In both of these varieties of eviction, 'sustainable development', and conservation are often the official justification used (Everett 2001; Türkün 2011). There have also been many controversial evictions in recent years associated with urban redevelopment ahead of 'mega-events' such as the Olympics and World Cup (Carrol 2012; Rolnik 2013). Sometimes evictions are legitimised simply through a discourse of urban 'beauty' (Harms 2012).

In many cases, however, the vulnerability associated with residence in an informal settlement is less visible but equally pernicious. Rather than serving as cheap housing for poor people, recent research shows that housing in these settlements can be surprisingly expensive (Gulyani and Talukdar, 2008). Moreover, the beneficiaries of these high rents are often political and economic elites who provide protection against eviction in return for rents, thereby undermining the incentive for government to support efforts to improve conditions (Fox 2014; Goodfellow 2013a). For example, in the infamous informal settlement of Kibera in Nairobi, one study showed that 57 per cent of landlords were government officials or politicians who rented out land and units informally, and whose interest was clearly in maintaining the status quo (Gulyani and Talukdar, 2008).

The vulnerabilities posed by the prospect of eviction and economic exploitation of the urban poor are compounded by physical vulnerabilities associated with the quality of housing units themselves. Housing in informal settlements often consists of makeshift shelters of corrugated iron or zinc sheets, scavenged pieces of cardboard and wood, and industrial scraps. Poor urban residents often settle on marginal land such as flood plains, riverbanks, steep slopes and reclaimed lands, which are particularly susceptible to natural disasters. Tenuous construction combined with a natural hazard can make for devastating consequences, and it is likely that such devastating events will become more common in the years ahead (see Chapter 6). Even those living in apparently more permanent structures often fall prey to the poor quality of their dwellings. For example, the poor building quality has been identified as the primary cause of the 40,000 lives lost in Bam, Iran during an earthquake in 2003 (ibid.: 136). Similarly, many of the lives claimed in the earthquake that devastated Haiti in 2010 were due to poor quality construction with modern materials – traditional forms of construction were frequently more resilient (Audefroy 2011).

Hazards associated with poor access to basic services are often even more pervasive than those posed by natural disasters. Dwellings in low-income areas are rarely incorporated into formal urban infrastructure networks, such as water, drainage and sewerage systems, for reasons explored in more depth in Chapter 5. Limited access to these basic services exposes poor urban residents to debilitating diseases. This creates ideal conditions for the incubation and transmission of infectious and parasitic diseases. Almost half of all urban residents in Africa, Asia and Latin America are victims to ill-health associated with poor water and sanitation facilities, such as diarrhoeal diseases and worm infections (WHO and UN-Habitat 2010). An estimated 88 per cent of all diarrhoea infections globally are attributed to unsafe water supply, unhygienic practices and a lack of sanitation infrastructure (Evans 2005). Recent cholera outbreaks in Zimbabwe, Senegal, Zambia and Ghana can be explicitly linked to unsanitary urban conditions (Sasaki et al. 2008; Drechsel et al. 2010). Children are particularly vulnerable and child mortality rates can be exceptionally high in informal settlements; a study in Kenya found that child mortality rates in some informal settlements in Nairobi were over twice as high as the national average, and the average for rural areas (Mitlin and Satterthwaite 2013).

Urban residents – and the urban poor in particular – are also susceptible to outdoor air pollution, which is concentrated in cities due to automobile use, industrial activity and the burning of waste (WHO and UN-Habitat 2010). Globally, the WHO estimates that such pollution caused over one million deaths in 2004 (ibid.), and the number has almost certainly risen since then due to significant increases in the number of automobiles on the roads in developing countries (see Chapter 5). Indoor air pollution, from the use of solid fuels for cooking and through occupational exposure, is thought to have accounted for roughly 2.5 million further deaths in the same year (ibid.).

All of these vulnerabilities associated with the physical environment can be compounded by the social and economic environment in cities. Kinship ties, social networks and connections with friends and neighbours often serve as a kind of informal social safety net for the poor, but they are also demanding of people. It is hard to reciprocate when you have nothing to share. Customary forms of managing health risks, economic insecurity, and other forms of vulnerability are placed under severe stress when low-income urban dwellers face hard times, and the struggle to survive in cities can lead to the breakdown of families and friendships. Borrowing and begging by destitute kin, with no possibility of repayment, drains limited resources and becomes problematic for relatives who themselves are already on the edge (Beall 1995, 2001). The strains of maintaining reciprocal relationships under conditions of persistent vulnerability can contribute to social fragmentation and alienation. In some cases, families may find themselves forced to sell what assets they have, to scavenge in the street or even to engage in illegal activities rather than borrowing from relatives and friends to survive hard times. The breakdown of social ties can be further exacerbated by crime and violence, which can undermine community trust and cohesion (see Chapter 7), and hence the informal social safety nets that so many of the poor rely upon to get by.

Finally, vulnerability in cities is fundamentally linked to peoples' reliance on a monetised economy. Urban residents need money to acquire such basic necessities as food, water, shelter, and clothing. Absolute reliance on a monetised economy is one of the key differences between urban as compared to rural poverty – rural residents can also rely on subsistence farming, or even foraging and hunting in lean times while urban residents cannot. Although there is a growing body of research on the possibilities for urban agriculture to contribute to livelihood strategies, particularly in peri-urban areas (Maxwell 1995; Redwood 2012), it is unlikely to make a significant contribution to the consumption needs of the majority of the urban poor. When agricultural output falls short of demand for food, urban dwellers are particularly vulnerable (see Box 4.4).

Box 4.4 Urban food riots in 2007–2008

During 2007 and 2008 the world witnessed spontaneous protests and riots in over 30 cities in countries across the world, including Guyana, Haiti and Bolivia in Latin America and Caribbean; Cameroon, Egypt, Ivory Coast and Senegal in Africa; Surinam, Bangladesh, South Korea and Uzbekistan in Asia; and Yemen in the Middle East. In Cameroon food riots resulted in 24 deaths, while in Haiti six died and the prime minister was forced to step down. In Ivory Coast anti-riot police used tear gas to disperse the predominantly female demonstrators and the president cancelled customs duties and cut taxes on basic household products after days of violent protests in the capital, Abidjan.

This wave of protest and violence followed a steep rise in prices for staple foods over just a few months – the prices for wheat and corn rose by 68 per cent in the first quarter of 2008. This price shock was the result of a confluence of factors, including high oil prices (which affect the cost of fertilisers and transport), rising demand for bio-fuels (which encourages farmers to divert production away from food and toward bio-fuel feedstock), and the growth of demand for food and other primary products from the expanding economies of Asia and Latin America.

The Executive Director of the World Food Programme warned that the world was confronting 'a new face of hunger' and that 'often we are seeing food on the shelves but people being unable to afford it'. Increasingly this is an urban face, as people have started to make their misery known on city streets. Governments and international development agencies have geared up to provide better support to agricultural development and food production. It remains to be seen whether they will pay similar attention to the plight of the urban poor.

This unrest in many large cities over rising food prices is the angry and desperate face of urban poverty, one that is evident in the history of advanced economies in the past, and may well persist globally into the future.

Sources: Patel (2008), Pilkington (2008), BBC News (2008).

Income, assets and sustainable livelihoods

Operating in a primarily monetised economy renders income, either from self-employment or wage labour, a centrepiece of urban livelihood strategies. Indeed, a study of poverty and governance across nine cities of Africa, Asia and Latin America found 'jobs and income earning opportunities...to be the most fundamental preoccupations of the urban poor' (Beall 2004: 54). In the 1950s and 1960s, urban employment was expected to grow with industrialisation, and urban development policies focused primarily on investments in infrastructure and housing for the formal workforce – especially government employees – and their families. By the 1970s however, it became clear that urban labour markets were failing to keep pace with urban growth and the discourse of international development was shifting towards the basic needs paradigm, with increasing attention being paid to such issues as access to basic healthcare, education and other essential services (Stewart 1985; Streeten et al. 1981). Then, in the wake of the economic crises of the late 1970s and early 1980s, neoliberal development prescriptions rose to the top of the agenda and encouraged a shift away from proactive government efforts to promote industrialisation and expand access to basic services. Instead governments increasingly embraced policies designed to help poor people help themselves, for example through small and medium enterprise development and microfinance initiatives.

The 'rolling back' of the state under structural adjustment reduced social sector spending in many countries, exacerbating in particular the already vulnerable position of the urban poor:

> The urban poor were faced with a price-income squeeze, as the effects of
> unemployment and downward pressure on wages were compounded by the
> marketization of public goods. The majority of new recruits to the labour market
> were left with underemployment in the informal sector as the only option left
> open to them.
>
> (Watt 2000: 103)

Moreover, the impact of job losses, the withdrawal of food and
transport subsidies and the introduction of user charges for goods and
services fell more heavily on some than others, with women being
particularly vulnerable to downward pressures on income and
consumption, and children engaging in income generating activities
in greater numbers (Moser 1996) – including hazardous activities
such as waste collection that pose substantial risks to health and
wellbeing (see Plate 4.2). In response to the negative social impacts
of macro-economic reform and a barrage of critiques from scholars
and development practitioners, new policies and programmes were
devised to ease the pain of adjustment (Cornia et al. 1987; Kanji
2002; Donkor 2002). These were essentially temporary remedial
programmes that sought to ensure basic needs were met, but which
ultimately depended on external aid and did nothing to address the
fundamental problem facing the urban poor: a lack of opportunities to
generate a stable income (see Jørgensen and Van Domelen 2001;
Tendler 2000; Beall 2005).

The failure of these kinds of policies to make a significant dent in
urban poverty led to a shift in thinking about how to deal with
disadvantage at the urban scale. A key critique of past approaches is
that they focused either on short-term income generation or the
maintenance of minimum consumption levels, neither of which
proved a sustainable way of lifting people out of poverty (Moser
2007: 87). Instead researchers have increasingly sought to expose the
complexity of urban livelihood strategies, and policymakers have
begun to address the challenges of establishing sustainable
livelihoods that reduce urban vulnerability (Rakodi and Lloyd-Jones
2002). Based on years of research on agrarian change and influenced
by Amartya Sen's analysis of poverty and deprivation (Chambers and
Conway 1992), the concept of livelihoods had come to the fore by the
late 1990s.

Livelihoods are not just about income earning but rather the wide
spectrum of capabilities, assets and activities required to make a
living. Commonly known as the 'sustainable livelihoods framework'

Plate 4.2 *Young waste-pickers in Bangalore, India*

Source: Jo Beall.

(SLF), this perspective adopts a broad conceptualisation of poverty and vulnerability and is associated with integrated approaches to development that include a focus on economic opportunity, social protection and a concern with environmental sustainability (Scoones 1998). The key insight of the SLF is that income from waged employment is only one dimension of the livelihood strategies of most individuals and households. Other factors, such as achieving material assets, good health, skills and knowledge, and maintaining social networks constitute equally important components of people's livelihood strategies. A livelihood strategy is deemed sustainable when it allows an individual or family to cope with and recover from stresses and shocks in the immediate term and in the long term, as well as maintain or enhance the capabilities and assets of a household and its members, both now and in the future, without undermining the natural resource base (Carney 1998: 4). There are a number of options available to low-income households in managing their livelihoods. These include reducing consumption, increasing labour or income-earning activities, investing in tangible and intangible assets, such as houses and social networks respectively. In order to ensure long-term security, households might substitute one asset for another, for example selling jewellery to purchase tools; or sacrificing assets for consumption to deal with short-term survival needs under conditions of stress.

The SLF has been taken up in different forms by national governments, international NGOs such as Oxfam and Care International, bilateral donors such as DFID and by UNDP. As a framework for analysing poverty, the SLF is useful because it invites a holistic approach, but as an operational tool it has proved more limited. SLF-informed initiatives frequently encompass fairly traditional projects and programmes, such as employment-generating infrastructure projects, microfinance programmes and the promotion of participatory approaches to development planning and practice. However, SLF-informed development strategies often combine such initiatives in creative ways that are more attentive than previous approaches to the challenges of generating sustainable improvements in livelihoods.

The SLF was first developed in relation to rural livelihoods but was subsequently explored in relation to urban risk and vulnerability, with a particular focus on asset accumulation (Moser 1996, 1998; Rakodi 1999). Caroline Moser has been particularly influential in this field with her research into poverty dynamics and mobility, including through intergenerational resource accumulation and loss. Drawing on the work of Anthony Bebbington (1999) and Amartya Sen (1981), she defines assets as a stock of financial, human, natural or social resources or capital endowments on which people build livelihoods, but which also give them the capability to be and to act (Moser 2007: 2). She shows how urban households manage complex asset portfolios over their life course, arguing that intergenerational strategies are not only about household survival on a day-to-day basis but about investment in long-term security (Moser 1996, 1998, 2007). Indeed, a key point about livelihood strategies in urban areas in particular is the propensity for city-dwellers to adopt what have been termed 'multiple modes of livelihood' strategies (Owusu 2007) – in other words, to adopt a range of different tactics to diversify risk and bolster their livelihood security.

While approaches such as the assets and sustainable livelihoods framework can provide useful principles by which to organise our understanding of social reality, they can also become rigid grids that

Box 4.5 Livelihood assets or 'capitals'

Human capital: The labour resources available to households both in terms of quantity, that is the number of members engaged in income-earning activities, and quality, referring to their education, skills and health status.

Social capital: The social resources on which people draw in pursuit of livelihoods, for example social networks and relationships based on reciprocity.

Physical capital: The basic infrastructure, equipment and tools of their trade that people need to pursue a livelihood.

Financial capital: The financial resources people rely on to pursue various livelihood options, such as savings, credit, remittances and transfers.

Natural capital: The natural resources used in the pursuit of livelihoods, for example, land, water and common pool resources.

Source: Carney (1998).

seek to codify complexity and do not easily accommodate the messiness and micro-politics of people's everyday home and working lives. A second weakness is that the livelihoods perspective tends to emphasise agency – what the poor can do for themselves – over the inalienable structural factors that condition and limit the extent of people's agency (Beall 2002: 75). Agency is clearly important, but so too is the broader context within which the urban poor seek to negotiate their livelihoods. This becomes clear when we consider the pervasive condition of informality in developing countries.

The informal economy

Anyone who has worked, travelled or done research in a city in a low- or middle-income country will probably be familiar with the following portrait of the vibrant economic life of city streets drawn from an International Labour Organization (ILO) report:

> The streets of cities, towns, and villages in most developing countries – and in many developed countries – are lined by barbers, cobblers, garbage collectors, waste recyclers, and vendors of vegetables, fruit, meat, fish, snack-foods, and a myriad of non-perishable items ranging from locks and keys to soaps and detergents, to clothing. In many countries, head-loaders, cart pullers, bicycle peddlers, rickshaw pullers, and camel, bullock, or horse cart drivers jostle to

make their way down narrow village lanes or through the maze of cars, trucks, vans, and buses on city streets.

(ILO 2002: 9)

From São Paulo to Nairobi to Cairo to Manila, millions of people rely on these types of activities for their livelihoods. For the most part they operate in the 'informal economy' (also called the informal sector, shadow economy, second economy or parallel economy). Keith Hart first introduced the concept of the urban 'informal sector' in a 1971 conference paper on Ghana, published two years later (Hart 1973). In the face of high costs of living and limited formal employment opportunities, Hart found that informal employment was a critical source of income for many urban residents and constituted a significant amount of urban economic activity. The key factor distinguishing formal from informal employment was 'whether or not labour is recruited on a permanent and regular basis for fixed rewards' (ibid.: 68) – i.e. self-employment versus waged labour. The ILO, which rapidly picked up on the idea and popularised it through a 1972 report on employment in Kenya, offered a more comprehensive definition of the informal sector, listing the following characteristics: (a) ease of entry; (b) reliance on indigenous resources; (c) family ownership of enterprises; (d) small scale of operation; (e) labour-intensive and adapted technology; (f) skills acquired outside the formal school system; and (g) unregulated and competitive markets.

Hart, and most researchers since, are careful to note the difference between 'legitimate' informal income opportunities and 'illegitimate' ones. Castells and Portes (1989) distinguish between formal, informal and criminal (or illegal) economic activities. This distinction is important because the overwhelming majority of informal economic activity in cities in low- and middle-income countries revolves around activities that, if recorded and regulated by governments, would be considered legal or of unclear legal status (having not been adequately legislated for) rather than strictly illegal.

Implicit in the ILO definition above, and in the work of many researchers studying informality, is the notion of a bifurcated or dual urban economy. On the one hand there is assumed to be a modern, capital-intensive, formally regulated, productive tier with relatively secure employment, and on the other a 'traditional,' labour-intensive, unregulated and marginally productive lower tier. Some trace this dualism back to the colonial origins of urbanism in low- and middle-income countries (Potts 2007; Drakakis-Smith 2000: 125), although research has shown that the conceptual distinction is often blurred in practice. Many studies have illustrated the interdependence and close linkages between informal and formal economic activity in industries as diverse as textiles, food service, and electronics (see Portes and Haller 2005), as well as labour mobility between the informal and formal sectors (Maloney 1999). Indeed, from early on the concept of a distinct informal sector was widely criticised in the academic community. Bromley (1978) argued that it reflected a crude, logically inconsistent dichotomy, providing no clear guidelines about how to classify most urban economic activities. Harriss (1978) likewise attacked the rigid dualism of the informal/formal distinction. The concept's popularity soared nevertheless, enjoying a 'meteoric career in the world of policy' (Peattie 1987). Recognising these problems with dualistic approaches that treated informality as a distinct sector, the term informal sector has largely given way to 'informal economy,' which can be defined as

> those economic activities that circumvent the costs and are excluded from the benefits and rights incorporated in the laws and administrative rules covering property relationships, commercial licensing, labor contracts, torts, financial credit and social security systems.
>
> (Feige 1990: 992)

This shift was partly a reflection of increased interest not only in whether enterprises were formally regulated but also the broader employment relationships and how these were structured in the absence of legal frameworks (Chen 2006). The relationship between the formal and informal economies is regarded by those on the left as exploitative and likely to perpetuate poverty, while those on the right stress the constructive entrepreneurial initiative of those operating in the informal sector – initiative that is put at risk by state interference in markets. Hernando de Soto's *The Other Path* (1989) was a particularly influential work in this regard. He argued that the informal economy was a product of over-regulation by states. In contrast to the dual economy perspective, which emphasised the failure of the modern sector to intervene to create jobs as generating informality, de Soto's neoliberal perspective emphasised excessive regulation, bureaucratic inefficiency and corruption as drivers of the informal economy, as people sought to escape the reach of the state. This latter view has been particularly influential within the World Bank, which now regularly gathers data on how difficult it is for an individual to start their own formal enterprise. Recent estimates show that the barriers can be formidable: in Haiti it could cost nearly two and a half times the average annual income, while in Venezuela it could take nearly five months (see Table 4.1). Confronted with high-costs to entry, many entrepreneurs may well choose informality. Indeed, a World Bank study of informality in Latin America found that self-employed workers and micro-enterprises *prefer* the informal sector, as it offers more autonomy, flexibility, stability and mobility (Perry et al. 2007).

The picture is different, however, for waged labourers in the informal sector. The same study found that these workers often aspire to the security of income and conditions of service afforded by formal employment that are simply not available in the informal economy. The study proposes an analytical framework whereby the first group choose to *exit* the formal economy due to the advantages of informality, while the second group suffer *exclusion* from the formal economy (ibid.).

More recently, the International Labour Organization has adopted a definition whereby informal employment 'refers to those jobs that generally lack basic social or legal protections or employment benefits and may be found in the formal sector, informal sector or households' (ILO, 2011: 12). This definition therefore includes not

Table 4.1 *Costs of starting a business in selected countries*

	Procedures (number)	Time (days)	Cost (% of income per capita)
Angola	8	66	123.5
Bolivia	15	49	64.4
Cambodia	11	101	139.5
China	11	31	0.9
DR Congo	7	16	30.0
Haiti	12	97	246.7
India	12	28	12.2
Indonesia	10	52	21.1
Kenya	10	30	42.7
Venezuela	17	144	49.9
Vietnam	10	34	5.3
Yemen	6	40	66.3
Japan	8	11	7.5
United Kingdom	6	6	0.3
United States	6	6	1.2

Source: World Bank (2014).

only individuals working in the informal sector, but also those working in the formal sector, with effectively informal terms of contractual engagement (i.e. no security or basic protections) (ILO 2013b). Using this definition, it is estimated that informal employment comprises more than half of all non-agricultural employment (which is primarily but not exclusively urban employment) in developing countries today (Vanek et al. 2014) leading some observers to claim that 'marginal is becoming mainstream' (Perlman and UN-Habitat 2014). In some regions informal employment constitutes the majority of employment, such as South Asia (82 per cent), sub-Saharan Africa (66 per cent) and East and Southeast Asia (65 per cent) (Vanek et al. 2014). In other regions it is somewhat less, but still very significant (see Table 4.2).

The global research-policy-action network known as WIEGO (an acronym standing for 'Women in Informal Employment Globalizing and Organizing') provides an even more nuanced definition by identifying six categories of informal employment: informal employers, informal employees, own account operators, casual wage workers, industrial outworkers/subcontracted workers, and unpaid but contributing family workers (Chen, 2012: 9). Not only does such an understanding highlight the heterogeneity of the informal economy, but it also emphasises the need to include unpaid work in analyses of the informal economy. Given that women are disproportionately likely to be primary care givers and carry out unpaid work, such an understanding touches upon the gendered element of the informal economy and the ways in which effects of informality are experienced very differently (see Box 4.6).

Table 4.2   Informal employment by region, 2004–2010

	Informal employment as a % of non-agricultural employment		
	Women	Men	Total
Latin America & Caribbean	54	48	51
Sub-Saharan Africa	74	61	66
Middle East & North Africa	35	47	45
Eastern Europe & Central Asia	7	13	10
South Asia	83	82	82
East & Southeast Asia (excluding China)	64	65	65
China (sample of 6 cities)	36	30	33

Source: Vanek et al. (2014).

Box 4.6 Gendered dimensions of informality in the Philippines

The informal economy in the Philippines accounts for over half of its GDP, but although the percentage of men and women in the informal economy is almost identical – 69.9 per cent and 70.2 per cent respectively – there are significant gendered differences. Filipina women in the informal economy not only tend to earn significantly less than men, but are also disproportionately focused in more vulnerable jobs with worse working conditions – typically domestic work, street vendors, waste pickers, and sex work. What is more, in the Philippines the brunt of the effects of the global financial crisis was borne by women: as joblessness among Filipino men increased , more women were forced into the informal economy. But, as mentioned, the jobs that most women found were those which tended to have the least stability and social protection.

The social context in the Philippines has particularly important impacts on the gendered nature of the informal economy. Filipina women are typically expected to carry out reproductive and domestic work – types of work often overlooked in analyses of the informal economy. This not only plays a role in the formation of employment niches, but it also means that they suffer from a particularly acute double burden. A further factor is the large number of Filipina women who migrate to find domestic work. Around 10 per cent of the total Philippine population works abroad, and through their remittances they produce almost 10 per cent of the annual GDP. An estimated 300,000 workers leave the Philippines each year to find employment, and the largest single group among these is domestic workers who make up around a third of all emigrants. Beyond the important remittances that these domestic workers send, their emigration also creates an extensive, yet 'informal', global care chain. Filipina migrants not only often work informally abroad, but their original care giving duties in the Philippines are then carried out, again often 'informally', by other female family members or migrants. As a consequence, this drives the demand for domestic work in the Philippines, and reinforces the gendered dimension of the informal economy – locally *and* globally.

It is also important to consider how statistics on the informal economy can render transgender and/or non-binary people invisible. This is especially problematic because the significant population of transgender and/or non-binary informal workers are among some of the most vulnerable in the Philippines: they suffer from discrimination, overwork, and underpay, and they are also disproportionately likely to be found in high risk forms of sex work. Analyses of the gendered dimension of informal economies must therefore move beyond a simplistic male/female binary.

Sources: Briones (2011), Casanova-Dorotan (2010), ILO (2013b), Milgram (2012).

The notion that informality is a response to overactive states and byzantine bureaucracy lent further credence to structural adjustment policies in the 1980s and 1990s. These were expected to reduce the size of the informal sector by removing the supposed distortions leading to informality. Instead, there is now a broad consensus that informal economic activity actually increased substantially during this period (Meagher 2003; Perry et al. 2007; ILO 2002). And just as globalisation has created space for 'world cities' to expand their role in the operations of the global economy, informal producers and traders have taken advantage of increased integration to expand into global markets (Meagher 2003; MacGaffey and Banzenguissa-Ganga 2000). Indeed, informalisation is seen by some as a globally connected process in which formal sector profits are protected by increasing recourse to informal production, with the tacit support of governments in low- and middle-income countries (Meagher 1995). It is also important to understand the political nature of informality in particular contexts, and how this serves to perpetuate it. For example, efforts to bring informal traders into the formal economy by introducing new regulations and taxes may be effectively sabotaged by politicians who, seeking political support, routinely overrule the bureaucrats who are trying to bring these new systems into force (Goodfellow and Titeca 2012).

While informality is not intrinsically bad, it certainly represents a second-best or least-worst solution from an economic standpoint. Operating in the informal economy may be perfectly rational from an individual's perspective, or it may be a result of deliberate exclusion. Either way, life in the informal economy carries significant risks for individuals and significant costs for society and there is growing consensus that the negative impacts of ignoring it justify policy intervention to encourage formalisation (World Bank 2005; Williams and Lansky 2013) and expand opportunities for 'decent work' (ILO 2013a).

When it comes to policies to promote the formalisation of employment, these can be conceptualised as either 'hard' or 'soft' (Williams and Lansky 2013). Hard approaches include efforts to deter individuals and firms from operating informally (e.g. through crackdowns and steep penalties) or positive incentives for compliance (e.g. easing the burden of compliance or directly supporting micro-enterprise development); soft measures focus on 'developing a culture of commitment to being lawful' (Williams and Lansky, 2013: 370) and supporting measures that improve conditions of employment and expand decent work opportunities (ibid.). Adoption of 'hard' approaches can be particularly problematic in developing countries given the scale of the informal economy, and among development researchers and practitioners there is increasing consensus that 'soft' approaches are not only more ethical, but also more effective (ILO, 2013a). 'Hard' approaches in the form of sporadic crackdowns are also frequently susceptible to being used as an opportunity by local enforcement officers to take bribes rather than actually enforce the formal rules (Goodfellow 2015a).

Contemporary policy efforts to promote formalisation have shifted away from the neoliberal deregulation approach (which appears to have been serially ineffective) and towards pro-poor, employee-centric strategies that place focus on the micro- and meso-level, instead of simply considering the role of employers and the macro-level (Chen, 2012; ILO, 2013a). By moving away from universalistic approaches, this focus not only allows for the creation of much more holistic, situated, and context specific policy, but is also in keeping with what the ILO calls a 'transition to formality' (2013a). Instead of attempting to 'solve the problem of informality' with a short, sharp shock, the idea of transition recognises that a much more gradual process is both preferable and more effective. Underpinning this transition are four broad goals for policy on the informal economy (Chen, 2012: 17–19): a) create more 'decent' jobs – preferably formal ones; b) register informal enterprises and regulate informal jobs; c) extend state protections to the informal workforce, especially the working poor; d) increase the productivity of informal enterprises and the income of the informal workforce.

It is important to recognise that informality is not merely a labour market phenomenon. As discussed above, a large proportion of the urban poor live in informal settlements where land and housing units are secured through undocumented exchanges that fall outside the reach of state regulation. Moreover, access to basic services such as water, electricity and even security and justice are also frequently negotiated through informal mechanisms and markets. Collectively, these create a condition of informality in which value and the terms of exchange and control are in a state of constant negotiation, which can exacerbate the vulnerability that characterises urban poverty (Roy and Alsayyad 2004). Indeed, definitions of what is informal in a given context can constantly shift depending on policy, or on the whims of political and bureaucratic actors with whom urban dwellers engage on a regular basis. Those who work, live and access services informally are thus subject to the constant negotiability and shifting boundaries between the formal and informal (Schindler 2013), such that they are perpetually 'walking the tightrope', to use Lourenço-Lindell's (2002) evocative phrase.

Conclusion

In order to effectively analyse, measure and address urban poverty it is necessary to go beyond aggregate statistical assessments and examine the complex of vulnerabilities confronted by the urban poor. Poor housing and limited access to decent basic services render the urban poor particularly vulnerable to debilitating and deadly diseases and natural disasters. Disease or injury can have knock-on effects by inhibiting individuals from earning an income, which is essential given the monetised nature of urban economies and the importance of money in securing basic necessities. But cash alone is not enough to ensure sustainable livelihoods, which depend upon the expansion of people's capabilities. Achieving this involves investments in intangible assets such as skills, education, social networks as well as material assets such as housing. While many make a living and secure necessities in the informal economy, there is a need to expand decent work opportunities.

Given the complex interdependencies between housing, health, income, assets and sustainable livelihoods, efforts to tackle urban poverty must operate at many levels. Economic growth and development are essential to create jobs and resources for investment, but governments clearly have an important role to play in channelling investment into goods that will ease the burden faced by the urban poor caught in a web of hazard and insecurity. As we will see in the following chapter, this requires planning, confronting the land question, delivery of critical urban services and investment in the infrastructure that shapes the urban built environment.

Summary

- Concepts of poverty have developed from a concern about subsistence to focus on basic needs and relative poverty. Measurements of absolute poverty tell us little about the relative deprivation within countries or the multidimensional nature of poverty.
- The Millennium Development Goals were a set of targets to be achieved by 2015. While influential, they have been widely critiqued for not addressing the causes of poverty or accounting for the specificities of urban poverty. They were replaced by a new set of Sustainability Development Goals in late 2015.

- While urbanisation is contributing to a reduction in global poverty, poverty is becoming increasingly urban. A product of this 'urbanisation of poverty' in the developing countries is the growth of informal settlements in urban areas.
- Inhabitants of informal settlements often experience social and institutional exclusion, as well as living in precarious environments. As such they experience high levels of vulnerability, linked to limited access to services, exposure to environmental hazards, poor building standards, and a reliance on a monetised economy.
- Throughout the 1980s, as a part of the project of neoliberalism, the rolling back of the state and reduction of social sector spending in many countries has further exacerbated the vulnerability of the urban poor. This has also had specific gendered impacts.
- More recently there has been a focus on sustainable livelihoods, highlighting that waged employment is only one factor in the livelihood strategies of the urban poor, as they seek to diversify risk and bolster security.
- In developing countries the informal economy is central to urban life, which a vast number of the urban poor rely on for their livelihoods. Informality does not equate to illegality.
- While many entrepreneurs benefit from the autonomy, flexibility and mobility the informal economy offers, waged labourers can often suffer from exploitation and a lack of stability. Governments may attempt to tackle informality through 'hard' and 'soft' approaches.

Discussion questions

1. Discuss the challenges of measuring poverty, with reference to the terms 'absolute poverty', 'relative poverty' and 'multidimensionality'.
2. To what extent have the Millennium Development Goals been successful at tackling poverty globally and in urban areas specifically?
3. What are the key characteristics of urban poverty and vulnerability?
4. What are the advantages and limitations of taking a 'sustainable livelihoods' perspective to development in an urban context?

5. How may involvement in the 'informal economy' benefit and disadvantage the urban poor?

Further reading

Alkire, S. and Santos, M. (2014) 'Poverty in the developing world: Robustness and scope of the Multidimensional Poverty Index', *World Development*, 59, 251–274.

Mittlin, D. and Satterthwaite, D. (2013) *Urban Poverty in the Global South: Scale and Nature*. London: Routledge.

Moser, Caroline (2007) 'Asset accumulation policy and poverty reduction', in Caroline Moser, ed., *Reducing Global Poverty, The Case for Asset Accumulation*. Washington, DC: The Brookings Institution, pp. 83–103.

Rakodi, Carole with Lloyd-Jones, Tony (eds) (2002) *Urban Livelihoods, A People-Centred Approach to Reducing Poverty*. London: Earthscan Publications Limited.

Redwood, M. (ed.) (2012) *Agriculture in Urban Planning: Generating Livelihoods and Food Security*. London: Routledge.

United Nations (2015) *The Millennium Development Goals Report 2015*. New York: United Nations.

Williams, C. C. and Lansky, M. A. (2013) 'Informal employment in developed and developing economies: Perspectives and policy responses', *International Labour Review*, 152(3–4), 355–380.

Websites

Eldis livelihoods page: www.eldis.org/go/topics/resource-guides/livelihoods

Eldis urban poverty page: www.eldis.org/go/topics/resource-guides/poverty/urban-poverty

International Labour Organisation (ILO): www.ilo.org

UN Millennium Development Goals Website: www.un.org/millenniumgoals

UN Sustainable Development Goals Knowledge Platform: sustainabledevelopment.un.org/topics/sustainabledevelopmentgoals

Women in Informal Employment: Globalizing and Organizing (WIEGO): www.wiego.org

5 Land, housing and urban services

- Planning and development regulation
- Land tenure and housing policy
- Critical urban services: water, sanitation and waste management
- Infrastructure and urban transport

Introduction

Underpinning both urban economic development and urban poverty reduction is the critical question of urban land: who owns it, how its use is planned and regulated, and how it is built upon, travelled across, and tunnelled under as cities grow and evolve. This chapter explores issues around land and urban planning with attention to housing, infrastructure and basic services. These amount to the most fundamental physical challenges of urban development but also have very important social, economic and political dimensions. Urban planning, land tenure, housing, services and infrastructure are also interlinked in complex ways. Understanding these aspects of the urban built environment in a holistic and integrated manner is therefore critical.

This chapter begins with a brief overview of the foundations of urban planning and the central position of land use controls within it. We explore some of the specific challenges associated with urban land use regulation in developing countries, highlighting the colonial origins of planning and regulatory tools 'imported' from high-income countries. Following this, we turn to land tenure and housing policy, examining the various ways in which governments and donors have attempted to overcome the enormous challenges of providing decent and affordable housing to their urban populations. We then turn to 'critical urban services' – sanitation, water supply and solid waste management. We consider why the delivery of basic services poses such a formidable challenge in developing countries, before reviewing different modalities of service provision and the shifting roles of the public and private sectors in recent decades. This leads to a discussion of the critical infrastructure that underpins services, with particular attention to transportation infrastructure and the broader challenge of providing adequate and affordable urban transport.

Planning and development regulation

As illustrated in Chapters 3 and 4, the size and density of urban settlements create myriad externalities and interdependencies that require active intervention to maximise benefits of agglomeration and minimise costs. Efforts to apply principles of design to urban form, to exert some regulatory control over the built environment and to plan for city-wide infrastructure have therefore been a feature of urban development since at least the Roman period. However, as a formal discipline and profession, urban planning is a relatively recent endeavour. Prior to the demographic transition and industrial revolution in Europe, cities tended to have relatively static populations once established. It was only when urbanisation began to gather pace in nineteenth-century Britain (and subsequently elsewhere) that the need for more proactive, ongoing interventions was recognised. Planning scholar Peter Hall explains how twentieth-century city planning was a reaction to the evils of the nineteenth-century city (2002: 7). Burgeoning slums, outbreaks of infectious diseases, and gross inequality alarmed the ruling and middle classes in European cities, instilling a combination of guilt and fear of revolution. The response was a wave of research, legislation and action to address the growing problems of urban poverty and inequality and create healthy and disciplined populations (Taylor 1998; Huxley 2006; Watson 2009a).

Many early urban planners combined radical normative ideas about creating a better society with attention to specific areas of urban development demanding professional attention, including housing, infrastructure development, public transport and the regulation of land and the built environment (Hall 2002). One of the central problems of urban development is how to govern the ownership, use and abuse of land (McAuslan 1985: 8); without establishing clear land institutions it is very difficult to effectively address questions of infrastructure, housing and services – all of which require access to land and effective coordination across urban space. Since urban 'master plans' became popular in the early twentieth century in Europe and the USA, land use, zoning and building regulations are the primary tools through which these grand visions have been implemented (Hall 2002). These tools have had important implications for how planning has impacted on developing countries, having commonly been exported from Europe and the United States to Africa, Asia and Latin America.

It is worth distinguishing conceptually between plans and regulations as specific instruments for guiding urban development. Plans can be thought of as sets of agendas, visions, policies, designs and strategies for physical development (Hopkins 2001), encapsulated in a 'two-dimensional layout of the physical form of the city' (Neuman 1998). Urban development regulations, meanwhile, are binding rules concerning 'what is built, where it is built, and when and how it is built' (Kaiser et al. 1995), and mostly take the form of land use regulations, zoning ordinances, and building codes. Land use regulation refers primarily to the 'user purpose' assigned to different parts of a city (e.g. commercial, high-density residential, industrial, etc.), while zoning is a particular form of this, which specifies not only land use in a given area but details such as maximum plot size, setbacks from the road and building height. Building codes (or construction regulations) are applied to the construction process itself, providing basic guidelines for building quality and safety.

These specifications are often contained in local ordinances, passed in accordance with whatever overarching plan is in place. Adopted by German cities in the 1870s, zoning ordinances subsequently became very widely used in the United States as cities expanded after 1910 (Fischel 2004). They generally have the force of law, unlike master plans, which city councils consult but are usually not bound by (Birch 2008: 142). Historically, in many cases regulations preceded formal 'comprehensive', 'master' or 'structure' plans, which imbued regulations with considerable legal power; master plans had to defer to existing zoning on any potential issues of legal dispute (Sullivan and Michel 2003).

These urban development tools were used extensively by colonists in relation to cities in colonised territories in the twentieth century (King 1976; Home 1990), with the 1932 English Town and Country Planning Act providing a model for Anglophone colonies (Okpala

2008: 11–12). In many colonies, the zoning regulations that accompanied city plans served to segregate the colonies racially in ways that were similar to apartheid South Africa in nature if not in extent, and legacies of this divisive zoning reach into the present day (Myers 2003). Alongside divisive regulations, by the middle of the twentieth century a decidedly modernist vision of city planning had spread across the globe. As countries achieved independence from colonialism this became especially popular; modernist planning ideals – characterised by the prioritisation of aesthetic appearance, large amounts of space between buildings, and clear separation of land use functions – were widely associated with 'development', 'being modern', and 'catching up with the West' (Watson 2009a). With this focus on the physical and legal aspects of modernism, many of the social concerns of urban planning pioneers gave way to top-down, universalising approaches that were overly rational and insensitive to social and cultural diversity (Allmendinger 2002). These were some of the very problems that led modernist planning to be widely rejected in wealthy countries across Europe and North America. However, in developing countries this paradigm has generally been slower to adapt (Watson 2009a; 2009b).

Despite the embrace of the modernist urban vision and its instrumental counterparts – the city master plan and accompanying land use, zoning and construction regulations – the actual development of many cities in low-income countries looks nothing like the modernists' regulated dream. Formal planning systems had been put in place in most countries by the mid-twentieth century, giving rise to static blueprints that were rigidly applied in cities across the world; yet the impact of these was often negligible in practice. With respect to Africa, for example, Mabogunje noted that 'the pervading impression is of the failure of governments […] to make any appreciable impact' on problems of urban degradation through plans and regulations (Mabogunje 1990: 121). Plans took so long to produce that 'most of what was implemented had not been planned, and informal development overwhelmed the assumptions and projections of the plans' (Taylor 2004b: 4).

On the rare occasions where planning did have an impact, planners often adhered to these outdated master plans even if a city expanded in ways and directions that bore no resemblance to what was represented on paper or anticipated by the plan. In the worst cases, modernist planning principles tinted with imperial ambitions and overt racism continued to produce cities designed to segregate and control particular communities. Some critics even conceptualised planning itself as being complicit in the uneven and highly unequal outcomes of mainstream development processes (Escobar 1995). Meanwhile, from the 1980s onwards, urban planning in general was subject to an ideological attack on the grounds that it was statist, an inefficient use of resources and interfered with market processes (see Chapter 8).

Land use and construction regulations largely remained in place despite planning's fall from grace, although with planning authorities increasingly stripped of resources, implementing plans became even more difficult. This is not to say that all developing countries have been ineffective at regulating urban physical development in recent decades; even among some of the world's poorest countries, there are striking divergences with regard to how urban land use planning and construction are used and whether they are enforced. Through a comparative study of Kampala (Uganda) and Kigali (Rwanda), Goodfellow (2012; 2013a) argues that these differences are less a consequence of dissimilar policy or varying levels of state capacity than crucial differences in the political incentives to implement regulations or override them. In contexts with a discontented urban population and relatively active political opposition, politicians may find that their best hope for building popular support is to informally overrule plans and regulations that impose costs on their urban constituents. This may gain them new supporters, but at the expense of planned urban development.

Dynamics such as these feed informality in the sphere of physical urban development, as the widespread overriding of formal regulations leads to increasing amounts of land use and construction that is either illegal or of ambiguous legal status. For instance, Roy argues that the reason India 'cannot plan its cities' is not due to planning incompetence but the fact that the planning system is itself informalised, with the line between what is 'legal and illegal, legitimate and illegitimate, authorized and unauthorized' constantly shifting (Roy 2009: 80). This arbitrariness not only renders city-dwellers highly vulnerable to the whims of politicians and officials, but places the state itself outside the law and renders planning a façade in the face of informal state practices. Planning and regulatory systems that are obviously out of date and inappropriate can be difficult to change, precisely because their presence provides power-holders with opportunities for political and economic gain (Watson 2009b; Fox 2014).

Given the way that land use planning and regulation is used and abused in this way, some scholars have argued that basically abandoning it altogether can have positive outcomes. Sims (2011) makes the point that many of Cairo's informal settlements, which have been left to grow organically and are largely beyond the reach of the state, are in many respects more functional and appealing places than the formal suburban areas promoted by the state. China has also experimented with various approaches to the regulation of land use at the rural–urban interface, including suspending regulations in certain places at particular times (see Box 5.1). In contexts such as this, where urban populations are growing fast, there are particular difficulties in trying to impose regulatory norms on communities accustomed to collective, local control over their land and what they do with it.

Box 5.1 The growth and transformation of urban villages in China

After the *hukou* (household registration) system was introduced in China in 1958, movement between cities and the countryside was strictly limited. From 1985, however, rural migrants were allowed to register as temporary residents in urban areas, which spurred massive movement to the cities. In cities such as Shenzhen, one of China's industrial powerhouses, this led to many rural migrants settling on farmland on the urban fringe. Before long, these areas were surrounded by the ever-expanding city. However, they were still classed as rural land (which in China is collectively owned by communities) rather than urban land (which is owned by the state). Consequently, these areas are known as 'urban villages'. By 2005 there were 241 such villages in Shenzhen alone.

The fact that the land in urban villages is classed as rural has important implications. Traditionally in China, rural areas have not been subject to development control, with village communities deciding how they wish to develop their land and acting accordingly. As the migrants living in urban villages are mostly poor (earning around half of what official city residents earn), the built environment in urban villages is densely packed, based around courtyards and with low-grade facilities, in contrast to more 'modern' looking high-rise development of officially urbanised areas.

The Shenzhen government tried to introduce some basic regulations for construction in urban villages, fearing chaos in the context of rapid in-migration. These limited the size of plots and the proportion of the plot that could be built upon. However, the more regulations they introduced, the faster villagers built on their existing plots, seeing it as their last chance to make use of their available land and earn money by renting out rooms to migrants. In the mid-1990s, the municipality took the radical measure of stopping the approval of any construction at all. Ironically, this just meant that people built bigger and taller because the existing size and height regulations had been rendered meaningless. These trends were exacerbated by the fact that urban villagers were cut off from the land so could not farm, yet were also relatively uneducated; their property was therefore their primary or only route to making a living. At the same time, villagers' collective land rights do not extend to the right to sell the land, so selling their land to property developers for money was not an option.

The municipality's next move was to change tack and legalise all the unauthorised housing but charge penalties on anything over a certain size or height. This encouraged yet further illegal construction, as the penalties paid were easily offset by the money they made from renting to migrants. In many cases across China, urban villages have now been expropriated by government and turned into urban land. Invariably, new urban villages have then appeared on the peripheries; indeed the level of demand for cheap, relatively low-quality housing is so great that a certain degree of informality is likely to continue as China's urban transition progresses.

Sources: Wang et al. 2009, Wu et al. 2013.

So are land use and other development regulations part of 'the problem' in cities in developing countries? Such legal and regulatory tools are indeed often criticised for imposing costs on the urban poor and restricting their livelihood opportunities, for example by requiring them to build to standards they cannot possibly afford or preventing them from conducting their business from particular parts of the city (Rakodi 2001; Brown and Lloyd-Jones 2002). In many cases this is arguably because approaches to zoning and land use regulation have been inappropriate to developing country contexts. Colonialism again looms large here; laws and regulatory codes were often rooted in the European experience and 'imported verbatim from the colonising power with little subsequent change' (Watson 2009a: 176). Others, however, argue that over-regulation of urban space is a problem in and of itself, regardless of colonial origins. This position was most famously articulated by Peruvian economist Hernando de Soto, who enumerated the costs faced by the urban poor as a consequence of legal and regulatory processes they are subject to (de Soto 1989). For de Soto, legal institutions in urban Peru had become irrelevant in the face of migration and urbanisation, serving only to push the poor into a vulnerable status of extra-legal (i.e. informal) existence.

While there are undoubtedly costs associated with land use regulation, the influence of de Soto and the ideological climate since the 1980s has led to a tendency to focus on the negative aspects of land use regulation rather than the benefits. The latter can also be substantial, as studies of Malaysia and Ghana have shown (Bertaud and Malpezzi 2001; Baffour-Awuah et al. 2014). Moreover, in the context of climate change and environmental threats to cities (see Chapter 6), the case for regulating the use of land and the quality of the built environment is arguably stronger than ever (Godschalk 2003; Pelling and Wisner 2009). In both rich and poor countries, land use planning and regulation have crucial roles to play in promoting sustainable cities (Bulkeley and Betsill 2005). Nevertheless, to be effective and pro-poor this needs to be pursued in a contextually sensitive manner that takes on board the different 'rationalities' of planners and urban populations, and works with rather than against the existing practices of the latter (Watson 2003; 2009b).

Land tenure and housing policy

The question of how to plan for and regulate the use of land cannot be divorced from that of land rights and land tenure. Land tenure refers to the set of rules governing how land is acquired, sold, used and transferred in a given context. Systems of land tenure amount to bundles of property rights which determine who can use which parcel of land for how long, and under what conditions. These are crucial issues for development generally, as the rights and relationships defined by land tenure systems shape not only economic development but also social interactions, cultural values and power relations (FAO 2002). Until recently the centrality of land tenure for understanding urban development was often overlooked, with attention focused primarily on the importance of land tenure for agriculture and rural livelihoods. Yet in urban contexts there are often multiple, conflicting land tenure systems within relatively small and densely populated spaces, which create numerous obstacles and opportunities for urban development. Indeed, urban land tenure systems are central to determining how planning systems work and how people access and hold onto the most basic of needs: housing.

Consequently, in the very late twentieth and early twenty-first century, tenure reforms that aimed to provide people with secure legal rights to their land became a central pillar of approaches to addressing the urban housing challenge, which is one of the most central and enduring problems of urban development. As described in Chapter 4, hundreds of millions of people across the world live in housing that is widely acknowledged as inadequate, and how to effectively address this problem has eluded policymakers for decades. Before the surge of interest in land tenure reforms as a solution to the housing question – an issue to which we return below – a wide array of different approaches to housing supply, settlement upgrading and housing finance have been attempted across the world.

One of the underlying challenges of developing effective housing policy relates to the nature of housing itself and the tensions between housing as a shelter need (or consumption good) and housing as an economic good – an asset with a market value. Housing can form an important part of a city's fixed capital, with a buoyant housing market helping to fuel economic growth in other areas. At the individual or household level, housing can be critical for avoiding impoverishment and improving fortunes. While for some people shelter represents little more than a roof over their head, for others even a simple dwelling can be their most valuable possession and its value may increase over time. When informal settlements are regularised, land and housing values generally increase. The same applies when urban services are extended to slums through upgrading schemes. A house can also improve the economic wellbeing of its owner or occupier in other ways. For instance, just having an address makes it easier to find employment. A house can be used as collateral for securing credit, as a site for home-based or small-scale enterprise and can be a potential source of income through renting out rooms.

Viewed over the long term, housing has been seen primarily as an economic good, and therefore something that people should invest their own effort and resources in. However, during the early development decades, leaders of many newly independent states were inclined to see housing as a shelter need and initiated large-scale housing projects. In some measure, this mirrored trends in the post-war welfare states of Europe at the time as well as the modernist planning paradigm discussed above, and in part it reflected a desire to redress the injustices of colonialism and to wear post-independence development as a badge of national pride. Yet these high-profile housing projects were never meant to house the legions of poor people who came to populate cities, given that at this time it was widely believed that urban growth could be prevented through policy interventions (see Chapter 1). When it became clear that the provision of low-cost housing was an imperative, it was assumed that affordable public housing could be delivered incrementally by the state. Typical of the housing provided at this time were four-storey walk-up apartment blocks – now often tenements – which can still be found in cities across Africa and Asia, from Cape Town to Cairo; Kolkata to Cochin. In the end though, this response was both inadequate to need and unaffordable for most, and construction of conventional housing of this sort never constituted a very large proportion of shelter options in cities of low- and middle-income countries (Jenkins et al. 2006).

By the 1960s it was becoming clear that low-income households were finding their own shelter solutions, constructing makeshift dwellings in informal settlements or crowding together in slums (Turner 1972; 1976; Mangin 1967). The creative solutions that poor urban dwellers developed autonomously and without the help or interference of bureaucracies started being cast in a positive light. Turner (1972) spoke of 'housing as a verb' and argued strongly that if the urban poor were given security of tenure and a plot of decent land they would incrementally achieve for themselves respectable housing. It was argued that the responses of squatters who engaged in self-help housing solutions were just as rational as those of the middle- and upper-income classes. As such, they required facilitation and support through cost reduction measures and the lowering of standards for housing, not evictions and demolitions.

Champions of self-help housing were bent on showing that the urban poor are the best managers of their own housing solutions, and policymakers were more than ready to accept such arguments. After all, governments needed to find ways to finance low-cost housing that were economically feasible. This signalled a major shift in mainstream approaches to housing policy towards the promotion of reduced standards and low-cost solutions in the face of rapid urbanisation (Jenkins et al. 2006). This generally took one of two forms. The first is sites-and-services schemes whereby governments make serviced land available, divide it into plots and sell it to people who are responsible for constructing their own houses. In well-serviced schemes plots will have access to roads, drains, water supply and electricity. Minimalist approaches can include nothing more than a marked out plot and a pit latrine.

In an extension of the sites-and-services idea, sometimes core housing is provided. This comprises a simple structure on a serviced plot, with the idea that the owner can in time build on it. This was a part of the strategy pursued by the post-apartheid South African government, which sought to redress the housing deficit among historically

disadvantaged populations through its Reconstruction and
Development Programme (RDP). This aimed to build 300,000 houses
a year, with a minimum of one million low-cost houses to be
constructed within five years through the vehicle of government
subsidies to low-income citizens. The programme largely succeeded
in terms of number of units delivered. However, the plots are often
too small for decent size extensions, and people also used the housing
in all sorts of ways that were not anticipated by the state, confounding
government efforts to assess whether the programme had achieved its
objectives (Charlton 2013). More generally, core housing schemes
are often beyond the reach of most poor people, with units commonly
ending up in the hands of middle-income people who rent them out.
Even minimalist sites-and-services schemes can end up marginalising
the poorest households. For example, people do not always have the
skills to construct their own dwellings to a standard that meets
government regulations and cannot afford to employ others to build
their homes.

As the problems with sites-and-services schemes became evident in
the late 1970s, the emphasis shifted towards the second dominant
mode of low-cost solution: slum upgrading (Jenkins et al. 2006).
Since the mid-1980s this has been the housing policy of choice in
many countries because it offers a number of important advantages.
Unlike moving people to new sites, it preserves existing economic
systems and opportunities for the urban poor, including proximity to
formal sector jobs and employment and entrepreneurial activities in
the informal economy. It also preserves the low cost housing stock
already in existence at its present location, protects community
structures that have been built up, and avoids the disruptive costs of
resettlement – including the potential of fuelling discontent and
political opposition (Martin 1983). Slum upgrading projects usually
include improving infrastructure and services such as water supply
and sewerage, building health clinics and schools, as well as
providing financial services and building assistance to individuals.
Valued by residents, upgrading is often resisted by urban elites,
especially those living adjacent to low-income communities slated for
upgrading. Opponents to upgrading projects often argue that
regularisation legitimates illegal settlements; that providing slums and
informal settlements with services raises the value of assets that
people have acquired by breaking the law through squatting; and that
upgrading gives these settlements permanence, potentially lowering
the value of houses nearby. Slum upgrading programmes can also

strengthen the rights of tenants against local 'slumlords', generating fierce resistance from the latter.

Slum upgrading was also criticised on the grounds that it provided insufficient attention to the legal and institutional foundations of the housing problem, focusing too much on infrastructure at the expense of attention to providing legal tenure (Werlin 1999). This chimed with the emerging consensus on the importance of property rights in development generally, and a focus on security of legal tenure as critical for incentivising people to invest in their own housing – ideas that came to fruition with the publication of Hernando de Soto's landmark book *The Mystery of Capital* in 2000. This argued that the central problem of development is that in most countries many people lack clear legal title to their property, which both discourages people from self-investing to improve their dwellings and constrains the ability of the poor to realise the value of their assets (e.g. by using their property as collateral for loans) (de Soto 2000). As a consequence of such ideas, land titling (or 'land tenure regularisation') schemes have been initiated throughout the world. It was also thought that land titling programmes would yield better results in urban than rural areas because a culture of land transactions is more developed in the former, and there is less likelihood of formalised land rights competing with existing informal systems (Sjaastad and Cousins 2009).

Experience with titling programmes has for the most part produced very mixed results and not borne out these high hopes. The introduction of tenure regularisation can lead to speculation and price increases that impact negatively on the poor (Payne et al. 2009; Briggs 2011), and while some studies indicate it can have poverty-reducing effects in the long term, it has not been found to facilitate access to credit (Galiani and Schargrodsky 2010). Even in Peru, de Soto's homeland and inspiration, researchers found that the issuing of land titles did little to help poor people gain either jobs or credit (Kagawa and Tukstra 2002). Payne (1997) also suggests that such schemes may encourage the development of new extra-legal settlements whose residents anticipate being granted titles to the land they occupy. Furthermore, the expected benefits of formal titling may not materialise if a well-regulated financial system is not already in place and able to provide credit for land or housing improvements (ibid.). The titling process may also provide an opportunity for well-connected individuals to seize land, having better access to the institutions and organizations managing the titling process.

Meanwhile, there is no evidence that titling improves access to basic services (Payne et al. 2009).

Contrary to the expectation that in urban areas formal land titling would obliterate other informal or traditional systems of land tenure, competing systems frequently remain in place. In the Mozambican capital Maputo, for example, a World Bank-funded formal land titling programme did not displace informal practices that involved payments to local authorities in exchange for other, extra-legal land documentation. The uneasy combination of the two systems ultimately made people even more insecure (Earle 2014). With this in mind, it is important to distinguish between legal tenure and security of tenure (see Box 5.2). The latter requires more than simply holding a title, necessitating the continuous long-term possession of land that provides a *de facto* sense of security in addition to the possession of a document (Bouquet 2009). Some have even argued that security might be best achieved through means other than full property titles; in a study of the *Favela Bairro* slum upgrading programme in Rio de Janeiro, for example, Handzic (2010) argues that securing a 'concession of right to use' their land rather than full tenure reduced the bureaucratic burden on the poor while also limiting the 'massive gentrification' that can arise when full title is awarded (Handzic 2010: 16). Evidence from Sri Lanka likewise suggests that governments should 'emphasise *de* facto property rights and avoid the expense of full titling programmes wherever possible' (Redwood and Wakley, 2012: 166). The crucial thing is to provide a level of security that can prevent the state from engaging in forced evictions, keep records of land use through incremental upgrading programmes and promote occupancy rights if not full land titling (UN-Habitat 2003b: 171).

Box 5.2 Slum notification in India and the difference between legal and *de facto* tenure security

In India, state and local government agencies have implemented slum notification, a type of tenure formalisation policy that provides certain levels of property rights. Households in so-called notified slums are legally protected from eviction for a specified period of time, and are entitled to infrastructure and services provided by local municipalities. Although slum notification is based on a law dating back to the 1950s, implementation varies enormously and the proportion of notified slums in Indian states and cities ranges from 83 per cent in Andhra Pradesh to 9 per cent in Delhi. Renewed attention to slum notification and the possible benefits of implementing the law has been spurred by the widespread belief within the international community that property rights encourage slum households to invest in their properties.

While research suggests that living conditions in notified slums are overall better than in non-notified settlements, recent slum surveys challenge the impact of slum notification on people's housing investments. Findings revealed that although the average amount of money spent for improvements was higher in notified slums, households in non-notified slums are more likely to invest in self-help housing constructions. The proportion of households that take action to improve their houses is estimated to be 3.4 per cent higher in non-notified slums.

These findings underpin arguments made by researchers and development practitioners who highlight a difference in legal and *de facto* tenure security (i.e. security acquired by means other than official legal status, and often through other forms of social and political protection). According to these studies, households will improve their living conditions despite the lack of a legal status as long as they feel secure. Perceptions of security often depend on informal political structures, patronage systems or networks with municipal authorities, which appear to be stronger in non-notified settlements. Similar observations have been made in marginalised neighbourhoods in Lahore, Pakistan, where access to services and political support seem to play a stronger role in household decision-making than the provision of property titles.

Sources: Nakamura (2014), Wajahat (2012).

De Soto's approach has also been criticised on the grounds that it amounts to a minimalist agenda with little actual redistribution of wealth or power towards the poor (Payne 2001; Gilbert 2002; Payne et al. 2009). Titling programmes may even provide justification for the reduction in government support to low-income households (Davis 2006: 71–72). Indeed, the focus on titling reflects a more general shift in housing policy over several decades that has seen governments moving from the role of provider (of housing units or sites and services, for example) to that of 'enabler'. Through reforms to legal frameworks and financial support mechanisms, the state is seen as providing a series of basic incentives on top of which 'entrepreneurship in the private sector, communities, and individuals can effectively develop the urban housing sector' (Pugh 1995: 67).

As the housing sector was reconceptualised as something in which the state should be just one among many players, the contributions of governments, markets and individuals to the development of housing markets were initially assumed to be separate and divisible. It soon became clear, however, that coordination and cooperation were necessary. By the 1990s, international housing policies had evolved towards a 'whole sector' approach (Pugh 1995; 2001), seen as involving the development of housing finance, targeted subsidies, regulatory audits and improved organisation in the building industry as well as infrastructural improvements and secure property rights. This was a challenging prescription through which different institutions and economic sectors involved in housing were supposed to become more effective in their own right and also in their systemic relationship to each other. Examples of relatively successful comprehensive approaches of this kind are rare, but might include Singapore and Chile, both of which mobilised compulsory social security savings as part of the route to developing housing finance (Pugh 2001: 412).

Most of the attempts to take a 'whole sector' approach focus on providing incentives for private investment in developing a range of housing options for different income groups, rather than concentrating only on housing the urban poor. In reality, the latter almost always get left behind. For example, an approach known as the ABC model – standing for 'Ahorro' (savings), 'Bono' (subsidy) and 'Crédito' (loan) – has proved relatively successful at bringing in private housing finance in several Latin American contexts, but this has mainly benefited middle-income residents (Bredenoord et al. 2014). The same problems have affected countries the world over: even with incentives in place, private developers cannot make adequate profits from housing low-income groups, who are also too poor for the usual mechanisms of housing finance – such as long-term debt finance through commercial mortgages – to be effective (UN-HABITAT 2005). In cases where a developmentally oriented state is willing to shoulder more of the initial burden and put particular effort into targeting housing towards the poor – as in the Ethiopian case in Box 5.3 – some of these problems can be mitigated; but the housing units still rarely end up in the hands of the poorest.

These problems of finance tend to be compounded by low levels of domestic savings, as well as broader governance problems, insecure tenure and lack of municipal finance to provide basic infrastructure. Given the lack of large-scale finance in many contexts, shelter microfinance and community-based shelter funds may be the most that the poor can access, which generally can only support incremental upgrading (UN-HABITAT 2005). Microfinance can never address the totality of housing needs, but community mobilisation around shelter finance and broader shelter concerns is beginning to challenge some of the market-based models that so often

Box 5.3 The condominium housing scheme in Ethiopia

Since 2005, the government of Ethiopia has been implementing its Integrated Housing Development Programme (IHDP), an ambitious government-led programme targeted at low- and middle-income city-dwellers. The programme's initial goal was to construct 400,000 housing units, mobilise domestic savings by requiring people to save for the initial down payment, create 200,000 jobs and promote the development of 10,000 micro and small enterprises to enhance the capacity of the construction sector. This programme was conceived against a backdrop of exceptionally poor housing quality in Ethiopia's urban areas, over 80 per cent of whom were living in areas classed as slums by UN-HABITAT.

The IDHP programme is based on a condominium model of housing, whereby people purchase an individual unit in a storied block and residents collectively own and are responsible for shared areas. In many respects the programme has been a success. By 2010, 171,000 units had been constructed. A lottery system was developed for allocating the units. Within this, a 30 per cent quota is reserved exclusively for female-headed households, and evictees from the sites of condominium development were automatically allocated a unit (provided they could afford the down payment). Cross-subsidisation was used to make sure that the one-bed studios were especially cheap, with these being targeted at the poorest.

Plate 5.1 *Public housing under construction in Addis Ababa*

Source: Tom Goodfellow.

Affordability has nevertheless been the main concern with how the project has performed, and the poorest are frequently unable to benefit as they cannot afford the down payment of 20 per cent. In response to this and other concerns, the government diversified the programme, offering multiple condominium schemes. A scheme aimed at the poor now only requires a 10 per cent down payment, and local authorities have to verify incomes to make sure only the poorest can register.

Unfortunately these measures have not proved entirely successful, with the units for the very poorest undersubscribed; even 10 per cent may be unaffordable to low-income groups, who also tend to have large families so need more than a one-bedroom flat. In reality, poor families who manage to win a unit commonly rent it out to higher-earning groups due to their inability to afford the monthly repayments. Social problems have also emerged due to the fact that people are unused to living in storied accommodation, and blocks have very few communal spaces for activities such as family cooking or slaughtering livestock, leading to competition and conflict. The newest blocks are also far from the city centre, often without easy access to employment, generating high transport costs.

Despite these problems, the scheme has delivered an impressive number of units for a very low-income country. Even though poor families generally rent these to wealthier ones, this still improves their economic situation, providing an income stream that allows them to move into better quality settlements than they were in previously. The scheme has also helped to build local enterprise in the construction industry and created a substantial number of jobs. Notwithstanding its ongoing challenges, this model is therefore becoming influential with regard to housing policy elsewhere on the continent.

Sources: UN-HABITAT (2011), Ejigu (2012), Ondakie et al. (2015), field research by Tom Goodfellow.

fail to meet the shelter needs of the poorest. The activities of Slum/ Shack Dwellers International, for example, are helping some communities to negotiate over regulatory frameworks, pressure governments to provide basic infrastructure, and shield individuals from market-based vulnerabilities by securing group, rather than individual, land tenure (Mitlin 2011).

The question of where resources are best concentrated in the effort to improve housing for the poor is still far from resolved. The explosive growth of slums in many regions of the world, as detailed in Chapter 4, is testament to the widespread failure of all the above approaches to deliver for the urban poor. Given this reality, observers are increasingly questioning whether the drive to produce societies of property-owners – which all of the above schemes ultimately aimed to do – has been misguided. Home ownership frequently does not have the poverty-reducing affects it is assumed to have (Lemanski 2011), and can even worsen peoples' situation: for example, urban renewal schemes in

Istanbul have given many poor households little choice but to relocate to flats they must purchase through mortgages with long maturities at supposedly 'affordable' rates. This has been interpreted as forced incorporation of lower-income residents into mortgage payments that may actually further impoverish them (Karaman 2013).

These problems have led to calls for increasing attention to be paid to the potential of rental housing (Bredenoord et al. 2014), which until recently has also been largely ignored by international development agencies. Informal rental markets exist all over the world and perform important roles, including generating income for low-income city-dwellers who rent out rooms, backyard shacks and even floor space to even poorer tenants (Crankshaw et al. 2000; Beall et al. 2002; Gilbert and Crankshaw 1999; Kumar 2001). When policy attention is paid to the rental housing option, landlords are frequently portrayed as exploitative, but many are poor and earn only a modest but vital income from their housing asset (Kumar 2002). Instead of promoting rental solutions, government strategies such as de-densification of overcrowded areas can undermine them. This hampers the drive to eradicate housing poverty while at the same time compromising an important livelihood strategy of many low-income urban dwellers.

Critical urban services: water, sanitation and waste management

The health and wellbeing of urban residents everywhere depends upon the effective, efficient and affordable provision of safe water supply, sanitation and the removal and disposal of solid waste or garbage. Billions of people across the world lack access to adequate services – especially sanitation: 2.5 billion are currently estimated to lack 'improved sanitation' (WHO and UNICEF 2014). While the proportion of the global population without access to improved sanitation declined from 46 per cent to 39 per cent between 1990 and 2008, in urban areas it increased slightly, which in real terms represents a massive overall increase in the number of urban dwellers without access to this vital service. In some countries, urban areas are particularly badly affected; for example in the Democratic Republic of Congo, rural access to sanitation increased substantially from 11 per cent to 33 per cent between 1990 and 2012 but urban access to sanitation declined from 32 per cent to 29 per cent over the same period (WHO and UNICEF 2014: 56). Moreover, poor delivery of such services has especially pernicious effects in cities, where population density and residential concentration multiply the risk of disease transmission, as discussed in Chapter 4. Women tend to be particularly adversely affected by poor sanitation provisions. In places such as the Dharavi slum in Mumbai, many are forced to defecate in the open and stigma prevents them from going in daylight hours so they have to wait until after nightfall or go before sunrise, at considerable personal risk.

There have been more substantial advances with regard to water access in recent years. Nevertheless, over 700 million people still depend on unimproved drinking-water sources (WHO and UNICEF 2014). There are also major regional disparities: while 89 per cent of the world's population is now considered to have access to improved water supply, in sub-Saharan Africa the average is 64 per cent (ibid.). The figures tend to be higher in urban than rural areas, with at least 80 per cent of urban populations having access to 'improved' water even in some of the poorest countries – but with some notable exceptions: in Sudan for example, the number of urban dwellers able to access 'improved' water declined from 86 per cent to 66 per cent between 1990 and 2012 (WHO and UNICEF 2014).

There is also reason to question whether what the WHO classifies as 'improved' water and sanitation really represents an adequate service (Bartlett 2003). Some observers have argued that the census and survey data on which improved water and sanitation figures are based commonly do not ask critical questions such as how many people share facilities, how long it actually takes to access facilities, and how regular and safe the supply of water is (ibid.; Hardoy et al. 2001). One report even suggested that the number of African urban dwellers without adequate water provision could be as much as three times the amount suggested in figures on 'improved' provision (UN-Habitat, 2003a).

In addition to water and sanitation, solid waste management (SWM), or the collection and disposal of garbage, constitutes the third critical urban service upon which managing the disease environment of a city depends. SWM is often viewed as the 'Cinderella' of urban services: neglected in favour of water supply, which is vital for life, and sanitation services that are more evidently linked to issues of health and dignity. Nevertheless, poor SWM can impact negatively on both water supply and sanitation by blocking drains and sewers. This in turn can lead to the build-up of domestic and human waste, polluting water sources and breeding disease.

Plate 5.2 *Exposed water pipes in a slum area of Kampala, Uganda*

Source: Tom Goodfellow.

In addition to the major deficits in water and sanitation provision, most large cities in low- and middle-income countries have inadequate waste-collection and disposal services. In the Zambian capital Lusaka, for example, at the turn of the millennium around 90 per cent of the 1,400 tonnes of municipal waste produced every day was left uncollected (Hardoy et al. 2001: 80). Where collection does take place, in many cities it is invariably in commercial centres and wealthy districts. Residents of high income areas are able to access regular municipal waste collection services through their political influence or by paying for private collection services, but this is not an option for residents of low-income areas. Here people are forced to burn rubbish or dump it in ditches, lakes, rivers, on pavements and by the roadside.

The effects of inadequate waste management on the poor are powerful indicators of, and contributors to, socioeconomic inequality. Utilising the concept of 'ecological distribution', Baabereyir et al. (2012) show how the domestic waste burden breeds environmental injustice in the Ghanaian capital, Accra. The inequalities in spatial distribution of waste collection services and waste disposal sites ultimately reflect the uneven distribution of power within Ghanaian society (ibid.). Even where waste collection is taken care of satisfactorily, the ultimate disposal of waste constitutes a real problem for many cities. In the 'new towns' surrounding Delhi, for example, the municipal government has placed considerable emphasis on waste collection, driven by the priorities of India's 'new middle class'. When it comes to processing and disposal, however, the system unravels due to inadequate prioritisation and regulation (Schindler and Kishore 2015).

Managed landfill sites are rare in low- and middle-income countries. Instead, rubbish ends up in open landfills that are often tens of metres high and in serious danger of collapse. Many are located near informal settlements, and in some cases people who make their livelihoods from retrieving and recycling waste materials deliberately form residential communities on or near city dumps. A stark illustration of inequality in relation to urban waste relates to higher volumes of garbage as a result of rising levels of affluence alongside levels of poverty that see legions of men, women and children who make a living from this waste. These are the scavengers who have become the leitmotif of urban poverty, living on the garbage dumps of Manila and Mexico City and rummaging through the bins and waste heaps of Kolkata and Karachi (see Plate 4.2). They are the most visible symbols of the paradoxical relationship between the environment and inequality: namely that affluence produces abundant waste, while poverty does not; that poverty encourages efficient reuse and recycling of waste materials, while affluence does not; and that urban livelihoods built on resource conservation and recycling, ironically and tragically, are predicated upon persistent inequalities in income and consumption (Beall 1997b).

For all critical urban services it is the poorest districts most in need of serious assistance that pose the greatest risks for those engaged in service provision. When local governments lack the resources or political will to invest in provision, private actors often fill the gap, naturally giving preference to wealthier neighbourhoods. In better off areas the terrain is usually more suitable for the installation of bulk infrastructure, the locations are more central and accessible, and the payment of user charges is more predictable. In slum areas, on the other hand, numerous risks abound. Informal or illegal settlements are often located far away from existing trunk infrastructure such as water mains and sewerage networks, where the costs of upgrading infrastructure and improving service provision are high because the terrain is inaccessible or dangerous and land tenure is uncertain or illegal, and where layouts are dense and hence difficult to access.

The case for improving essential urban services is incontrovertible, for reasons of equity, to improve the health and capabilities of the urban poor, and to ensure the efficient functioning of cities and urban populations more generally. Better critical urban services have positive knock-on effects on household incomes, school attendance, family and community relations and many other (often gendered) outcomes. To take an example related to education, it is common for children – especially girls – to skip school in order to wait for water deliveries from private vendors (InfoChange News 2005). Meanwhile a study in Addis Ababa found that, among other benefits, improved water supply meant the time spent collecting water was reduced from 6–8 hours to 5–20 minutes, resulting in much less fatigue for women. The incidence of eye diseases and postnatal infections was also reduced, alongside improved menstrual hygiene, increased self-esteem and more time available for family and community activities (WaterAid 2001).

Given the urgency of the urban services challenge, it is difficult to understand why there are such gaping deficits in their provision. Part of the explanation lies with the scale of urban growth, discussed in Chapters 1 and 2; yet growth alone does not adequately account for failures to meet the challenge. Obstacles to more effective basic service provision are multiple, but can be categorised in terms of a simple trichotomy: supply constraints (for example the cost and technical complexity of providing good water and sanitation infrastructure), demand constraints (such as low willingness to pay for services), and institutional constraints (such as horizontal governance failures where neighbouring municipalities fail to cooperate sufficiently to manage services effectively) (Duflo et al. 2012). Problems such as urban sanitation provision are also subject to collective action failures at the community level, although community-driven improvements to sanitation also offer important opportunities and have led to considerable successes, particularly in South Asia (see Box 5.4). It is through this kind of 'micro-solution' that the most significant gains might be made in the absence of funds or political will for major infrastructure investment. As Duflo et al. (2012: 15) point out, 'The urban services challenge is not purely a technical one'. In addition to technology and infrastructure there is considerable scope for improving critical urban services through better pricing and revenue models, attention to breaking down barriers to collective action and creating incentives for politicians to direct adequate resources towards services.

Box 5.4 The challenge of community-driven urban sanitation improvement

The 'right to sanitation' is rising up the international development agenda, with universal access now embedded in the Sustainable Development Goals. Given difficulties in covering the cost of widening access through conventional means, finding low-cost solutions is increasingly urgent. Much attention has turned to community-driven sanitation projects, but these also face considerable challenges, especially in urban contexts. McGranahan (2015) identifies four major challenges to community-led efforts to widen access to sanitation in deprived urban areas.

The first is the challenge of local collective action. Collective demand for better sanitation is often not converted into changed behaviour or investment because the benefits any individual receives depend on the action of others – only if everyone improves sanitary behaviour will local communities enjoy improved public health. One approach to this problem, especially in rural areas, has been 'community-led total sanitation' (CLTS) programmes. These involve discussing the state of sanitation publicly, to 'ignite' collective disgust, leading to agreements about the need for collective investment and maintenance, enforced through community self-regulation. This, however, has limited purchase in urban areas due to the need for infrastructure investment and additional resources as well as behaviour change.

The second challenge is that of co-production: the idea that public services are best provided by a combination of government agencies and organised citizens. Communities have a comparative advantage in managing low-cost systems within the community, while public agencies can better develop technologically sophisticated systems outside it. Co-production can be successful when, for example, communities organise effectively to operate a toilet facility while government takes responsibility for removing and disposing of human waste. However, achieving effective cooperation can be hard where levels of 'social capital' are low both in the public sector and informal settlements.

Third is the challenge of affordability versus acceptability. In practice, sanitation considered acceptable by authorities is not affordable to people on the lowest incomes, particularly when subsidies are removed in order to scale up pilot projects. International donors are also reluctant to accept standards that could be taken to condone unsafe or degrading sanitation facilities. Both locally and internationally, there are strong pressures to prioritise improvements that meet some minimum standard over ones that have the potential to reach everyone in need.

The fourth challenge is ensuring that tenure issues do not undermine the incentive to improve sanitation. For example, utilities and local authorities may not officially be allowed to provide sanitation and water services to unauthorised settlements. Tenure insecurity can also reduce the incentives for people to invest in their own toilet facilities. Although there may be incentives for local politicians to improve sanitation, issues of tenure can undermine maintenance of facilities.

Two successful attempts to overcome these challenges have been in South Asia – the Orangi Pilot Project in Karachi, and the Alliance in Pune and Mumbai. Both of these depended on coproduction (for example the division between 'internal' and 'external' infrastructure, with lane residents responsible for the former and a public provider for the latter). They also both implicitly distanced themselves from 'rights based' approaches by privileging affordability over conventional standards of acceptability. Crucially, they also started out with agendas much broader than sanitation, with sanitation emerging as a priority through lengthy dialogues with low-income communities. This helped to address the collective action challenge.

Source: McGranahan 2015.

It is important to understand some of the ways in which approaches to service delivery have evolved in the context of the turn towards market-led approaches since the 1980s, and why these have often failed to deliver in the developing world. Central to this was the shift towards privatisation of services, in full or in part. The various different arrangements by which a state can transfer responsibility to private providers are summarised in Box 5.5, specifically in relation to water and sanitation. The case for private provisioning of urban services was made primarily on economic but also environmental grounds. First it was argued that precious commodities such as water have an economic value that should be reflected in prices. Second, when the public sector provides scarce goods such as water at subsidised rates, it was argued with some credence that they are over-used. When produced by private operators, and with the full costs of production passed on to consumers, the belief was that this creates incentives to use such scarce resources more parsimoniously. The argument was also made in terms of equity, on the grounds that markets are more likely than states to deliver the greatest good to the most people over time.

Whether we categorise particular urban services as natural monopolies, public goods or economic goods can greatly influence the organisational arrangements for delivery of that service, with some services lending themselves more naturally to private sector provision than others. For example, Batley (1996) placed sanitation close to the public goods end of the spectrum, arguing that because poor sanitation has huge negative externalities, there is a strong case for keeping it public, except for specific works that can be contracted out to private firms. By contrast, he identifies waste *collection* as an

---

Box 5.5 Types of contracts for water and sewerage projects

Service contract: The most basic and short-term arrangement, whereby a private company is contracted to perform a discrete activity such as fixing pipes or collecting bills.

Management contract: A more involved transfer of responsibility where the management of certain operations is handed over to the company but investment and expansion remains with the state.

Concession: The private contractors manage the whole utility at their own commercial risk. They are also required to invest in the maintenance and expansion of the system.

Build-own-transfer (BOT): Similar to concessions, BOT contracts are usually used for new projects where the private contractor is responsible for constructing infrastructure from scratch.

Divestiture: A model of full privatisation where a private company purchases the assets from government and takes over their operation and maintenance as a commercial business and on a permanent basis.

Joint ventures: An arrangement rather than a contract whereby private investors form a company with the public sector, which then takes on a contract for utility management.

Sources: Budds and McGranahan (2003), UN-Habitat (2003a).

ideal area for private provision as one can exclude those who do not pay for a door-to-door service. Waste *disposal*, on the other hand, has the properties of a public good because the negative impacts of poorly managed waste dumps and landfill sites affect everyone.

There are difficulties, both theoretical and practical, in seeing water as a private economic good, and water privatisation has been especially contentious. Water, like sanitation, possesses all the characteristics of a natural monopoly: it depends on large investments in a unified system of infrastructure capable of delivering a functional service at a city-wide scale. In addition, water is a public good, being necessary for the life and health of all citizens. Together these characteristics suggest that water should be provided by public agencies. There was, however, an aggressive scramble for contracts and concessions that ensued in low- and middle-income countries during the era of neoliberal reforms. The water sector worldwide is dominated by a small number of multi-national corporations, which together control the majority of the privatised water and sewerage market. However, by the late 1990s, water companies themselves were publicly admitting that they had been ill equipped to meet expectations in cities of low- and middle-income countries, in which it was very difficult to make their operations profitable (Joy and Hardstaff 2005: 16). From the perspective of many poor people, the privatisation of urban services has rendered them unaffordable. In some cases, prices for water rose dramatically: for example 500 per cent in Conakry, Guinea over five years and 200 per cent in Cartegena, Colombia over eight years (ibid.: 17). Simply put, the regulatory theories on which privatisation was based were not suitable for the institutional context in most developing countries (Estache and Wren-Lewis 2009).

Needless to say, the failure of privatisation to improve services for the majority of urban dwellers and the accompanying price hikes has led to substantial disillusionment and public protest. When privatisation has led to improvements, it has usually been at the expense of universal coverage, with low-income areas being excluded because they proved difficult or expensive to reach. When neither public nor private agencies are providing services in these areas, low-income people often have to rely on purchasing water from informal vendors, which can be extremely expensive. In some cities it is possible to find someone without access to a tap paying up to 50 times more for clean water than a resident living a stone's throw away in a fully serviced neighbourhood (see Table 5.1).

Table 5.1 *The cost of accessing water in selected Asian cities*

	Cost of water per cubic metre (US $)			Difference between price of house connection and price of private vendor
	House connections	Public tap	Water vendor	
Bandung	0.38	0.26	3.60	847%
Bangkok	0.30	–	28.94	9,547%
Chonburi	0.38	–	19.33	4,987%
Dhaka	–	0.08	0.84	–
Kathmandu	0.18	0.24	2.61	1,350%
Manila	0.29	–	2.15	641%
Mumbai	0.07	0.07	0.50	614%
Port Vila	0.42	0.86	8.77	1,988%
Seoul	0.25	14.13	21.32	8,428%

Source: UN-Habitat (2003a).

Because private contractors are unwilling to take the risks involved in investing in the poorest parts of cities, governments are forced to take on a compensatory role, coping with the most difficult and least profitable aspects of service delivery – or risk public outcry. Although the service they provide may be very poor, it is difficult for governments to evade responsibility for service delivery entirely in these unprofitable areas. This is how the 'hidden costs of privatisation' are passed on to government (Batley 1996: 749). As a consequence of these problems, the late 1990s saw the high water mark of privatisation (especially in water), and despite some early reformers' visions of fully privatised services, public providers still play a crucial role today. The role of public investment is particularly important with regard to the crucial factor underpinning all effective service delivery: infrastructure.

Infrastructure and urban transport

The most critical urban infrastructure systems are those underpinning water, sanitation, waste collection, energy and transportation, but infrastructure networks are central not only to service provision but growth and development more broadly. The challenge has always been how to develop urban policy that works across all these and other linked systems to provide an integrated infrastructural basis for urban governance (Marvin 1992). A recent wave of academic interest in urban infrastructure was sparked by Graham and Marvin's 2001 book *Splintering Urbanism*, in which they argued that the recent selective privatisation of networks for services such as communications, transport and water supply was causing the 'unbundling' of infrastructural networks in ways that fragment the urban social and material fabric (2001: 33). Under these circumstances the state, private sector and communities may each compensate for the service delivery shortfalls of the other, but with no clear mechanisms of accountability or co-ordination structures in place. As noted above, private firms cherry-pick the wealthier cities or regions (or the better off parts of a city), enabling them to 'bypass' poorer areas while more closely inter-connecting 'premium networked spaces', thus exacerbating divisions between rich and poor. These interpretations of infrastructural change have led to increased attention to the political nature of infrastructure, which in the past has all too often been conceptualised in narrowly technical concern (McFarlane and Rutherford 2008).

Many empirical studies have borne out Graham and Marvin's concerns; for example, one study of Cusco, Peru, found that multiple modes of organising infrastructure coexisted in the city and the urban poor were integrated only in ways that amplified their livelihood' vulnerabilities (Crawford and Bell 2012). While hugely influential, however, the 'splintering urbanism' thesis has been challenged from a number of angles. Some authors argue that the so-called 'modern infrastructural ideal' of fully integrated systems emerged in the 'global North' and was never a reality in the 'global South' anyway, even during colonialism (Legg and MacFarlane 2008). This perspective calls attention to the need for a more 'globally informed' understanding of infrastructure provision ideals and realities. Others have argued that there is no necessary link between unified, integrated infrastructure and the kind of cohesive social fabric that Graham and Marvin assume accompanies this. For example, in Santiago de Chile the connection of virtually all residents to essential networks has facilitated and legitimised policies of spatial segregation as well as political fragmentation (Pflieger and Matthieussent, 2008). More generally, it has been argued that the 'modern infrastructural ideal' has not collapsed simply because of externally driven factors like privatisation but because of its widespread historical failure to sustain universal and affordable services (Coutard 2008).

Regardless of who provides it, there is clear evidence that access to critical infrastructure makes an enormous difference to urban dwellers' wellbeing. The effects are particularly pronounced for women, who are especially disadvantaged by poor infrastructure. One study in Indian slums found that access to water and sanitation infrastructure was associated with a 66 per cent increase in education among females, as well as enhancing income by 36 per cent (Parikh et al. 2015). Legal access to electricity – which 40 per cent of the world's urban poor lack – is also hugely significant. Like solid waste, this tends not be prioritised as highly as water and sanitation in upgrading programmes as it is not so essential for life, but it is strongly linked to better educational outcomes, improved productivity, access to information and improved health linked to indoor air quality (Baruah 2010). An anecdote from De Boeck and Plissart's (2005) study of Kinshasa, one of the most deprived cities in the world in terms of infrastructure, is revealing about the power of an electricity connection to spur a diverse range of activity. They describe how a sole lamp post, powered by the generator of a nearby gas station, turned a quiet street corner with little movement after nightfall into a major meeting point and transport terminal with a large number of bars, shops and a cyber-café in a matter of months.

The challenge of equitable infrastructure provision is complicated by the question of appropriate standards. There is a danger in setting the bar too high if it means that provision will not progress beyond the easiest or most profitable areas: UN-HABITAT argues that 'it is better to provide a whole city's population with safe water supplies by means of taps within 50 metres of their home than to provide only the richest 20 per cent with piped water to their homes' (UN-Habitat 2003a: 3). At the same time, there are normative questions about whether it is acceptable, especially in middle-income countries, to deliberately provide a lower standard of infrastructure to people living in certain areas. Bond (2000) argued against South Africa adopting excessively low standards of infrastructure for poor areas on the grounds that it is not only inequitable but also inefficient. It is extremely difficult, after installing yard taps rather than in-house reticulated water and 5–8 AMP rather than 20 or 60 AMP electricity, to incrementally upgrade this infrastructure. This can result in 'permanently segregated low-income ghettoes' (Bond 2000: 54). Hence solutions based on lowered standards may seem appropriate in the immediate term, but can have undesirable consequences in the long run.

The most visible form of critical infrastructure is that associated with transport. Like other forms of infrastructure, roads and train tracks offer direct and indirect benefits for economic growth, including employment and contracts for local firms. Good road networks are also appealing to political leaders for numerous reasons. Walsh and Amponstira (2013) explore how in Naypidyaw, the new capital of Myanmar, transport infrastructure offers the opportunities for efficient governance and political control as well as linking the city to economically important nodes. Meanwhile in the Cambodian capital Pnomh Penh, transport infrastructure is less concerned with internal integration than linking up with cross-border markets (Walsh and Amponstira 2013).

The many economic and political advantages of transport infrastructure – including the very visibility which can make it a vote-winner – explain why it is often a major priority for domestic policymakers. Yet its usefulness in appealing to the masses does not mean that when installed it necessarily produces efficient or equitable outcomes. The consequences of transport infrastructure investments can be deeply exclusionary. In the Nicaraguan capital Managua, for example, there has been huge investment in new high-speed roads and roundabouts connecting parts of the city inhabited or used by elites. While these areas are now tightly interlinked, those areas deemed insignificant or dangerous are systematically neglected (Rodgers 2004). As a result, the city has become safer for the rich – both because of the quality of the roads and because they have increasingly little contact with the poorer and more threatening parts of the city – while the urban space available to the poor has become increasingly circumscribed, as 'large swathes of the metropolis' are carved out 'for the sole use of the urban elites' (Rodgers 2004: 123).

Meanwhile, the dramatic physical expansion occurring in many cities in low- and middle-income countries has led to new infrastructural demands as well as people commuting longer distances than ever before. In Jakarta, Indonesia, there has been an explosion of peripheral settlements in the wake of commercial property development in the city-centre and the demolition of more proximate informal settlements. In response, poor urban residents have moved to peripheral areas in search of cheaper land and housing and a lower cost of living. This can lead to commutes of up to 120 kilometres (Jellinek 2000: 272). The vast numbers of commuters travelling in and out of cities of this size can overwhelm the infrastructure, resulting in traffic gridlock. Yet there is a vicious cycle at play, whereby new roads inevitably result in new settlements near them, expanding the urban territory along new highways and rapidly placing further demands on roads that were built to try and meet existing demand. Transport infrastructure is thus linked to two of the greatest curses of contemporary urbanism: sprawl and congestion.

It is an unfortunate irony that as transport infrastructure proliferates and more people have access to vehicles, mobility – defined as the ability to move at an acceptable speed and travel time, and at acceptable levels of comfort, convenience and safety – is actually declining across most cities of the developing world (Gakenheimer 1999). This is partly a consequence of demand for motorised transport far outstripping supply, the high dependence on often very poor public transport systems, and the incompatibility between existing urban form and increased motorisation; for example, in China, street space is around 10 per cent of the city surface as compared to 25 per cent in the average city in Western Europe (ibid.: 673). Other key factors that render urban transport more challenging than in most high-income contexts are the scarcity of capital, poor road maintenance, lack of transport planning expertise, weak driver discipline, and huge variation in vehicle performance. As public transport is so bad, those who can afford it are desperate to acquire cars. This makes congestion worse for everyone, with particularly negative effects on the poor who still depend on ever-slower bus services. There is therefore another vicious cycle in which 'the declining quality of public transport reinforces private auto use and vice versa' (Davis 2006: 131).

The long commutes caused by peripheral settlement and traffic congestion have substantial impacts on national economies, with many developing countries losing an estimated 2–5 per cent of GDP to road congestion (Lohani 2010: 3). They also affect the lives of urban dwellers in multiple ways. For example, the urban poor in developing countries often spend 20 per cent or more of their household income on public transport (Godard 2013), and research has shown long urban commute times to be significantly correlated with poor health (Christian 2012). Despite these problems, challenges relating to urban transport and mobility have long been ignored in discussions about poverty, alongside the more general neglect of urban poverty (Godard 2011).

Given the issues identified above, a key question is why public transport systems remain so poor in most developing countries. As with other urban services, there is a complex combination of causal factors, encompassing far more than infrastructure. Public transport systems have varied (and often highly political) histories in different countries, but there are a number of common trends. In the context of rapid population growth in urban centres across the developing world, formal urban public transport systems were often set up by governments after decolonisation. Yet these state-provided transport services commonly collapsed within a few decades due to weak fiscal capacity, fares that were kept low for political reasons, and scarce investment (Cervero and Golub 2007). This resulted in failure to maintain the public transport fleet, eventually putting many public companies out of business (Kumar and Barrett 2008).

Meanwhile, rapid urban growth meant that demand was intensifying as supply was collapsing. Informal 'paratransit' services in the form of minibuses, microbuses and motorcycles proliferated to fill the gap, offering services that could respond quickly and flexibly to changing markets, had relatively low costs and were able to enter cramped informal areas without adequate roads (Cervero and Golub 2007: 456). While the extent of informalisation of the sector varies widely even among low-income countries, in general at least 30 per cent and as much as 95 per cent of the public transport journeys undertaken in the cities of Asia, Latin America and Africa are taken using informal 'paratransit' services (Godard 2011). Types of 'paratransit' vary considerably, but small scale (and hence minimal capital outlay) is always a feature. Paratransit varies from 12–24-seater minibuses to smaller vans or pick-ups (sometimes termed 'microbuses'), three-wheel 'tuk-tuks', motorcyles, push-bikes, rickshaws and even horse-carts (Cervero and Golub 2007).

This transformation in public transport rarely resulted from conscious decisions to deregulate; rather 'it was an indigenous response to growing demand and commercial opportunity' that operated in a 'regulatory vacuum' initially by default (Kumar and Barrett 2008: xi). In Nigeria, for example, reduced public expenditure meant a decline in municipal bus services. At the same time, public sector job cuts saw former government employees purchasing and running fleets of minibus taxis to make a living, taking advantage of the public transport gap (Oyeniyi 2007). De Soto (1989) described how in Peru, pressures for informal buses to formalise led them to be bogged down by the costs of regulation, with the effect that a new wave of 'pirate' operators stepped into the gap by the 1970s. Even if unregulated transport initially began by default, however, liberalisation and deregulation of the sector was being actively encouraged by the World Bank by the mid-1980s, following the experience in the developed world (Gómez-Ibáñez and Meyer 1993), and many countries officially adopted this approach in the 1990s.

Plate 5.3 *Paratransit traffic jam in a narrow street in Jodhpur, India*

Source: Tom Goodfellow.

In some parts of the developing world, demand for affordable urban paratransit has been met through motorcycle-taxis. Very heavy traffic congestion caused these to soar in popularity due their agility and capacity to dodge traffic. However, the severe hazards associated with this form of transport mean that in many countries they are prohibited – albeit with highly variable degrees of success. In Uganda, for example, the Kampala city government has attempted to control, tax or completely ban this form of transport several times, with no effect whatsoever; between 2003 and 2010 the number of motorcycle-taxis – known locally as *boda-bodas* – increased by 1,000 per cent from 4,000 to 40,000 in just seven years (Goodfellow 2015a), and is now thought to be as high as 100,000. The demand for these services, as well as the potential for huge profits and, significantly, the collective power of motorcycle-taxi drivers as a 'vote bank', mean that political will to enforce regulations is scarce (ibid.). Powerful cartels can come to dominate the informal transport sector, limiting or even obliterating effective competition, as well as efforts to implement regulations and introduce larger, more economical bus services (Kumar and Barrett 2008; Goodfellow 2012). In some cases this results in battles between cartels with significant violence and fatalities, as has for example happened in South Africa and Nigeria (Dugard 2001; Albert 2007).

While affordable and convenient, informally provided public transport is often unregulated and unsafe. In the minibus sector, competition between private operators can be immense resulting in tariff wars, overloading of vehicles and speeding to maintain a competitive edge, with the consequence that, for example, 93 per cent of all fatal accidents in Dar es Salaam in the average year involved minibus-taxis (Rizzo 2002: 144). Globally, road crashes are the second leading cause of death among people aged 5–29 years, and the third leading cause of death among people aged 30–44 years. Traffic injuries are therefore a 'neglected public health epidemic' (Nantulya 2002). The situation is significantly worse in developing countries (see Figure 5.1) and getting worse over time: in 2004, the WHO estimated that without immediate action to improve road safety, road traffic deaths would increase by 80 per cent in low- and middle-income countries by 2020 (WHO 2004). In cities, it is generally not those in vehicles but those on foot who are at the greatest risk, and this invariably means the least well off. For example between 1977 and 1994, 64 per cent of fatalities from traffic accidents in Nairobi were pedestrians (UN-Habitat 2006: 131).

For all these reasons, despite the 'solution' that informal paratransit offered in the context of collapsing national public transport services, it has also ushered in a host of new problems. The substantial capital and maintenance costs involved in transportation, alongside entrenched interests that resist reform, mean that while many cities have exhibited leadership with regard to transit service innovation, few have actually succeeded in substantially improving the transport situation (Gakenheimer 1999). There are some cases in which these difficulties have been overcome, however. Curitiba, Brazil, provides a paradigmatic example, generating numerous innovations in bus transportation since the 1970s. The city has a centralised and integrated system rather than a series of competing routes, although private bus companies operate them. Dedicated bus-ways were established from the outset so that road capacity was given over to buses, and speed is facilitated by passengers being able to pay for their trips before boarding the bus, which they do very rapidly. Curitiba thus created the first 'Bus Rapid Transit' (BRT) system in the world (Lindau et al. 2010), with buses able to move people from one area to another as swiftly as the New York subway but at a fraction of the cost (Tannerfeldt and Ljung 2006: 101–02). Moreover,

Figure 5.1 *Traffic deaths per 100,000 people in 2010*

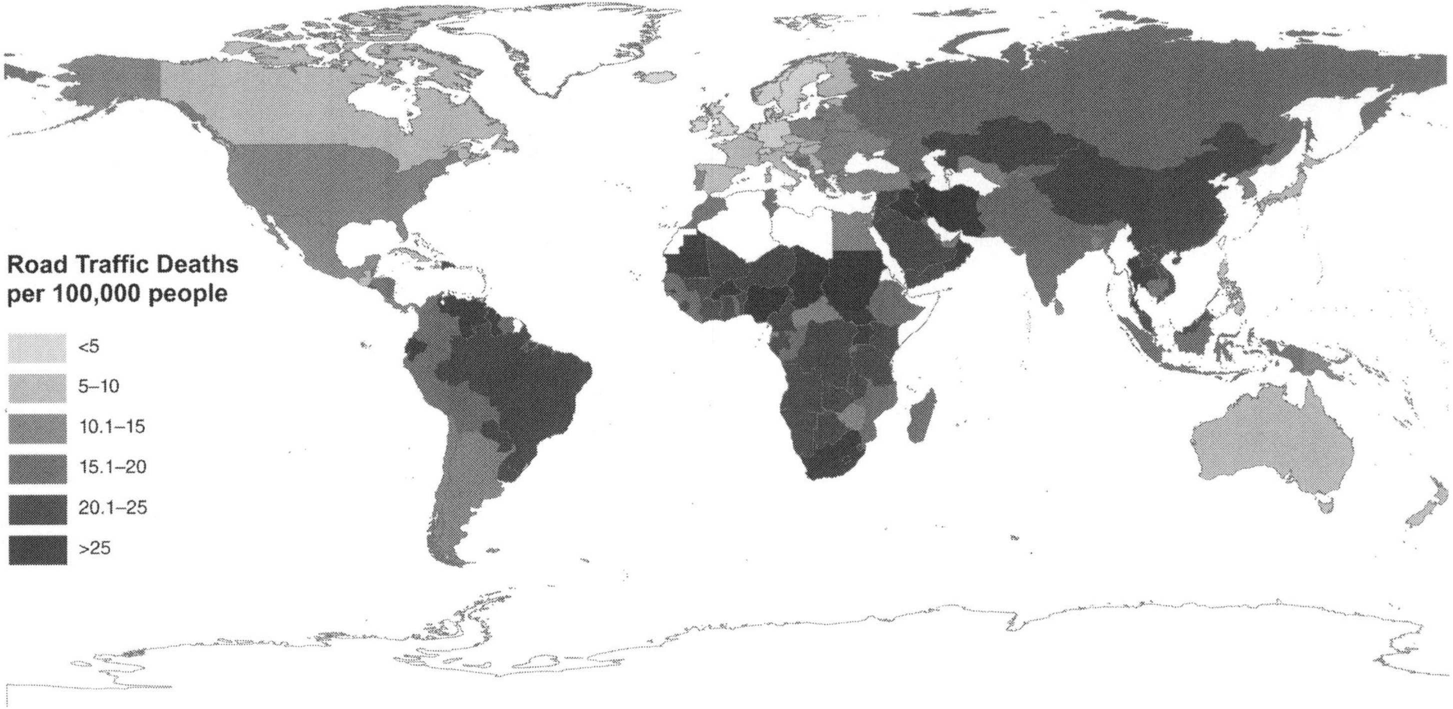

Road Traffic Deaths
per 100,000 people

<5
5–10
10.1–15
15.1–20
20.1–25
>25

it has made substantial innovations in sustainability, introducing its Green Line in 2009, which includes the operation of 100 per cent bio-diesel articulated buses (Lindau et al. 2010).

Another success story is Colombia, which has innovated in several of its large cities. Bogotá, the capital city, introduced a BRT system (known as the *TransMilenio*) by way of a public–private partnership in 1999, which offers frequent and affordable transport across the capital. The BRT system traverses rich and poor parts of the city alike, with free feeder buses ferrying passengers to the 114 TransMilenio stations, also served by 250 kilometres of cycle paths. This has improved congestion and quality of life in the city dramatically (Hidalgo et al. 2013). An average trip now takes a third of the time it used to, and this combined with service frequency means that the buses are 95 per cent full as opposed to less than half full in the past. Accidents have decreased by 80 per cent and air pollution has dropped almost 40 per cent. The determination and political will of two former mayors of Bogotá – Enrique Peñaloza and Antanas Mockus – and the political coalitions they forged around them were central to the improvements in planning in transport.

In some cities, topography and urban form may not facilitate BRT due to narrow roads, steep hills or a dense built environment. Colombia has again innovated in such circumstances, for example by using ski-slope cable car technology to improve access to dense and hilly low-income informal settlements (Brand and Dávila 2011). However, the combination of factors behind these Colombian successes – mayors with considerable devolved power, substantial financial resources and a favourable political coalition for sustained urban development – are relatively rare (see Chapter 8). Hence these achievements have been difficult to replicate in most lower-income countries.

Conclusion

While Chapter 4 explored the vulnerabilities of the urban poor, this chapter has illustrated how intimately linked these are to the urban built environment and how efforts to physically develop cities aim to ameliorate these vulnerabilities. Planning, housing, infrastructure and service provision are some of the most fundamental challenges for improving urban living conditions, and are also critical for urban economic development. Providing these goods at the scale needed is extremely problematic, however. Central among the underlying challenges is the question of land – both in terms of how it is allocated and secured, and how it is regulated. The intrinsic economic value of urban land means that it tends towards uses that benefit high-income groups at the expense of the poor, who can frequently only afford to access it illegally, reinforcing their vulnerability.

Delivering housing, infrastructure and urban services at scale is also constrained by a range of co-ordination, governance and finance challenges, all of which are exacerbated in conditions of rapid urban growth. Unable to access the benefits of formal housing and infrastructure, the only way in which the urban poor can make a home and access basic services is often through informal provision, which can put them at risk and be very expensive, or community-based solutions, which are limited in scale and can be gender-biased and exclusionary. Yet the challenges raised in this chapter are not only concerns in terms of how they affect the urban poor: land use, infrastructure and services – especially transport – are also central to the relationship between cities and the natural environment, and the urgent question of urban sustainability. It is to cities, climate change and the environmental challenge that we now turn.

Summary

- Planning and development regulation have been very limited in many cities in the developing world, and since the 1980s have been widely critiqued for being out of touch with reality, as well as imposing costs and unreasonable standards on the urban poor.
- Despite this, efforts to plan cities and regulate urban land remain very important but need to be adequately context-sensitive and avoid simply 'exporting' models of planning from the developed world.
- Effectively housing the urban poor is one of the key challenges of urban growth. Land tenure security is now acknowledged as central to addressing this problem, but often difficult to implement due to the high value of such land and pressure from elites.
- Widespread public housing delivery in the mid-twentieth century gave way to 'self-help' based approaches including site-and-service schemes, core housing provision, slum upgrading, and

tenure titling programmes – all with specific challenges and mixed results.

- Water supply, sanitation and solid waste management are considered the critical urban services and, along with energy and transport, require sound infrastructure to deliver.
- Billions of people still lack adequate sanitation, and the management of solid waste – an often neglected urban service – is a stark indicator of inequality, given that most waste is created by the wealthy but is recycled by the poor, often negatively affecting their health.
- Attempts to meet service and housing needs exclusively through the private sector have been problematic as companies struggle to make profits in the poorest areas.
- Improving transport infrastructure has important economic and health benefits. However, some urban infrastructure projects exacerbate inequality and encourage urban sprawl.
- Urban transport has been increasingly provided by private and informal actors, contributing to rising congestion, pollution and road deaths. However, there have been some success stories of integrated public/private-funded 'Bus Rapid Transit' systems.

Discussion questions

1. What are some of the problems with 'exporting' European, North American or East Asian planning and infrastructure design to cities in the developing world?
2. How does the provision of property titles benefit the urban poor, and why is security of tenure so difficult to achieve?
3. 'The urban poor are the best managers of their housing solutions': discuss with reference to recent planning approaches to informal settlements.
4. Why has it proven so difficult to rectify deep inequalities in the delivery of critical urban services?
5. Why is it so difficult to overcome the obstacles to good urban public transport in developing countries?

Further reading

Batley, Richard and Larbi , George (2004) *The Changing Role of Government, The Reform of Public Services in Developing Countries.* Basingstoke and New York: Palgrave Macmillan.

De Soto, Hernando (2000) *The Mystery of Capital.* New York: Basic Books.

Graham, Stephen and Marvin, Simon (2001) *Splintering Urbanism: Networked Infrastructures, Technological Mobilities and the Urban Condition.* London: Routledge.

Hall, P. (2002) *Cities of Tomorrow* (Third Edition). Oxford: Blackwell

Jenkins, P., Smith, H. and Wang, Y. P. (2006) *Planning and Housing in the Rapidly Urbanising World.* London: Routledge.

UN-HABITAT (2013) *Planning and Design for Sustainable Urban Mobility: Global Report on Human Settlements 2013.* Nairobi: UN-HABITAT.

Watson, V. (2009) '"The planned city sweeps the poor away…": Urban planning and 21st century urbanisation', *Progress in Planning,* 72(3), 151–193.

Websites

Cities Alliance: www.citiesalliance.org

Global Land Tool Network (GLTN): www.gltn.net

Shack/Slum Dwellers International (SDI): www.sdinet.org

UN-HABITAT Mobility pages: unhabitat.org/urban-themes/mobility

Water Supply and Sanitation Collaborative Council (WSSCC): www.wsscc.org

World Water Council: www.worldwatercouncil.org

6 Cities and environmental change

- The challenge of sustainable urbanism
- Urbanisation, economic development and climate change
- Urban environmental risk, 'ecological security' and climate migration
- Adaptation, resilience and climate governance

Introduction

Of all the aspects of the global urban transition that have attracted interest in recent years, few have raised as much concern as the relationship between cities and the environment, and especially the nexus between urbanisation and climate change. This relationship is multifaceted, complex and contested. Cities have been blamed for climate change as well as hailed as the answer to sustainable living, while their residents have variously been demonised for their consumption patterns, valorised for their capacity to recycle and reuse resources, and feared for due to their vulnerability to the effects of global temperature rise. This chapter explores key evolving debates around cities, sustainability and environmental change and how the increasing urgency of these issues has impacted on ideas and policies aimed at addressing the environmental challenge.

The global environmental movement and constantly evolving academic discourses on the environment have generated a bewildering array of concepts and debates, which can be challenging to navigate. With the aim of keeping our focus squarely on cities, and how the global environmental crisis intersects with urbanisation and urban living, this chapter is divided into four sections. The first is concerned with ideas of sustainability at the urban level, and how this relates to ideas about the 'metabolism' of cities and their 'ecological footprints'. This leads into a section on the relationship between cities and global climate change, which weighs up the evidence about cities' contribution to our changing climate, and considers the various sources of carbon emissions in cities at different levels of development and how these might be mitigated. We then turn to the question of the environmental risks to which cities are subject and the all-important question of vulnerability. It is increasingly recognised that urban dwellers in low-income countries are the world's most vulnerable people when it comes to deleterious impacts of climate change, but equally important is the highly differential vulnerability of different categories of urban dwellers across cities in developing countries. Here we also consider ideas of urban 'ecological security', and the perceived risks to urban security posed by 'climate migration' into urban areas. Finally, we turn to the question of how cities should respond, and are responding, to climate change. We consider the challenge of adapting to the reality of a changing climate and its effect on specific locations – and the currently much-debated concept of 'resilience' – as well as how cities can help to limit climate change, both individually and by working collaboratively across international borders.

The challenge of sustainable urbanism

Sustainability has risen to the top of the global development agenda, to the extent that it has been built into the foundations of the 'post-2015' development vision embodied in the 'Sustainable Development Goals'. The concept of sustainable development represents an effort to pull together the agendas of environmental protection, social justice and economic growth into a unifying 'grand narrative' for global development. The term 'sustainable development' was first used in its current sense in a landmark report on *The Limits to Growth* (Meadows et al., 1972) by a group of scientists at Massachusetts Institute of Technology. It only came into common use, however, after the World Commission on Environment and Development (WCED) (also known as the Brundtland Commission) produced its report, entitled *Our Common Future*, in 1987. It was in this report that sustainable development was defined as 'Development that meets the needs of the present, without compromising the ability of future generations to meet their own needs' (WCED 1987).

Inevitably, a concept with such broad scope has been open to a very wide range of interpretations, and today 'sustainable development' as a concept is almost as contested as the idea of development itself. Claims of sustainability have become part of the standard rhetoric of virtually any policy or enterprise, and sustainability has become a form of 'comfortable rhetoric' which, much like 'good governance', is highly ambiguous, serving the various interests of those who use it (Connelly 2007). For some, it is about scaling back human impacts on the planet; for others it is about improving stewardship of the earth's resources for the benefit of future human generations. For others still it is about overconsumption in the developed world and the need to promote global socioeconomic equality (Wheeler and Beatley 2014: 88).

The relationship between cities and sustainability has become a focus of growing interest amid awareness of the challenges arising at the intersection of two of the defining trends of the twenty-first century: threats to the natural environment and dramatic increases in the numbers of people living and working in cities (Alusi et al. 2011). Notwithstanding ongoing debates about the actual impacts of urban living on climate change (discussed below), the importance of reimagining urban planning, housing, economic activity and leisure in ways that actively promote sustainability has generated fertile debate around ideas of 'sustainable urbanism'. This term, along with 'urban sustainability', is often taken to refer to a desired end-state, while 'sustainable urbanisation' and 'sustainable urban development' refer to the process of striving to achieve this state (Shen et al. 2011: 19). In general, however, all of these terms reflect a concern to redirect urban economies in ways that are restorative rather than exploitative of the environment, such that they do not rely on ever-growing consumption of material products and long-distance trade (Wheeler and Beatley 2014).

Environmental change caused by urbanisation is inevitable. Throughout history, cities have changed the environments around them locally, regionally, and ultimately internationally through the intensification of global trade linked to urban productive activity; urbanisation is not only a demographic but an ecological transformation (Wackernagel and Rees 1996). Consequently, despite the recent interest in climate change, some observers argue that the most pressing aspect of environmental change linked to urbanisation is biodiversity loss and the fact that nitrogen pollution exceeds safe planetary limits (World Bank 2010a). One of the ways of understanding environmental change linked to urbanisation is through analysis of the resources used and waste produced by city-dwellers. A long-standing thread in the debate on cities and sustainability has thus involved analysis of the 'ecological footprint' of cities, meaning the combined use of resources by urban dwellers beyond the physical boundaries and 'carrying capacity' of a city (Rees 1992).

Ecological footprints are calculated by converting resource needs and pollution into the equivalent land area that would be required to offset or produce these. Like 'a cow in its pasture', this approach asks of any given city how large a space would be needed to produce all its 'food' and absorb all its waste (Wackernagel and Rees 1996: 378). London's ecological footprint, for example, is estimated to extend to around 125 times its surface area of 159,000 hectares; in other words, to nearly 20 million hectares, equivalent to a figure of 2.8 hectares per person (Dodman 2009: 185).

The sources of such enormous 'footprints' are multiple. For example, meeting the needs of urban dwellers for firewood can speed up deforestation, while cities can use up the lion's share of limited water sources in a region. Similarly, poor urban sanitation can contaminate

Plate 6.1 *Visible air pollution in Shanghai, China*

Source: Tom Goodfellow.

rivers with the consequences being felt downstream and hundreds of miles beyond the urban edge. The damage done is not only to people's health but also their livelihoods. Sea and river fish yields can decline as a result of water pollution from activities in nearby cities. Air pollution has a huge regional impact, with acid rain often falling hundreds of kilometres away from the source of pollution, devastating agriculture by ruining soil and vegetation. The expansion of cities onto previously rural land can push agriculture into less suitable areas, while the commercialisation of agriculture stimulated by the need to feed cities renders many small farms redundant. The natural demands of cities increasingly shape the environment of their surrounding regions; indeed, while many cities first developed as market centres to serve the farms and farming households around them, urban hinterlands now increasingly work to serve the city. The case of water supply in Mexico City provides a stark example of how regional resources can be drained in a context of rapid urban growth – but also, importantly, how unsustainable agricultural water usage can feed urban crises (see Box 6.1)

Another growing strand of debate on cities and resource use has focused on the idea of 'urban metabolism'. This involves the study of inflows of materials and energy into a system, the stocks and internal flows within the system, and outflows in the form of exports, waste and pollution into other systems (Pincetl et al. 2012). Marking a resurgence of interest in bio-physical analogies in urban planning and urban geography, the idea of urban metabolism draws our attention to critical urban infrastructures and the relationship between technologies, space and society. Interest in urban metabolism has also found practical application in the form of material flow analysis as a method for interpreting energy demands and consumption in urban areas (Pincetl et al. 2012). Approaches to this vary, with some focusing on energy equivalences and others more broadly on flows of water, materials and nutrients (Kennedy et al. 2011).

As with many of the concepts in the field of urban sustainability, metabolism has different meanings among different groups of users of the term, and there are distinct intellectual traditions marking the study of urban metabolisms. For example, the tradition of 'industrial ecology' has tended to view nature as an external source of materials for urban metabolism and a destination for its wastes, while urban political ecology approaches have challenged this separation of society and nature, often drawing on Marxian political economy and emphasising the role of humanity in creating and reshaping nature.

Box 6.1 The water crisis in Mexico City

Water scarcity has become so severe in Mexico in recent years as to reach crisis proportions in its capital city, where much of the water supply is now piped in from a neighbouring province. The water shortage is often described as an issue of national security, and some observers have even suggested that by 2020 there will be a 'mini-revolution' in the city as a consequence of anger about increased scarcity and the costs of acquiring fresh water: Mexicans already consume more bottled water than any other country. The origins of the problem are multifarious, which contributes to the complexity of finding solutions. Mexico City is highly dependent on underground aquifers, but these are not being adequately replenished and are in danger of collapsing. As a side effect, this is resulting in the sinking of the city, which has sunk by an estimated ten metres over the last century. Meanwhile, thousands of small leaks in the piping system across Mexico City result in huge leakages: the city loses 1,000 litres of water per second as a consequence of this – 40 per cent of the water travelling through the system. Many city-dwellers have turned to harvesting rainwater as a solution to shortages, though the cost of installing the necessary technology is prohibitive for most families.

Some have partly blamed this situation on Mexico's system of restricting politicians to a single term in office, with the effect that they are more concerned with using their office to secure future careers than providing public goods to ensure re-election. The roots of the problem are decades old, however, going back as far as the 1920s, when developmental imperatives in the wake of the Mexican revolution led to the unregulated and unsustainable pumping of groundwater for agrarian purposes. Successive warnings from engineers in the twentieth century were largely ignored. Ironically, one of the effects of growing water shortages was that agricultural livelihoods became unviable, adding fuel to the demographic shift towards urban areas. The current water crisis in Mexico City is therefore a clear example of how past decisions and present circumstances can interact to produce particular environmental challenges. It also serves as a reminder that the use of water for agricultural purposes can impact on urban water security, sometimes decades later.

Sources: Wolfe (forthcoming) and https://nonprofitquarterly.org/policysocial-context/25153-water-water-nowhere-mexico-city-endures-severe-shortages.html.

Such approaches seek to emphasise how flows of materials such as water play a role in urban socioeconomic structuring as well as resource use (Gandy 2004). In this view, urban metabolisms characterised by fragmented technological landscapes have important political-economic as well as environmental implications.

The growing interest in sustainable urbanism is reflected not only in conceptual debates and academic discourses but in policies at national and local level. In China, for example, a national energy and resource-saving policy adopted in 2006 means that every city now has a financial incentive and punishment system in place to encourage water saving (Wu 2013). In many countries around the world ambitious 'eco-cities' have been proposed, though often with limited success (see Box 6.2). At a more everyday level, policymakers frequently adopt packages of sustainability indicators to assist with goal-setting and policy evaluation (Shen et al. 2011). This has proved

Box 6.2 The trials and tribulations of 'eco-cities'

The idea of developing 'eco-cities', which rapidly gained momentum in the 2000s, refers both to initiatives to make existing cities more sustainable and the creation of brand new cities predicated on principles of sustainability. One survey of 79 eco-city initiatives across the globe found that initiatives under the banner of 'eco-city' development ranged from car-free tourist resorts in Europe to free trade zones using renewable energy in Africa (Joss 2010). In 2010 the World Bank launched its 'Eco2 Cities' programme, which aims to provide an operational framework that can be customised to the particular context of any city to help integrate infrastructural frameworks and optimise the efficiency of resource flows while establishing platforms for collaborative design and decision-making (Suzuki et al. 2011). The vogue for eco-cities has also coincided with an interest in the idea of 'smart cities', and there is a significant overlap due to the proposed use of sophisticated ICTs in eco-city initiatives to more efficiently managed complex systems from traffic to the electric grid in order to reduce emissions and conserve resources (Alusi et al. 2011).

Eco-cities are not without their problems, however. Some grand proposals, such as those for the highly publicised Dongtan city in China, fail to get off the ground due to funding difficulties despite considerable investment in planning and design and apparently significant political commitment. Among the most ambitious and famous ongoing eco-city projects is Masdar in the United Arab Emirates, located in the desert ten miles from Abu Dhabi. The original proposal was extremely ambitious, proposing a city that would be zero carbon, being powered by renewable energy, car-free and producing net zero waste. The proposal was scaled down to reducing carbon emissions to 50 per cent of 'business as usual' in Abu Dhabi, and the original proposed completion date from 2016 to 2021–2025. The city is a special economic zone that allows 100 per cent foreign ownership and zero taxes (Alusi et al. 2011). Despite its innovative and ambitious designs, the project has been extensively criticised, primarily on the grounds that its foundations are strongly rooted in economic concerns, which the environmental dimension serves to mask: a phenomenon sometimes called 'greenwashing'. Critics suggest that the primary motivation for the city's creation is not sustainability but the effort to promote Abu Dhabi as a global clean-tech hub and attract investment (Cugurullo 2013). Moreover, Masdar will ultimately rely on real estate sales and rentals to repay banks and other capital providers for the enormous cost of its construction, an imperative that will likely override the city's environmental and social goals (Alusi et al. 2011: 17).

Sources: Joss (2010), Suzuki et al. (2011), Alusi et al. (2011), Cugurullo (2013).

controversial, not least due to tensions between citizen-led and expert-led models of urban sustainability. The tendency towards long lists of indicators based on technical conceptions of what is 'unsustainable' can obscure local understandings and values regarding what it is that should be sustained. This reflects a more general problem in balancing the social aspects of sustainability with economic and environmental concerns. While economic growth, social justice and environmental protection are generally considered the 'three poles' of sustainability, in many policy discourses the environment is prioritised at the level of rhetoric while the economy remains the 'bottom line', leaving the social aspects marginalised. Indeed, in many respects social sustainability is the most difficult of the three poles to define, and the most subjective (Shen et al. 2011: 19).

Despite all the interest in urban sustainability, until very recently there was no clearly articulated 'universal' agenda for sustainable *urban* development to match the broader critical moments of the 1972 *Limits to Growth* report, 1987 Brundtland Commission, 1992 Earth Summit in Rio de Janeiro or the adoption of the Millenium Development Goals in 2000 (Wheeler and Beatley 2014). This however, changed in September 2015 with the introduction of Sustainable Development Goal no. 11: 'Make cities and human settlements inclusive, safe, resilient and sustainable.' There are a wide range of specific targets associated with this goal, including expanding access to affordable housing and services, expanding participation in planning processes, strengthening heritage preservation initiatives, reducing vulnerability to natural disasters, and reducing the environmental impact of cities with regard to pollution and waste (UN 2014). Despite the problems of combining so many diverse issues into one goal, this recognition of the centrality of cities to sustainable development is long overdue.

There are undeniable ecological changes that accompany urbanisation, but it does not make much sense to 'blame' cities or the urban transition for negative impacts on the environment. As Tannerfeldt and Ljung (2006: 66) have observed, it is increasing affluence and the production and consumption associated with economic growth rather than the spatial configuration of cities that leaves the footprint. Thus while economic growth depends on cities, it is important to distinguish between the effects of that growth and the environmental effects of city life itself. In fact, the relationship between urbanisation and our changing global climate is highly contested. Far from cities being a problem, there are many reasons for believing that they are a major component of the solutions we so urgently need.

Urbanisation, economic development and climate change

There are clear reasons why cities are associated with, and sometimes blamed for, climate change. Since the industrial revolution at least, cities have concentrated industries, transport infrastructure and construction activities – as well as concentrating households – all of which result in the generation of greenhouse gases. When badly managed in environmental terms, cities can contribute overwhelmingly to carbon emissions. The top 10 greenhouse gas emitting cities in the world produce more emissions than the whole of India, despite having a combined population only 15 per cent the size of India's (World Bank 2010a: 18). Particular cities' per capita emissions often outweigh the average for the population of the country in which they are located; Rotterdam in the Netherlands, for example, emits some 29.8 metric tonnes of greenhouse gases per capita – over twice the average for the whole of the country, which is 12.67 (World Bank 2010a: 25). Some developing countries too have cities that emit disproportionately; Beijing emits 10.1 metric tonnes per capita, almost three times the overall Chinese figure of 3.4 (ibid.: 25).

There are all sorts of ways in which cities contribute to greenhouse gas emissions, beyond the obvious industrial output aspect; indeed, in most developed countries it is not industrial activity that generates these gases so much as activities such as transport, heating, cooling and lighting. Also critical are changes to land use that result from urbanisation; creating impervious surfaces, filling wetlands and fragmenting ecosystems, results in heat emission and poor water storage, which can generate further emissions (UN-HABITAT 2011a). In Bulkeley's words, 'climate change is actively being produced through the urban condition' (Bulkeley 2013: 229). Overall, there is no escaping the fact that while only around half of the world's population live in urban areas, cities produce about 80 per cent of greenhouse gas emitted around the world (World Bank 2010a: 15). This is unsurprising given that cities also produce the majority of the world's GDP and consume the majority of energy in the process. Yet the causes of high carbon emissions differ enormously between cities. In Chinese cities overall, 65.1 per cent of emissions came from industrial energy consumption, with only 10 per cent coming from transportation and 7.7 per cent from household consumption (Wang et al. 2012: 6197). This differs substantially from the situation in Los Angeles, for example, where ground transportation contributes significantly more than industry, electricity use and heating combined (see Figure 6.1). In Brazilian cities, meanwhile, many of the emissions come from solid waste rather than transport or industry (Dodman 2009: 190).

Linking emissions at the city level to responsibility for climate change is extremely problematic, not only due to the porousness of city borders but because of the question of whether responsibility for emissions should sit primarily with the production of goods that generate greenhouse gases or their consumption. In a global context of industrial outsourcing, many cities that now have high registered per capita emissions do so because they produce large numbers of goods for export, the demand for (and consumption of) which lies elsewhere – sometimes on the other side of the world. Many of the 'dirtiest' links in global value chains are also those requiring large amounts of cheap labour, leading them to be outsourced to cities in Asia, for example. Meanwhile, richer countries deal with the higher value-added and cleaner elements of production, as well as being sites

Figure 6.1 *Composition of CO₂ emissions sources (selected cities, 2005)*

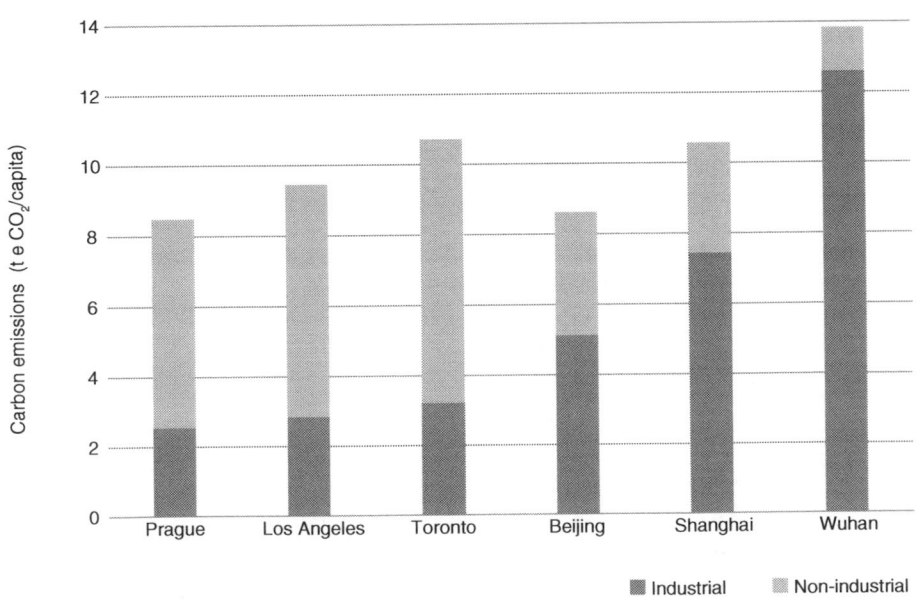

Note: Data are from Wang et al. (2012).

of consumption. Should those cities saddled with the most polluting productive activities, by virtue of their relative poverty, be held responsible for emissions created in the production of export commodities? As Walker and King (2008: 199–200) note:

> Many of the countries in the western world have dodged their own carbon dioxide emissions by exporting their manufacturing to [...] China. Next time you buy something with "Made in China" stamped on it, ask yourself who was responsible for the emissions that created it.

This, of course means that reducing emissions cannot be tackled by thinking only about production, and requires sustained attention to the geography of consumption. Complicating matters is the fact that as countries such as China become rich though industrial production, they too develop consumption habits that are more environmentally damaging. It is telling, for example, that air conditioning units in China were increasing by a rate of about 25 per cent per year in the 2000s (Wang et al. 2008). Cities in Asia and other parts of the developing world too are rapidly increasing their energy consumption, not least because of the steep increases in vehicle ownership due to the severe inadequacy of public transport systems (Kumar 2013).

It is crucial to understand that despite cities producing a disproportionate share of global carbon emissions relative to their population share, this fact is completely skewed by the emissions produced in certain cities of the developed world and a few in emerging economies. In contrast to the examples of Rotterdam and Beijing above, in most countries cities actually produce far fewer carbon emissions per capita than the national average, both in rich and poor countries. People in Oslo produce less than a third of the national average, and in Kathmandu the figure is around a twelfth. Concentrating on urban emissions per capita also only presents part of the story, given the very large differentials in emissions that exist within urban centres. Residents of Dharavi slum in Mumbai, for example, are emitting a tiny fraction of the per capita emissions of Mumbai's rich neighbourhoods (Satterthwaite 2008).

In fact, there is no direct relationship between the level of a country's urbanisation and the level of its carbon emissions. This is shown in Figure 6.2, which plots levels of urbanization (x-axis), carbon emissions per capita (y-axis), and GDP per capita (bubble size) for 20 countries from across the globe. The United States, with an estimated 80 per cent of its population living in urban areas, emits more carbon per person than any other country in the sample. However, there are several countries with higher levels of urbanisation and yet lower emissions per capita including South Korea, Japan, Sweden, Brazil and Singapore. The figure also shows that wealth and carbon emissions do not necessarily go hand-in-hand: Sweden and the USA have very similar levels of wealth, but the Swedish people emit about one-third of the carbon that Americans do per person.

While it is true that wealthier countries do, on average, have higher levels of carbon emissions per capita than very poor countries due to higher levels of energy and materials consumption, this depends on the 'energy intensity' of a country's economy. In theory, richer countries with advanced technologies should be able to generate national wealth at a lower level of energy use – hence Sweden has lower emissions per capita than industrialising emerging economies such as China, Malaysia and South Africa. But, as the case of the USA demonstrates, this is not always the case if a country is highly dependent on road transport and entrenched vested interests prevent a shift to less energy-intensive industrial and agricultural practices.

Figure 6.2 *Urbanisation, carbon emissions and GDP per capita*

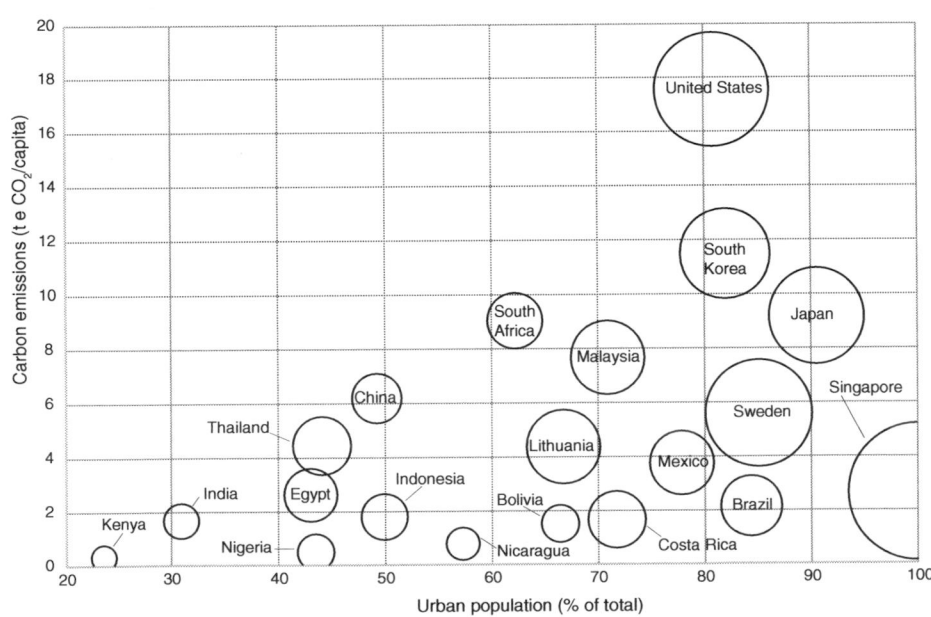

Notes: Data are from the World Bank World Development Indicators, accessed April 2015. Bubble size represents PPP adjusted GDP per capita in constant 2011 international dollars.

The composition of economic production, variations in energy demand associated with climate, and urban geography also help to explain why countries with similar levels of urbanisation and income can have very different emissions per capita. The distribution of people in space can exert a very significant effect on emissions, not least due the effects of this distribution on transport use. As Figure 6.3 illustrates, cities with higher population densities (x-axis) produce far less emissions (y-axis) than their low-population density counterparts, regardless of income level (bubble size). For example, Los Angeles and Barcelona have similar levels of income per capita but Barcelona packs in 19,500 citizens per kilometre squared, while Los Angeles accommodates just 1600 in the same space. The former emits less than half the amount of carbon per person per year than the latter: 4.2 vs. 13 tonnes equivalent of CO_2 per person per year. The levels of emissions in the USA reflect the fact that in the country as a whole, people are more than fifteen times as likely to drive to work than to take public transport – though certain cities such as New York and some on the West Coast provide notable exceptions (Glaeser 2011: 208–10). Indeed, it seems that the highest contributors to greenhouse emissions on a per capita basis are rural-dwellers and suburbanites who rely heavily on individualised automobile transport for daily activities such as commuting and shopping (Glaeser and Kahn 2009).

Recent research indicates that we could leverage an 'urbanisation wedge' by promoting energy efficient urbanisation, which – if implemented in Asia especially – might reduce energy use by over 25 per cent compared to a 'business-as-usual' scenario (Creutzig et al. 2015). This is based on an analysis of a dataset of 274 cities of different sizes across the world, which demonstrates that economic activity, transport costs, geographic factors, and urban form explain 37 per cent of urban energy use and 88 per cent of energy use from transport (Creutzig et al. 2015: 1). Crucially, however, the kinds of policies needed to produce energy-efficient urbanisation differ substantially by city type. For 'mature' cities in affluent regions, increasing taxes on petrol and encouraging compact urban form (such as greater mixed-use urban design) will be critical, while in developing countries with nascent infrastructure there is greater potential to prevent unsustainable urban form at an early stage through strategic urban transport planning (Creutzig et al. 2015).

Figure 6.3 *Urban population density, CO₂ emissions and GDP per capita in selected cities*

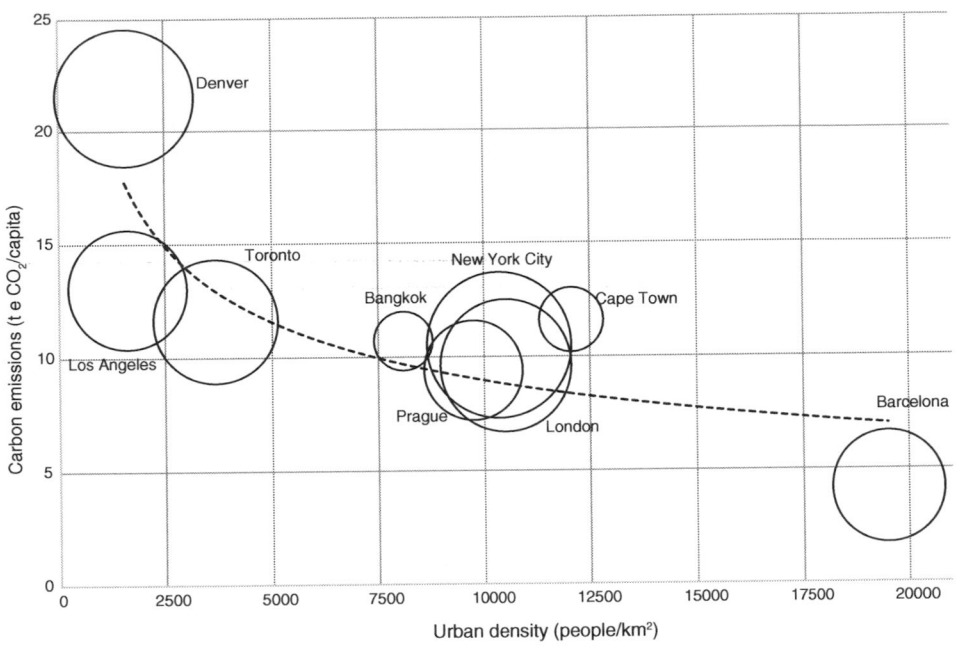

Notes: Bubble size represents PPP adjusted GDP per capita in 2005 $US. Data are from Kennedy et al. (2011).

A greater awareness of the possibilities for low-energy urban living – which holds the greatest promise in developing countries not yet bound by their urban form to high-emitting private transport patterns – reminds us that there is no fundamental reason why cities should be especially damaging to the environment. On the contrary, urban living actually presents many opportunities for minimising such damage. Proximity and agglomeration make it cheaper to provide the infrastructure needed for good environmental management, the prospects for waste-to-energy generation plants are greater in cities, and the relative proximity of home and work can (in theory) encourage walking and cycling and mass transit over automobile use (Dodman 2009). The association of cities with climate emissions is therefore not 'natural' or necessary. Brazil, for example, is one of the world's most urbanised countries, with 85 per cent of its population living in cities, which is higher than the USA; yet Brazil has a relatively low level of carbon emissions per capita, with urban dwellers in its major cities emitting lower still (World Bank 2010a: 25). Indeed, most of Brazil's emissions come from rural emissions relating to activities such as deforestation and cattle-raising (Dodman 2009: 190).

Realising the potential for cities to reduce rather than increase emissions can be seen as dependent on three dimensions: improvements in planning (including better transport and building upwards), advancements in technology (for example in the construction and transport sectors) and changes in behaviour (for example with regard to waste and use of heating). Innovative institutional arrangements can also make a significant difference, and cities such as Santiago in Chile and Delhi in India have demonstrated the ability to 'bundle' together emissions reductions with other issues such as cost savings and the generation of additional finance, improving the prospects of mitigation (Heinrichs et al. 2013: 1871). In many cases, it has been cities and their mayors that have taken the initiative, often acting as 'first movers' in response to climate change and placing it on their agenda as early as the 1980s (Bulkeley 2010). Even some low-income urban informal settlements, such as Khayelitsha in Cape Town, have been involved in projects that reduce energy use while also addressing energy poverty and helping to create jobs (ibid.). In Glaeser's words, 'Anyone who believes global warming is a real danger should see dense urban living as part of the solution' (Glaeser 2011: 201).

Urban environmental risk, 'ecological security' and climate migration

While the association of cities with global warming and generally being part of the 'problem' goes back some way, it is only more recently that concerns have been seriously voiced about the vulnerability of cities to climate change and what can be done to address this. The ways in which urban areas, and especially the urban poor, are rendered vulnerable to environmental change (including but certainly not limited to climate change) are multifarious. The impact of natural disasters, in particular, is intricately linked to the conditions of urban life; indeed urbanisation itself has been interpreted as one of the main causes of the dramatic increase in disaster-related deaths and economic losses in recent decades (Pelling 2003). This increase, which saw the number of people affected by natural hazards each year quadruple between 1975–84 and 1996–2005 (Emergency Events Database 2007, cited in Brecht et al. 2013), actually has a lot more to do with population concentration in particular areas than it does with climate change itself (Bouwer 2011; Neumayer and Barthel 2010; Brecht et al. 2013).

To understand how urban dwellers are subjected to the effects of environmental change it is important to unpack the various aspects of risk, which is a multifaceted concept that involves several elements. In their attempt to build a 'global urban risk index', Brecht et al. (2013) build a model that conceptualises risk as constituting three key elements: hazard, exposure and vulnerability. If hazard is about the likelihood of a future adverse event (such as flood, cyclone or earthquake) occurring, exposure refers to the elements (including people, buildings, communities and infrastructure) that are in harm's way as a result of this hazard, and vulnerability to how well the elements affected will be able to cope with disaster. Any increase in natural hazards occurring as a consequence of climate change contributes to the environmental risk faced by populations. Urbanisation itself can be considered a risk factor, as noted above, due to issues such as population and building density which increase the exposure of populations to potential hazards. The risks associated with urbanisation can, however, be managed where the resources and systems are in place to do so, either by limiting exposure or reducing the vulnerability of those exposed.

For example, the earthquake that struck Chile in March 2010 was one of the ten most powerful earthquakes in the last century, releasing 500 times more energy than the one which devastated Haiti in the same year. In Chile, however, only 521 people died, while the death toll in Haiti is estimated at over 230,000. The main reason for this difference was the fact that people in Chile were less vulnerable as the buildings were built to higher standards, in accordance with building codes and subject to regular inspections, which was not the case in Haiti (Brecht et al. 2013: 3). To take another example, there was a 20-fold difference in mortality between the Philippines and Japan when they were hit by typhoons of the same intensity (Satterthwaite 2013). The greater exposure and vulnerability of less developed countries in the face of hazards is also reflected in the fact that disaster-related economic losses are twenty times greater as a proportion of GDP in poor countries relative to rich ones (World Bank 2006).

Risk profiling and ranking of cities is an increasingly widespread practice, typically using three sets of indicators: the climate hazards a city faces, the demographic and geographic features of the city that affect its exposure to those hazards, and some indicators of the city's vulnerability or capacity to withstand and adapt to hazards (Adger et al. 2005; Heinrichs et al. 2013; Solecki et al. 2013). In low- and middle-income countries as a whole, the hazards themselves are astonishingly high, with approximately 1.1 billion people at risk of floods, 560 million at risk of earthquakes and 660 million facing the risk of landslides. When looking at risk levels for specific urban areas, certain cities rank especially high on multiple fronts, including Metro Manila, Tehran and various cities in India and Bangladesh. Meanwhile, Addis Ababa alone accounts for 31 per cent of Africa's earthquake mortality risk (Brecht et al. 2013: 22). According to a climate change risk index produced by a global risk analysis firm, the top ten cities facing extreme risk as of 2013 are all in South Asia, South-East Asia or Africa, with Dhaka topping the list (Maplethorpe 2013).

Other cities particularly vulnerable to floods and sea level rise include Mumbai and Shanghai, with parts of both being only one to five metres above sea level; Lagos, Alexandria, Port Harcourt and Banjul in Africa; New Orleans in the US and the coastal towns of the Caribbean states, where between 20 and 50 per cent of the population resides close to sea level (ibid.: 24). The most dramatic case in terms of absolute numbers is China, which alone has more than 78 million people living in vulnerable low elevation cities – a figure that is increasing annually at 3 per cent (McGranahan et al. 2007). In addition, populations living in flood plains in China have increased by 114 per cent and in cyclone-prone coastal areas by 192 per cent (UN-ISDR 2011). Even when the storms, floods, droughts, heat waves and landslides are not directly attributable to climate change, the extent of urban destruction they can cause is proof of how vulnerable many city populations will be in the face of inevitable increases in events that are climate-change induced (Satterthwaite et al. 2007: vii). Particularly vulnerable are urban centres located in low-elevation coastal zones less than ten metres above sea level. These areas, which comprise less than 2 per cent of the world's land area, contain 10 per cent of its total population (over 600 million people) and 13 per cent of its urban population (around 360 million people). Moreover, almost two-thirds of the world's cities with more than five million inhabitants fall at least partly in this zone, and low- and middle-income countries have nearly twice the proportion of their urban populations in this zone as compared to rich countries (Satterthwaite et al. 2007: 22).

The way in which different aspects of vulnerability overlap in urban areas, which are also exposed to complex hazards, has led some to call for cross-disciplinary research in pursuit of a more concerted, urban-focused sphere of 'vulnerability studies' (Parnell et al. 2007). This would also help to unpack the important intra-urban differentials with regard to the vulnerability of urban populations. In any city, those most at risk tend to be low-income households because they live in poor quality dwellings located in areas most vulnerable to threat (see Chapter 5). They are also those least able to escape disaster, cope with hazards or recover from these and other impacts of climate change. As well as their location and the quality of their housing, the capacity of different populations to cope with natural disasters is affected by their overall health, the financial resources available to them, the access to shelter after a disaster and the quality and inclusiveness of governance structures and community organisations in which they are embedded (UN-HABITAT 2011a: 13). Higher income groups therefore tend to be significantly less vulnerable even when in the same city and exposed to the same hazards. In many cities there are also significant gender dimensions to vulnerability to climate hazards (see Box 6.3).

Importantly, it is not just extreme weather events that render city-dwellers vulnerable; cities are also subject to incremental change, and to changes in the *average* environmental conditions that in the long term are just as devastating (Dodman and Satterthwaite 2008). One of the more gradual effects that climate change is having on cities is the exacerbation of urban heat islands. Cities are already hotter than their surrounding suburban and rural areas due to the absorption of heat by building materials and the effect of built form on natural evaporation and ventilation (Solecki et al. 2013). As a consequence, cities can be up to 5–11 degrees centigrade hotter than their surrounding areas, with urban sprawl exacerbating the effect (Campbell-Lendrum and Corvalan 2007). While climate change will intensify summer heat waves and exacerbate air pollution in urban areas, urbanisation, urban growth and increased population density in urban areas exacerbates exposure and vulnerability. Higher average precipitation intensity is also likely to overwhelm drainage systems even as the melting of mountain snow and glaciers impacts on water availability for urban dwellers who depend on these sources for water (Solecki et al. 2013). Often these risks can overlap in unexpected ways. Living in an urban settlement increases the risk of 'concatenated hazards', meaning a situation where a primary hazard such as a heavy storm leads to secondary hazards such as water contamination caused by flooding (UN-HABITAT 2011a: 13).

Again, slum conditions and overcrowding result in heightened exposure to these hazards and increased vulnerability. Vector-borne diseases thrive in situations where water services are limited and

Box 6.3 The gender dimensions of climate vulnerability in Dhaka

While many policy initiatives on adaptation to climate change make reference to gender, in reality these often just mean making generalised provision for women, without attention to the complex power relations that render both women and men vulnerable in different ways. Factors such as the gender division of labour and particular cultural gender roles, for example, can contribute to different kinds of vulnerabilities.

One study of Dhaka, Bangladesh highlights the gendered vulnerabilities people are subject to in the context of one particularly notable effect of climate change affecting the city in recent years: exposure to intense heat over long periods of time. The study found that male dominance in decision-making about the built environment resulted in a high priority being placed on privacy and security for women, with the effect that many rooms designed for women lacked windows and ventilation, subjecting them to extreme heat conditions in hot periods. At the same time, however, men's more frequent presence in outdoor and public spaces can increase their exposure to other heat-related hazards, particularly where they are involved in work such as rickshaw-driving or construction, providing them with few ways to control their exposure to direct sun.

Attention to these gender-based climate vulnerabilities is critical if we are to better understand adaptive capacity at the community level. The study found that overall, the dominant patriarchal system in which women are significantly under-represented in decision-making means they are less well-equipped to build their adaptive capacity, though this is starting to change as women become increasingly involved in income generation. The shifting balance of power may have significance for overall community resilience, given that women are thought to be more aware of incremental changes to the climate due to their reproductive and household roles.

Source: Jabeen (2014).

during floods, which are a common occurrence in informal settlements lacking adequate drainage (Solecki et al. 2013). These impacts of climate change are also going to hit many countries in the context of the epidemiological transition they are undergoing as they urbanise and develop, including exposure to higher levels of pollution and diseases linked to increased sewage and demands on water and sanitation (McGranahan 2013). In settlements where public health challenges have not been managed well thus far, the impacts of climate change on health are likely to be felt particularly harshly (ibid.). Urban heat islands in themselves generate new health risks including the spread of disease, and the poorest are most vulnerable. One study found that while heat waves led to fatalities in London and Delhi, in London mortality fell within a few days of temperatures dropping but in Delhi the increase in mortality continued for weeks afterwards as diseases took their toll (Hajat et al. 2005).

Plate 6.2 *Water pollution in Kampala, Uganda*

Source: Tom Goodfellow.

Many of the risks facing cities in the context of environmental change are now framed within a discourse of security – an aspect of the 'new security agenda' in the twenty-first century that reflects fears of terrorist attacks on critical infrastructure layered onto concerns about the effects of natural disasters on infrastructure and energy sources. In the urban context, an emerging discourse of 'urban ecological security' is emerging, guided by the principles of 'secure urbanism and resilient infrastructure' and concerned with ensuring urban ecological and material reproduction (Hodson and Marvin 2009). This builds on urban political ecology approaches that emphasise the entanglement of human and non-human, socioeconomic and material aspects of cities, and their mutual systemic vulnerability to risk (Heynen et al. 2006). It also draws on the broader, nebulous concept of 'human security' that has been popular since the 1990s (Heinrichs et al. 2013; Barnett and Adger 2007), and feeds into ideas about the 'socio-technical transitions' necessary to ensure urban energy security into the future (Rutherford and Coutard 2014).

A further emerging concern that can be linked to ideas of urban ecological security, particularly with regard to developing countries, involves the relationship between climate change and migration. In particular, the climate-related changes already occurring in many rural areas of the world – including worsening droughts, increased flooding, degradation of farmland and heightened storm risk – can induce population migration with significant knock-on changes for cities and ecological security. For obvious reasons, migration and increased mobility are responses to environmental change, as they have always been through history. This prospect has raised considerable concern in recent years, especially given the figures that are now frequently cited: it is estimated that by 2050, the number of people forced to migrate primarily because of climate change will be 200–300 million (Myers 2005; Christian Aid 2007). Cities are, in many cases, likely to be the places to which people migrate, even though many of those fleeing extreme weather events will also be coming from cities, for reasons discussed above. Even for those cities that have not yet felt significant impacts of climate change, migration is the route through which they may be most imminently affected as they face increasing pressure to accommodate refugees from other cities and rural hinterlands (Biermann and Boas 2010; While and Whitehead 2013).

Despite these very real concerns, however, there is a risk of an alarmist, oversimplistic, negative and poorly evidenced narrative of a 'human tide' of climate migrants descending on the shores of relatively less-affected parts of the world (Tacoli 2009). The idea of the 'environmental refugee' first entered into debate in the 1970s, around the time of neo-Malthusian pessimism about population trajectories that was epitomised by the *Limits to Growth* report (see above). On the whole, however, migration and environmental change were not much linked in the literature on either subject until a marked increase in interest in the 2000s. Climate change was inevitably going to reignite interest in this issue, and a growing body of evidence suggests that climate change is indeed playing a role in migration (Jäger et al. 2009; Warner 2010). Some have argued that available evidence indicates that 'the climate refugee crisis will surpass all known refugee crises in terms of the number of people affected' (Biermann and Boas 2010: 61). While numerical projections are fraught with methodological difficulty, it seems clear that by far the largest group likely to migrate is those at risk from sea level rise, which Myers has estimated at 162 million, relative to 50 million refugees projected to migrate due to drought and other climate change impacts (Myers 2002).

Linked to the difficulty of estimating numbers with any accuracy are the huge complications in determining who is rightly considered a 'climate migrant'. Some effects of climate change are incremental and may be difficult to distinguish from other environmental and social changes, and people often have multiple reasons for migrating (Tacoli 2009; Warner 2010; Biermann and Boas 2010). The question of whether people have to have been 'forced' to leave by climatic conditions to qualify as climate migrants is also contested (Reuveny 2007; Warner 2010). More generally, the whole discourse of 'climate refugees' is predicated on the assumption of a direct causal link between environmental change and migration, which rests more on 'common sense' ideas about how people behave rather than detailed research on the multiple factors that impact on migration decisions (Tacoli 2009: 516; Parnell and Walawege 2011).

If we accept that migration linked to climate change is likely to increase, there is a further question regarding how this affects rural–urban dynamics, particularly with respect to a possible acceleration of urbanisation trajectories in developing countries already urbanising apace. Barrios et al. (2006) find that climate change is indeed affecting urbanisation in Africa, with declines in rainfall driving a certain amount of rural–urban migration, though they find that this link is not evident in other parts of the world. A report by the International Organisation for Migration likewise suggests that an expansion of cities and towns in Africa could result from global environmental change (Laczko and Aghazarm 2009), and there are good reasons for thinking that increasing strain will be evident in many of Africa's rural areas as climatic conditions lead to land degradation and reduced livelihood options (McGrahahan et al. 2009: Annez et al. 2010). Research in Bangladesh, meanwhile, has found that the impact of climate change on livelihoods is a central driver of rural–urban migration, with claims of 500,000 rural migrants arriving in Dhaka's slums every year. Some rural districts are said to be becoming 'ghost lands' as migrants make the decision never to return to places now devastated by swollen rivers and typhoons (Meagher and Farhana 2012). Based on survey work and interviewing, one researcher argues that 'climatic factors have never affected any other group so severely at both their origin and destination as they have affected poor climate-induced migrants' in Bangladesh (Adri 2014).

While specific cities such as Dhaka may undeniably be feeling the effects, a more general relationship between climate change and rural–urban migration is harder to establish. The available evidence and literature provide a very inconclusive picture of the dynamics of migration in Africa, for example, even before the factor of climate change enters the picture (Parnell and Walawege 2011). Establishing evidence on migration patterns is rendered especially difficult by long-standing practices of circular and seasonal migration patterns in Africa (Potts 2010; Tacoli 2009). While it may be the case that people begin to settle more permanently in urban areas due to climatic factors, it is equally possible that the effects of climate change in poorly managed cities drive people to depend more heavily on their rural linkages (Parnell and Walawege 2011: S18). We do know, however, that the most common migratory responses to environmental disasters are likely to be within national borders or across proximate borders, rather than from the 'global South' to cities of 'the North' as warned by those who fear an influx of climate migrants in developed countries (Annez et al. 2010).

Regardless of uncertainty over whether climate change is likely to be an independent cause of rural–urban migration, the combination of increased population movement due to climate change with more general urban growth trends is raising security concerns. A 2007 Christian Aid report entitled 'The Human Tide' argued that migration due to climate change has the potential 'to de-stabilise whole regions where increasingly desperate populations compete for dwindling food and water' (Christian Aid 2007: 2). In these rather alarmist accounts, Africa has been a particular concern, with commentators arguing that climate-related migration could tip 'fragile states into socio-economic and political collapse' (Brown et al. 2007). However, while speculative scholarship has attempted to trace the links between climate change-induced migration and potential violent conflict (Barnett and Adger 2007; Reuveny 2007), robust empirical evidence to support these links is very scarce. Reuveny (2007) finds that in a number of cases there were conflicts of medium intensity following incidences of environmentally induced migration into urban areas, but the causal link is unclear.

It is also important to consider how anyone benefits from such negative framing of climate-related migration, when mobility is (and has always been) an important strategy alongside income diversification to reduce vulnerability to environmental, social and economic shocks (Wisner 2003; Tacoli 2009: 514). There is a need to guard against the possibility of a socio-environmental agenda focused narrowly on perceived 'threats', leading to the legitimisation of a renewed authoritarian politics of urban social control (While et al.

2010; While and Whitehead 2013). What is required instead is the development of national and international governance regimes that support climate refugees, accompanied by a broader global governance framework to facilitate both sustainable resource use and adaptation to climate change (Biermann and Boas 2010). It is to the question of adaptation and climate governance that we now turn.

Adaptation, resilience and climate governance

As concerns about climate change began mounting in the 1980s and 1990s, efforts to address the problem were initially conceptualised in terms of mitigation: in other words measures to prevent further climate change or limit the extent of climate change. The First Assessment Report of the Intergovernmental Panel on Climate Change (IPCC) in 1990 was a typical example, focusing on emissions reduction with very little attention to adaptation: the question of how to readjust life to the reality that a certain amount of climate change will inevitably occur. Later IPCC reports in 1995, 2001 and 2007 said progressively more about the need for adaptation, though only the latter really emphasised the urgency of this aspect of the climate challenge (Pelling 2010: 9). The now widespread acceptance that a certain amount of climate change will occur has placed adaptation firmly on the agenda. The debate about how much to emphasise mitigation versus adaptation is still far from settled, particularly given that placing too much emphasis on the latter might risk slowing momentum on the former, at great human and financial cost. Indeed, the 2006 Stern Report on the economic consequences of climate change suggested that dealing with climate change through adaptation would cost in the order of five to 20 times as much as containing climate change through mitigation (ibid.: 11).

At a global level, the mitigation problem is especially relevant for developed countries, while adaption is especially urgent in low- and middle-income countries which, as noted above, face the greatest risks and are affected by higher vulnerability. This distinction has tended to bolster the tendency for the developed world to largely ignore adaptation, at least until recently. Setting targets for mitigation, and monitoring progress, is a relatively simple way for cities in high-income nations to be seen as taking actions to deal with climate change (Satterthwaite and Dodman 2013: 294). There is also a tendency in the policy world to frame mitigation as a global challenge, due to the collective action challenges associated with achieving binding commitments to emissions reduction, while adaptation is a local effort that impacts on (and should be organised at) more local scales (World Bank 2010a: 11). This, however, is oversimplistic, given the importance of high-level legal frameworks and voluntary agreements to support local adaptation, and of local pressure on national and international institutions to pursue both mitigation and adaptation (Pelling 2010: 7). In other words, developed countries with the vast majority of the world's resources need to be playing leading roles in terms of adaptation as well as mitigation.

Even at the level of cities in developing countries, mitigation is often seen as a strategic city planning concern while adaptation is viewed as something that largely involves very local, household-level interventions (Parnell 2015). In reality, given the severe constraints on municipal government resources in most developing countries, it is likely that much of the investment and activity for adaptation will indeed have to come from individuals, households, communities and firms. Yet whether these investments and activities succeed depends in large part on what local governments support and prevent, in terms of both regulation and the provision of infrastructure and services (IPCC 2014).

There are also multiple meanings of the term 'adaptation' and varying approaches to adaptive interventions in terms of scope, scale and timing (Smit and Wandel 2006; Bulkeley 2010). In simple terms it is possible to distinguish between 'hard' adaptation associated with infrastructure and changes to the built environment (such as flood defences) and 'soft' infrastructure, relating to capacity building, awareness and information systems. In practice the latter tend to be rather neglected, receiving only around 10 per cent of the aid related to urban adaptation (IPCC 2014). Nevertheless, increasingly central to the discussion on adaptation is the idea of the 'adaptive capacity' of particular places, encompassing both the ability to prepare for risks and opportunities associated with climate change (proactive adaptation) and to cope with potential and actual negative effects as they unfold (reactive adaptation) (Smit and Wandel 2006; Heinrichs et al. 2013). In relatively early contributions, Carter (1996) distinguished between autonomous (automatic, spontaneous or passive) and planned (active and strategic) adaptations, while Burton et al. (1993) distinguish between adaptations that prevent loss, spread loss, change use, activity or location, or engage in restoration.

Adaptation is thus clearly more complex than it might first appear. In concrete terms, examples of adaptation interventions at different scales include the following: exploring new groundwater sources for mains water supply in the event of reduced rainfall; government programmes to re-employ people migrating from drought prone regions in China (Li 2013: 41); dredging and widening of drainage canals and conducting of topographical and fluvial surveys in Cartagena, Colombia (Stein and Moser 2014: 178); and exploring new construction materials to reduce 'heat island' effects in Santiago de Chile (Barton 2013: 1926). Adaptation efforts do appear to be yielding important results in many cases. In Bangladesh, for example, efforts to decrease vulnerability to tropical cyclones after a devastating hurricane in 1991 led to an early warning system and public shelters for evacuees, which were tested by another cyclone in 2007 of almost equal force. The 32-fold reduction in death toll in the second of these cyclones, in which 4,234 people died as compared with 131,000 in 1991, suggests that adaptation had made a very substantial difference (Paul 2009). Most highly prized are adaptation efforts that also deliver mitigation co-benefits, as some authors have argued carefully planned greenbelts and transport policies can do (Viguié and Hallegatte 2012).

As attention to adaptation has escalated, a related concern with *resilience* has rapidly grown in momentum in recent years, drawing on longer intellectual trajectories within the fields of ecology and systems thinking and emphasising the flexibility and responsiveness needed to ensure effective adaptation (Folke 2006; Tyler and Moench 2012). The interest in the idea of resilience is not exclusive to thinking on climate change and has proliferated across a range of issues and disciplines in response to new global insecurities. One study found that the number of articles on resilience increased by 400 per cent between 1997 and 2007 (Swanstrom 2008). This is likely to have further accelerated since, especially given the popularity of the term in the wake of the 2008 financial crisis. A typical definition of resilience refers to the ability to 'absorb disturbance and reorganize while undergoing change so as to still retain essentially the same function, structure, identity and feedbacks' (Folke 2006: 259). Resilience in the context of climate change implies a capacity not only to withstand shocks and stresses but also to fully recover, including from events that may be entirely unexpected (Satterthwaite and Dodman 2013: 295). It also draws on a longer history of research on disaster risk reduction (Blaikie et al.

1994) to encompass the ways in which systems may become more robust through the experience of shocks, for example through 'building back better' (Lyons 2009).

As highlighted earlier, cities in developing countries are likely to suffer particular difficulties in achieving resilience in the face of climate change due to their often extreme vulnerability. The assessment or measurement of vulnerability itself usually encompasses a measure of adaptive capacity (Heinrichs et al. 2013; Jabeen 2014). It is, however, important not to assume that urban dwellers in poorer countries are necessarily always the most ill-equipped to achieve resilience. Households, small businesses and communities in places particularly exposed to climate change are in many cases already doing a great deal to invest in their own resilience in the absence of substantial economic resources (Stein and Moser 2014). In some cases, the informality associated with cities in the developing world can actually serve to bolster resilience. Comparing Hurricane Katrina in the USA with major floods in Mumbai, Anjaria (2006) argues that while the former were met with social disorder and violence, the latter were met with widespread acts of altruism and generosity. Anjaria attributes this to the strong informal networks and sense of community prevalent throughout much of the city, which are often lost in the formal areas of the city. It can also be said that the urban poor contribute substantially to resilience through activities that effectively adapt and mitigate simultaneously; the huge number of people engaged in livelihoods linked to solid waste management and recycling in Asian cities, for example, considerably reduces greenhouse gas emissions (Kumar 2013: 1462).

Despite its soaring popularity, the idea of resilience remains controversial. Critical appraisals of the concept caution against assuming that resilience-building necessarily leads to positive outcomes, especially if it serves to entrench problematic systems and existing inequalities (Orleans Read et al. 2013). The question of who is expected to adapt and how, and who has to become more resilient, is critical; it is here that politics enters the equation and without attention to this, adaptation is unlikely to have socially just outcomes (Barton 2013). Smith and Stirling likewise highlight the need to reflect on what is being made resilient, and for whom that resilience is good or bad (Smith and Stirling 2010: 11). There is also a concern that the focus on building resilience to shocks and ignoring long-term stress 'may lead to robustness which inhibits adaptability and transformability' (ibid.: 11). In a landmark contribution to the adaptation and resilience debate, Pelling (2010) argues that approaches to adaptation have been overly technical, ignoring the role of politics at the centre of adaptation, which is ultimately a matter of environmental and social justice. Climate change and the threats it poses are, in this view, symptoms of a deeper 'social malaise', and to focus on technical responses is to ignore the fundamental challenges posed by climate change to the way societies are organised.

Pelling poses three distinct adaptation pathways with different ultimate purposes, of which resilience is the first and least radical. This requires a focus on achieving functional persistence in the face of change, and necessitates social learning and organisation. However, like some of the authors cited above, Pelling argues that since resilience is basically about maintaining the status quo, it can allow unsustainable or unjust practices to persist; for example, service delivery systems that are dependent on precarious and risky livelihoods for the poor can persist and be 'resilient' to environmental change without actually *reducing* the vulnerability of the poor. A second pathway is characterised as moving beyond resilience to achieve 'transition', which is about incremental change in systems and requires going beyond social learning and functional persistence to take into account governance arrangements and the need for socio-technical transitions. Pelling's third adaptation pathway is conceptualised as 'transformation', which goes further still, involving radical change and incorporating the need to fundamentally revise or replace existing social contracts, addressing the root causes of vulnerability and promoting new understandings of human security (Pelling 2010). Without a more transformative approach, the most vulnerable in society may not be able to benefit from systemic resilience at all, because for them improved security and opportunity means going beyond mere functional persistence and requires changes in social systems (Satterthwaite and Dodman 2013: 291).

Transformational approaches in many respects bring adaptation full circle back towards mitigation, in the sense that they are about adapting 'with' rather than 'to' climate change, which involves shifts towards the low-carbon lifestyles necessary to achieve ecological justice in cities (ibid.; Pelling 2010). For Pelling, the re-linking of mitigation and adaptation is an echo of the earlier 'sustainable development' debate, which 'manifestly failed' to effectively challenge dominant forms of development (Pelling 2010: 2). The new attention to power and justice in discussions of adaptation has led scholars to emphasize the political pressure and organisation needed to 'accumulate' resilience (Satterthwaite 2013), and to call for 'transformational resilience thinking' that pushes the boundaries of what we understand by responding to climate change (Bahadur and Tanner 2014). The way in which we think about who the key actors are with respect to adaptation and resilience also needs to evolve. For example, the role of businesses and the private sector is critical, especially with regard to insurance companies and how they should be facilitating adaptation (Pelling 2010).

As is obvious from the earlier discussion of cities' contribution to climate change, mitigation must remain firmly on the agenda. Building adaptive capacity needs to be accompanied by a fundamental transformation in the way we live our lives and the way in which the quest for economic growth in cities is pursued. The critical emphasis in recent debates is on how to think about mitigation and adaptation in tandem, rather than as two separate processes. This is important not least because mitigation and 'green growth' have themselves become an industry, and in trying to combine mitigation with growth, attention to the needs of the most vulnerable can easily fall by the wayside. The overriding concern with economic growth in most countries of the world also has a tendency to impede local agency with respect to transformative approaches to climate change. In China, for example, any policy initiative needs to be built into an economic growth framework and will not even be considered unless it is shown to be consistent with increased competitiveness and continued growth (Li 2013). This raises concerns about how to build political will through changes that may not be directly linked to achieving growth but have very important human impacts.

In reality, initiatives that combine both mitigation and adaptation elements with continued commitment to urban economic development and growth are likely to be the most appealing to local and national policymakers (as well as private firms) and consequently the most effective. There is an increasing 'marriage of political and economic interest' in low-carbon urbanism (Bulkeley et al. 2012) that must be capitalised on. An example of an environmental initiative that could be seen as a 'win-win' in both economic and social terms is the improvement of water pipes to increase water supply efficiency while also widening access to water and reducing the vulnerability of the poor (World Bank 2010a: 12). Individual initiatives will also need to be supported by new ways of thinking about urban governance and planning that bring together urban designers, public health experts and transport engineers to develop innovative ways of creating compact and diverse communities, where multiple goals of adaptation and mitigation can realistically be pursued in tandem (Blanco et al. 2009).

In this respect, it is important to explore the opportunities cities present for climate action rather than focusing too much on the threats they pose (While and Whitehead 2013). Cities can and do serve as 'laboratories for climate-change action' (Solecki et al. 2013), and urban-level initiatives are increasingly taking place across the developing as well as developed world. While there are administrative obstacles to effective action, there are also opportunities for municipal-level bureaucratic innovation, as explored by Aylett in the case of Durban, for example (Aylett 2013). Urban initiatives such as city-level carbon trading schemes are also proceeding in parts of the world, with potential for mitigation gains that may be more challenging to achieve at the national level (see Box 6.4). Bulkeley et al. (2013) argue that the most interesting examples of city-level climate governance are taking place not through policies and plans but through experimentation, which is happening across the public–private divide and has been particularly innovative in relation to energy use and carbon control.

Meanwhile, cities are coming together across borders to explore urban-level opportunities for climate action. The C40 Cities Climate Leadership Group is a network of 'megacities' and 'innovator cities' across the world that aims to demonstrate that 'ending climate change begins in the city', showcasing examples of effective adaptation and mitigation from which others can learn. Of the almost 60 cities in the network, over 4,700 climate change actions are currently in effect. There has also been broader progress in terms of urban international collective action. In December 2011, the Climate Summit for Mayors was hosted in Durban to highlight the role of cities in both mitigation and adaptation, while Local Governments for Sustainability (ICLEI) has brought together more than 1,100 city and local government authorities committed to promoting sustainability through local action. The importance of high-level 'buy-in' and particularly having mayors that really champion the cause is evident from a range of cases of successful climate action (Solecki et al. 2013). Unfortunately, however, in the least developed countries it is hard to replicate some of the successes of cities such as New York, both due to the lack of resources and the constrained autonomy of mayors and city governments discussed in Chapter 8.

Box 6.4 The Tokyo Emission Trading Scheme

A greenhouse gas emission trading scheme (ETS) is a market tool that attaches a monetary cost to emissions in the form of carbon credits (1 credit = 1 tonne of CO_2 equivalent). Carbon credits can be traded, and this provides a profit motive for facilities to reduce emissions and sell their excess credits. Despite optimism for ETS policy in the wake of the 1997 Kyoto Protocol, early schemes (e.g. the multi-national EU ETS) achieved minor emission reductions.

With the largest city population in the world, and the highest greenhouse gas emissions of any city, Tokyo is a vast metropolis. In 2006 the Tokyo Metropolitan Government committed to reduce emissions by 25 per cent (relative to 2000 baseline level) by 2020. A city-wide ETS was launched in 2010 as the main policy to achieve this target. The Tokyo ETS is restricted to the regulation of downstream electricity usage from large commercial and public facilities. The 2019 reduction target for compliant facilities is 17 per cent below 2000 baseline levels. However, by 2014 total emissions were already 23 per cent below baseline, with 90 per cent of compliant facilities bettering their 2020 target.

The Metropolitan Government's initiatives have been an important driving force behind these impressive results. First, a number of policies paved the way for the introduction of the ETS by testing the suitability of particular methods, such as mandatory electricity reporting by facilities, which was being enforced before the ETS was in place. Second, the government is implementing measures to motivate acceptance of the ETS policy. In particular, public engagement meetings provide a platform for addressing public concerns and identifying policy areas to develop, and a certification scheme provides incentive for facilities to cut emissions beyond that required under the ETS.

Tokyo is viewed as a model for city-scale ETS development worldwide, with a number of cities following suit. However, it remains to be seen whether the transfer of this policy will achieve equivalent emission cuts, or whether the results from Tokyo are a testament to the particularly concerted effort made by local government in the city.

Sources: IETA (2015), TMG (2015), Niederhafner (2013), Rudolph and Kawakatsu (2012).

Support from higher tiers of government and an international architecture attuned to the need for (but also the intrinsic political complexities of) adaptation is therefore essential (Bulkeley 2010). Solutions cannot be left to local government, even large metropolitan authorities. Urban institutions can take actions to provide services, conserve water, recycle waste and reduce greenhouse gas emissions, but they cannot be held responsible for reducing climate change risk beyond their jurisdictions. Recognition of the complexity of dealing with climate change issues that might at first look like 'city level' problems has meant that climate change response is one of the areas in which 'multi-level governance' is being most thoroughly discussed and theorised (Bulkeley 2010; 2013). As well as regional and national support, these activities also need to be supported and promoted internationally, with recognition that neither effective adaptation, resilience or transformation are likely to be effective without supportive global processes. For a start, the costs are potentially infinite. Oxfam estimates that a minimum of US$ 50–80 billion annually is necessary for low- and middle-income countries to adapt to climate change (Oxfam 2007). Low- and middle-income countries and their urban centres do not have the power and resources to address the profound global inequities that caused climate change in the first place, even if they are home to those most vulnerable to its impact. This lack of resources is leading to experimentation, innovation and entrepreneurship on the part of urban authorities, though capacity constraints remain an enormous challenge (Anguelovski and Carmin 2011).

Conclusion

This chapter has explored a number of dimensions of the relationship between cities and environmental change, aiming to locate the urban within a range of evolving debates and processes concerning sustainability, mitigation, adaptation, resilience, transformation and climate governance. Understanding the relationship between cities and the natural environment in all its complexity is crucial, because it is very clear that both the challenge of climate change and our best hope for meeting it are intimately linked to how the world's cities and urban areas evolve. While policymakers are finally waking up and responding to climate change, the urban dimensions of the environmental challenge remain fairly low down the agenda of aid donors and international agencies. This is something that needs to change – and fast – if further catastrophic climate-related disasters are to be avoided and the potential for mitigation maximised.

Of central importance is the need to pursue strategies for mitigation and adaptation in tandem wherever possible. As Satterthwaite et al. (2007: 50) have argued, these 'are not alternative strategies but complementary ones that need to be pursued together ... Failure to mitigate sufficiently in high-income nations will create ever more adaptation failures, mostly in low- and middle-income nations.' There is also thus a major global imbalance that lies at the heart of the climate change challenge, and will continue to provide a major hurdle in international negotiations as well as potentially leading to unforeseeable political conflicts in the future. An important avenue for overcoming some of these differences is for cities to work together internationally, in recognition of their common interest in a low-carbon, climate-resilient global urban future that is also economically productive and socially equitable.

Summary

- The importance of sustainable urbanism is rising up the global development agenda, although the concept of sustainability remains debated and contested.
- Cities can be understood to have 'ecological footprints' and their effects on the environment are complex and multifaceted. The idea of 'urban metabolism' can help understand the relationship between cities and their surroundings.
- Emissions from cities are significant, but the global connections between production and consumption must be considered as well as inter and intra-city differentials in emissions.
- While cities have long been conceptualiscd as an environmental 'problem' due to their contribution to climate change, they can also offer important environmental solutions. This relies upon improved planning, advances in technology and changes in behaviour.
- The environmental risks facing cities are determined by levels of hazard, exposure and vulnerability.
- Risks facing urban populations vary substantially between and within cities. The top ten cities facing extreme risk are in developing countries, and low-income urban dwellers are especially vulnerable to acute and long-term effects of climate change.
- An agenda of 'urban ecological security' is emerging, linking environmental and security issues. Concern is growing about the effect of climate change on rural–urban migration, but the relationship between the two processes is contested.
- Sustainable urbanism requires a combination of mitigation, adaptation and increased resilience at both the local and global levels. To achieve such goals, cities themselves are starting to take the lead – often across borders.

Discussion questions

1. What does sustainability mean in an urban context?
2. In what ways can cities be considered part of the solution to climate change?
3. 'Urban populations in developing countries face the same hazards as those in developed countries, but their exposure and vulnerability is higher.' Discuss.
4. How do urban environmental and political concerns intersect to frame discourses of urban ecological security?
5. What is the relationship between mitigation, adaptation and resilience, and how do these apply to the ideas of transition and transformation?

Further reading

Bulkeley, H. (2013) *Cities and Climate Change.* London: Routledge.

Dodman, D. and Satterthwaite, D. (2008) 'Institutional capacity, climate change adaptation and the urban poor', *IDS Bulletin*, 39(4), 67–74.

Pelling, M. (2010) *Adaptation to Climate Change: From Resilience to Transformation.* London: Routledge.

Satterthwaite, D. and Dodman, D. (eds) (2013) Special Issue of *Environment and Urbanization:* 'Towards resilience and transformation for cities', 25(2).

Sclar, D. E., Volavka-Close, N. and Brown, P. (eds) (2013) *The Urban Transformation: Health, Shelter and Climate Change.* Abingdon: Earthscan.

UN-HABITAT (2011) *Global Report on Human Settlements 2011: Cities and Climate Change.* United Nations Human Settlements Program, Earthscan.

World Bank (2010) *Cities and Climate Change: An Urgent Agenda.* Washington DC: The World Bank.

Websites

Arcadis Sustainable Cities Index: www.sustainablecitiesindex.com

C40 Cities group website: www.c40.org

ICLEI (Local governments for sustainability) website: www.iclei.org

Sustainable Cities Collective website: www.sustainablecitiescollective.com

UN-Habitat Climate Change pages: unhabitat.org/urban-themes/climate-change

World Resources Institutes Cities site: www.wricities.org

7 Violence, crime and insecurity

- Conflict, violence and development
- Conceptualising urban violence and insecurity
- Crime, endemic violence and urban development
- War, terrorism and urban reconstruction

Introduction

Crime, endemic violence, war and terrorism are among the most dramatic and disturbing manifestations of vulnerability in cities, which are uniquely prone to these threats given their economic, political and cultural significance. Historically, many cities were designed and planned with this vulnerability in mind. From the ziggurats of Mesopotamian city-states to medieval fortress towns in Europe to fortified colonial outposts, the impetus for city building has often been inspired by a dual desire to concentrate and project power, as well as to defend it against perceived external threats. However, as spaces where political and economic power is concentrated, where diverse actors converge and where inequality is highly visible, threats just as often emerge from within as without. This too has not been lost on city builders: the grand boulevards of Paris were not designed for promenading, but rather military parading, to emphasise the power of the state and ensure that military order could be maintained in the streets if insurrection should materialise.

Despite the intrinsic links between cities, city-building and conflict, the relationship between these factors has received only scant attention in the context of development. Yet crime, violence, war and terrorism present very real and very serious obstacles to development in cities and beyond. We begin this chapter by exploring the relationship between conflict, violence and development more broadly, setting out key definitions and reviewing the changing nature of global security challenges, particularly as they pertain to developing countries. This is followed by a discussion of the challenge of urban violence specifically, and how this has been conceptualised and analysed in the context of growing attention to the topic and interest in the relationship between different forms of conflict and violence. We then turn to the most pervasive sources of urban insecurity – crime and endemic violence. We explore how this challenge maps out across different world regions, as well as how it impacts on urban development through the creation of urban geographies shaped by fear. Finally, we consider how cities are affected by large-scale armed conflict and terrorism, and the challenges of urban reconstruction as well as the potential cities hold for cultivating peace and building a concrete image of a society's collective aspirations.

Conflict, violence and development

The relationship between conflict, violence and development is one of the perennial themes in Development Studies. Sometimes it is treated as a contextual factor or engaged with only implicitly, but periodically it has risen to the surface as a central focus of debate. The period since the 9/11 attacks in the United States, the 'Global War on Terror' and the preoccupation with 'fragile states' marks one such phase in which conflict and violence have been particularly salient with regard to how development is understood and development interventions prioritised. In this period there has been a growing tendency to link global security concerns to international development prospects, with war, terrorism and organised crime depicted both as evidence of development failure and obstacles to development. Yet the relationship is much more complex than is suggested by the idea that conflict and violence are bad for development and therefore prioritising security should be paramount. In fact, there is a well-established stream of literature suggesting that conflict and violence are central parts of the development process, a perspective that has been used on the one hand to reject the very idea of development, and on the other to justify violence.

Some basic definitions are useful here. First, it is important to be clear that conflict broadly defined is something endemic to human societies. The interests and values of diverse individuals and groups in society are bound to conflict, but most conflicts can be resolved through non-violent means. Violence itself, on the other hand, can be defined as the use of physical force to cause injury, harm, deprivation or death (Fox and Beall 2012). This too is something that all societies face, but is much more problematic. It is generally considered a fundamental obstacle to development, destroying economic assets and social capital as well as human lives. A very influential strand in recent scholarship has argued that development has, throughout human history, been driven by efforts to contain the problem of violence – albeit with highly variable success and through different institutional forms (Bates 2001; North et al. 2009).

Two further caveats regarding the conceptualisation of violence are in order here. The first is that it can be useful to distinguish between violent acts by individuals (interpersonal violence), and organised group violence (collective violence) – whether perpetrated by gangs, armies, governments or terrorist groups. The second is that definitions of violence often subsume within them not only the use of physical force but the *threat* to use such force. If violence is considered to include the threat of force, then this renders its relationship to development rather different. Weber famously defined the state as an organisation holding a monopoly of legitimate violence (Weber 1994 [1919]); on this basis, the threat of force that underpins modern states, even if rarely exercised, is what makes most development possible. Indeed, coercion and the threat of force are 'as much a part of everyday life as are markets and economic exchange' (Bates 2001: 50), and often what we consider to be 'order' in society is sustained by the strategic use or threat of violence to manage conflict (Kalvyas et al. 2008).

Some critical scholars stretch the definition of violence much further, for example distinguishing between direct or personal violence – visible violent acts performed by a clearly identifiable agent – and systemic or 'structural' violence: violent outcomes resulting from impersonal forces for which no particular human agent can be deemed ultimately responsible (Galtung 1969; Žižek 2009). Some definitions of structural violence do not even involve physical injury as a necessary outcome, using the concept to encapsulate 'the violence of poverty, hunger, social exclusion, and humiliation' (Scheper-Hughes and Bourgois 2004: 1). The 'post-development' trend in scholarship has long taken the view that structural violence is central to the development project; for these authors, development is essentially about the enforcement of an idea of modernisation which necessarily involves large-scale displacement and discards or marginalises those who stand in its way (Escobar 1992; 2004).

Despite these important contributions to theoretical debates, to avoid confusion in the remainder of this chapter we use violence in the narrow sense to mean intentional use of force (or threat of such use) to cause physical harm or material damage.

The challenge of understanding the relationship between violence and development depends as much on our definition of development as our definition of violence. If we think of development solely in terms of particular interventions designed to bring about specific outcomes (i.e. development as a practice or a goal) then it is hard to imagine any way in which violence could be viewed as being positive or constructive. If, on the other hand, we consider development as the underlying processes through which societal structures evolve then it is hard to deny that violence has historically played a central role, albeit at enormous human cost. The work of Charles Tilly (1985, 1992), which has recently become popular with some scholars seeking to understand processes of conflict in the contemporary developing world (Leander 2003; Herbst 2004; Taylor and Botea 2008) has emphasised the centrality of war in the formation of states that subsequently enabled development in Europe (see Chapter 2). The most developed societies today are arguably those in which systems predicated on violence have evolved into ones where conflict is primarily managed non-violently; where over the *long durée,* conflicts have produced institutional innovation capable of sustaining development, rather than reproducing cycles of destructive violence.

These considerations serve to underline the complexity of the relationship between conflict, violence and development, and the fact that while conflict has been (and will always be) central to processes of development, the overriding concern of all those concerned with development as a practice must be to take 'a course of least violence' (Parfit 2013: 1191). Unfortunately, however, recent decades have been characterised by changing patterns of violence that pose enormous challenges for development both as a practice and an objective or goal. Violence rose to prominence as a development issue in the 2000s following a tumultuous decade of civil wars in many developing countries fuelled in part by the shifting geopolitics of the immediate post-Cold War era (Kaldor 1999) and growing concern about the threat of terrorism emerging in 'failed states' (Rotberg 2002). For some scholars, the new millennium will be characterised by persistent urban violence (Moser and McIlwaine 2014), which for some represents 'a normal feature of late modernity' (Hagedorn 2007: 298).

There can be little doubt that violence has devastating consequences on human development outcomes, regardless of whether particular forms of violence sometimes perform state-building functions in the much longer term. Violence limits human freedoms and erodes trust and social cohesion, as well as diverting resources away from productive investment and destroying productive assets. It has been estimated that the average economic cost of civil war is equivalent to more than 30 years of GDP growth for a medium-sized developing country (World Bank 2011: 5). On average, a country that has experienced major violence over the period 1981–2005 also has a poverty rate 21 percentage points higher than a country that has not (ibid.: 5). There are reasons to debate the direction of causality when it comes to the relationship between violent conflict and poverty, and a large body of cross-national quantitative research has argued poverty is itself a fundamental cause of civil war (Fearon and Laitin 2003; Collier and Hoeffler 2004). Regardless of ongoing debates about the causes of civil war, however (see Sambanis 2004; Blattman and Miguel 2010; Cramer 2006), few would dispute that significant violence of any kind also has profoundly damaging consequences for human wellbeing.

Human insecurity and vulnerability to violence is thus widely seen as 'a primary development challenge of our time' (World Bank 2011: 1). The prevalence of contemporary violence is certainly alarming, with one and a half billion people thought to live in areas affected by the spectre of violence (ibid.: 1). Collective violence is also becoming more of a recurrent problem, with countries that have previously experienced a violent conflict much more likely to be tipped back into conflict than in the past. Ninety per cent of the civil wars that began in the 2000s were in countries that had experienced conflict in the previous 30 years – a much higher percentage of repeat conflict than at any other time in the past half century (World Bank 2011: 3).

It is not, however, civil wars that are the primary source of violence today, and in fact the death toll from civil wars in the 2000s was around a quarter of what it was in the 1980s, though it increased again in the early 2010s. Numbers of both inter-state and civil wars have been in decline since the 1990s (Newman 2009; Blattman and Miguel 2010), while the soaring fatality rates from civil war and terrorism since 2010 are largely confined to a few conflict-ridden states in the Middle East and North Africa (GPI 2015). However, in many parts of the world, other forms of low-level instability and conflict are on the increase, and globally violence in non-war situations is responsible for far more deaths than civil war (Harbom and Wallensteen, 2009; Fox and Hoelscher, 2012). As Table 7.1 shows, interpersonal violence (i.e. homicides) is responsible for roughly four times as many deaths per year as collective violence such as armed conflict, terrorism and state repression. Some regions are particularly badly afflicted. Many Central and South American countries, which saw a significant decline in domestic armed conflicts in the late twentieth century, now suffer from some of the world's highest homicide rates. Indeed, a central emerging theme in the literature on conflict, violence and development is the degree to which, in some parts of the world, violence in one sphere of life often rises as other forms of violence decline (Beall et al. 2013; Moser and Rodgers 2012; Moser and McIlwaine 2014).

The shift from organised conflict in the form of interstate and civil wars towards more prosaic forms of everyday violence has occurred alongside a reconceptualisation of the idea of security. While in the Cold War era security largely meant the sovereign integrity of states, since the 1990s the UN has popularised a new concept of security – human security – which places people, not nation-states, at the centre of the agenda and has been used to justify humanitarian interventions that undermine state sovereignty (Kaldor 2007). In some respects paralleling the idea of human development (see Chapter 1), human security is a holistic concept that embraces a wide range of concerns, including economic security, food security, health security, environmental security as well as personal security (UNDP

Table 7.1 *Regional comparison of the burden of collective vs. interpersonal violence*

	Deaths per 100,000 people			
	Collective violence		Interpersonal violence	
	2000	*2012*	*2000*	*2012*
Caucasus & Central Asia	0.8	0.3	8.1	5.3
Eastern Asia	0.2	0.2	2.4	1.0
Latin America & the Caribbean	0.6	0.2	27.4	27.4
Oceania	0.0	0.0	5.5	5.0
South-eastern Asia	1.1	0.4	8.0	6.1
Southern Asia	1.3	1.3	5.3	5.3
Sub-Saharan Africa	11.7	1.5	15.2	14.5
Western Asia	0.9	34.2	3.7	3.6
World	2.0	*1.7*	7.1	7.1

1994: 24–25). While arguably too broad to be of much analytical use, the emergence of this concept reflects the degree to which vulnerability to violence has become perceived as part of a bundle of interrelated vulnerabilities, all of which constitute critical development challenges.

The focus on human security has also emphasised the degree to which security issues pay increasingly little heed to international borderlines, as international terrorism proliferates and civil conflicts spill across multiple borders. In an increasingly globalized world, it is argued, the effects of poverty and insecurity are no longer strictly confined by national boundaries, and it is impossible to be fully isolated from the effects of insecurity and conflict in other parts of the globe (Liotta 2007: 12–13). Valid though this concern is, the increased focus in rich countries on the effects of conflicts in poor ones has raised concern among development scholars and professionals who fear that conventional development goals, such as reducing poverty and inequality, will be eclipsed by apparently more pressing national security concerns (Beall et al. 2006). Yet there is a real and growing divide between a group of countries that is experiencing unprecedented levels of peace today and a set of developing countries that are descending further into violence (GPI 2015).

Conceptualising urban violence and insecurity

As the global urban transition proceeds, understanding the dynamics of violence, security and development in an urban context has become a critical concern in both policy and research circles. Between 2009 and 2014 there have been at least four major journal Special Issues on the topic (see Further Reading at the end of this chapter) and two significant policy reports (World Bank 2010b; Muggah 2012). The urban dimensions of conflict and insecurity have also become a growing concern within military circles, as armies increasingly find themselves fighting long-term conflicts in urban areas (Graham 2004; 2010) and a preoccupation with 'fragile states' gives way to interest in 'fragile cities' (Muggah 2014). From a human development standpoint, urban violence is seen as having multifarious and devastating effects on the urban poor. Beyond the obvious threats of injury and death, violence can also have profoundly negative impacts on social cohesion, trust, livelihoods and economic productivity within communities, as well as gender relations and the social contract between governments and citizens (Meth 2003; Moser and McIlwaine 2006; Fox and Beall 2012; Muggah 2012).

Urban violence is a broad category that encompasses a wide range of phenomena, from isolated crimes to outright warfare. Box 7.1 provides an example of one of the many efforts to develop a basic descriptive categorisation. While typologies vary, most focus on categorisation in relation to the primary *actors* involved in perpetrating violence (e.g. violence perpetrated by armed combatants versus that perpetrated by terrorists, gangs or domestic partners), the degree of *organisation* of the violence (e.g. violence linked to organised crime versus 'anomic' crime), and/or the *motivation* for the violence (see below). Unsurprisingly, given the diversity of forms of violence affecting cities, it is difficult to establish underlying causes that apply generally to all. However, cross-national evidence suggests that poverty, unstable political institutions and socioeconomic inequality are all correlated with violence (Fox and Hoelscher 2012; Muggah 2012) and therefore considered particular risk factors. Despite the common perception that urbanisation and rapid urban growth are drivers of violence, there is little systematic empirical evidence that this is the case (Fox and Hoelscher 2012; Urdal and Hoelscher 2012). The increased salience of urban violence does not, therefore, map neatly onto countries with faster growing cities, younger populations or more rapid urbanisation rates, though these may combine with some of the above risk factors to feed violence in specific contexts.

In seeking to better understand the causes of urban violence, many observers have focused on the motivations behind violent acts, generally making a distinction between economic, social and political impulses. Economic violence is usually considered to be motivated by material gain and associated with criminal activities such as drug trafficking, car hijacking and kidnapping, through to mugging and armed robbery, small arms dealing and petty theft. Social violence is more difficult to isolate. At one extreme it includes domestic violence in the form of physical or psychological abuse, and at the other, interpersonal violence in public places such as drunken brawls, arguments getting out of hand and road rage. Political violence also takes many forms, from riots to rebellions to terrorist attacks. A further category of institutional violence is sometimes used to refer to violence perpetrated by state institutions as well as institutionalised non-state organisations such as vigilante groups (Winton 2004; Moser and McIlwaine 2006).

Box 7.1 Typology of urban violence

Anomic crime	*Organised crime*
Main actors: Individual criminals, state security/police forces. **Organisational features:** Ad hoc acts of violent crime and delinquency, usually economically motivated. **Impacts/outcomes:** Sporadic murder, assault, gender-based violence, robbery/theft.	**Main actors:** Drug cartels, human trafficking networks, arms smugglers, state security forces (intelligence) and police officers. **Organisational features:** Command structure, often transnational, limited territorial control, mainly economically motivated. **Impacts/outcomes:** Targeted killings, kidnapping, extortion, systematic sexual abuse, human trafficking and enslavement, small arms proliferation.
Endemic community violence	*Open armed conflict*
Main actors: Urban gangs, vigilante groups/community defence organisations, ethnic militias, state security forces and police officers. **Organisational features:** Widespread/routine violent crime in the context of failed public security, limited command structure and territorial control, primarily economically motivated. **Impacts/outcomes:** High rates of gang/police/civilian casualties, unlawful killings, recruitment of 'urban child soldiers', social cleansing, gender-based violence, inter-gang warfare and police shoot-outs, kidnapping, trafficking, robbery/theft.	**Main actors:** Rebel groups, paramilitaries, state military forces. **Organisational features:** Struggle for territory in interstate or civil war context occurring in cities, usually large scale, political/ideological/identity motivated. **Impacts/outcomes:** Significant civilian casualties, mass population displacement, war crimes and crimes against humanity, genocide, terrorism, humanitarian crises, gender-based violence, recruitment of child soldiers.

Source: humansecurity-cities.org (2007: 16).

These various forms of violence can often overlap and their boundaries are very fuzzy, rendering a simple typology of the motivations for violence problematic. For example, the contested 2008 election result in Kenya resulted in clashes between ethnic groups supporting opposing candidates. When violent political clashes fall along ethnic lines, is this political violence or social violence? When people took to the streets to riot about rising food prices in Uganda in 2011, which was arguably due both to global commodity prices and mismanagement of the economy by the government (Goodfellow 2013b), was this economic or political violence? Even when urban riots and insurrections have local causes rooted in economic concerns they can have wider political implications (Hobsbawm 2005). Moreover, violent community mobilisation and vigilantism in contexts such as South Africa often involve significant collusion or even covert organisation by the police, confounding the divide between institutional and social violence (Winton 2004; Cooper-Knock 2014).

The predominant manifestation of violence in a given society may also change over time while the underlying causes remain fairly constant. Rodgers has argued that rising urban violence in Central America is not a qualitatively new form of conflict but a continuation and relocation of earlier – and ostensibly more political – forms of violence motivated by discontent surrounding gross inequalities in wealth. Revolutionary struggles have given way to endemic urban violence, which he argues represents the continuation of past political struggles in a new spatial context (Rodgers 2009). Indeed, one of the arguments recurring in recent scholarship on urban violence and development is that much or even most urban violence can be considered political in nature, resulting from perceptions of injustice and contentious power relations between state and society or among different societal actors (Moncada 2013: 219). There is also evidence that some of the structural causes of social violence (e.g. homicides) may be similar to those underpinning political violence (e.g. armed conflict) (Fox and Hoelscher 2012).

The limitations of traditional typologies have led to various attempts to conceptualise urban violence – and its relationship to other forms of conflict – in new ways. For example, Beall et al. (2013) explore the widely observed global decline in civil wars relative to other forms of violence and instability through a distinction between 'civil' and 'civic' conflict. The former term refers to violent conflicts between two or more armed groups (organised primarily within sovereign boundaries) that have publicly stated political objectives, at least one of which seeks to take control of all or part of the state. 'Civic' conflict, on the other hand, refers to the reactive expression of grievances vis-à-vis the state or other actors, which may reconfigure power relations but does not generally involve an effort by some group to take control of formal structures of power. When it becomes violent, civic conflict comprises a broad range of expressive, grievance-driven actions from gang warfare, terrorism and religious or sectarian rebellions to violent protests and riots over state failures to provide basic services. One thing that all these forms of violence have in common is a tendency to play out in urban areas, unlike many of the civil conflicts of the twentieth century (Beall 2007). This focus on civic conflict accords with other research that identifies the roots of many outbreaks of urban violence in acts seeking 'popular justice' (Moncada 2013), which may be a response to political marginalisation by the state (Goldstein 2004) or triggered by alarm at the erosion of traditional mechanisms for the provision of social order (Buur 2008).

Another trend in current research has explored how different forms of violence bleed into, feed and substitute for one another. This has been explored in relation to how increases in urban civic conflicts often occur after national-level civil wars end (Rodgers 2009; Beall et al. 2013; Branch 2013; Moxham and Carapic 2013), but also has relevance for the relationships between different forms of violence that occur within city boundaries either simultaneously or successively. Moser and Rodgers introduce a concept of 'violence chains' to explore how urban violence frequently operates systemically, involving a range of interconnected processes (Moser and Rodgers 2012). In Patna, India for example, although many city-dwellers' perceptions of security have increased in recent years due to targeted policing of organised crime, overall crime rates have actually continued to rise. This paradox is explained by the fact that violent conflict has largely been geographically displaced to the city's slums, where violence has intensified (ibid.: 7–8).

Interventions targeted at one form of violence can thus have a knock-on effect on others, and focusing on just one form of violence (e.g. organised crime) in isolation from others (e.g. violence in the home) presents an incomplete picture (Moser and Rodgers 2012; Moser and McIlwaine 2014). Indeed, a crucial and often overlooked element of urban violence emerging in recent research is the way in which gender-based violence (GBV) and violence against women and children (VAWC) relates to other forms of violence beyond the home (Moser and McIlwaine 2014). On the one hand, violence in the home can be seen to contribute to the normalisation of violence in wider society, while on the other, the exposure of men to high levels of violence in the public sphere can cause fear, humiliation and crises of masculinity that manifest in high levels of violence against partners and children in the home (Meth 2009; Moser and Rodgers 2012; Winton 2014). Some of the impacts of gender-based violence, and responses to it, are explored in the context of urban Pakistan in Box 7.2.

Box 7.2 Gender-based violence in urban Pakistan

Levels of domestic violence are very high in many parts of Pakistan. In one study of Karachi, a shocking 76 per cent of women reported physical abuse, along with 12 per cent reporting sexual abuse and 100 per cent verbal abuse. In the vast majority of cases the abuser was their husband, and other studies in Pakistan have found that similarly high proportions of men admit to perpetrating these forms of abuse against their wives. The Karachi study found the mean duration of abuse was 11 years. This abuse has been shown to have grave psychosocial consequences on women, including depression, attempted suicide and substance abuse as well as negative impacts on physical and reproductive health. To make matters worse women are also vulnerable to crime-related violence in cities such as Karachi.

It is common for domestic violence to follow a sequential pattern, beginning with verbal abuse that then escalates into anger and explodes into various kinds of violence. Ninety per cent of women in the Karachi study initially respond to this abuse with silence or efforts to minimise provoking violence, though 58 per cent eventually resisted and either answered or fought back, with around a quarter leaving or attempting to leave their husband. Increasingly, women's groups are organising to respond to the problem of gender-based violence, drawing both on arguments about the interpretation of Islam and international human rights discourses. In the city of Lahore, an NGO called Dastak ('Knock on the Door') was established by four women with legal expertise in 1990 to provide crisis interventions and safe havens for women and children, as well as legal aid, counselling and resettlement services. The organisation has grown progressively, employing outreach staff and psychologists, and had served over 5,000 women by 2008. Dastak's efforts extend far beyond service delivery, however, encompassing lobbying and advocacy to gain public support for women's basic right to safety and security and the use of litigation to ensure that existing laws establishing rights for women are properly interpreted and enforced. The organisation is fighting to change the prevailing notion of 'protection' for women, epitomised by state-run shelters that isolate them and constrict their movement rather than strengthening and liberating them by challenging societal norms.

Sources: Rabbani et al. (2008), Critelli (2010).

These analyses of violence in different but interlinked urban arenas reflect a more general interest in questions of space and scale in the study of violent conflict and insecurity. Winton, for example, explores the role of prisons as key places that create alternative organisational spaces for gangs, which then feed new forms of gang organisation and activity in wider urban society (Winton 2014). Recent research has also highlighted the role of access to urban space and land in fomenting violence widely assumed to be about ethnic identity, religion, or other socioeconomic divisions. In Karachi, for example, different informal systems for regulating urban space in adjacent unplanned neighbourhoods have fomented violence by exacerbating frictions between different groups of migrants while simultaneously undermining the capacity of the state to govern space (Gazdar and Mallah 2013). In more concrete terms, the spatial divisions engendered by fear and insecurity have become a pervasive feature of many cities across the developing world, as explored in the following section. Drawing on Galtung's 'structural violence', some observers have suggested that the carving up of urban space in ways that connects certain groups while excluding others can be considered a form of 'infrastructural violence' (Rodgers and O'Neill, 2012).

In addition to this focus on space, 'multi-level' explanations of urban violence, which consider dynamics between city-, regional- and national-level institutions, are increasingly deployed. For example, differential patterns of urban violence in Nairobi and Lagos have been explained with reference to contrasting relationships between central governments and urban militias (LeBas 2013). Studies of India (Weinstein 2013) and Brazil (Hoelscher 2015) also find different forms of urban violence to be linked to the nature of relationships between municipal governments and state governments. Another scalar aspect of urban violence relates to migration across borders and links among diaspora groups abroad. McIlwaine (2014) argues that Colombian migration around the world is partly driven by urban violence, while remittances sent back home can feed the kind of spatial division that perpetuates fear and social exclusion, further feeding cycles of violence. The multiple scales on which much contemporary violence operates has led some observers to describe it as an 'unbound phenomenon' that challenges our understandings of violence as strictly social or political, urban or rural, national or local (Moncada 2013: 222). This has important implications for the question of which authorities and institutions should be expected to deal with the problem of violence. Indeed, it implies that to achieve and maintain human security in cities requires governance strategies that involve cooperation from the very local to global levels, engaging communities, city authorities, national governments and international agencies.

Crime, endemic violence and urban development

Although generally overlooked in development research and scholarship until relatively recently, crime is an important development challenge in urban contexts that is linked directly to inequality, perceptions of injustice and political dynamics (Meth 2014). The very concept of crime is societally specific. Much like the idea of informality, it involves politico-legal decisions about what kind of behaviour is desirable, preventable and punishable in a given society, which often reflects prevailing power relations and political strategies. Indeed, many of the urban poor are criminalised automatically through residence in informal settlements officially designated as illegal. Because acts defined as criminal in a particular society span the range from petty crimes or misdemeanours punishable through fines to severe crimes punishable with long prison sentences or death, the relationships between crime and urban development are diverse and complex. This section therefore focuses specifically on violent crime, and the related concept of endemic violence: interpersonal violence that is common and frequent in a given society, in contrast to the acute, episodic and often group-based violence of wars, terrorist attacks or ethnic pogroms. Violent crime and endemic violence are not synonymous, especially as not all forms of the latter are criminal (e.g. domestic abuse is legal in many places). There is however a significant overlap and both undermine basic human rights to safety and security.

In a sense, violence is endemic to all societies, but this should not obscure the fact that it is much more endemic to some than others. Figure 7.1 illustrates this using country-level estimates of homicide rates, which are a convenient proxy for violent crime or endemic violence and generally considered to be less subject to classificatory confusion and manipulation than other violent crime data (Fox and Hoelscher 2012). By this measure, the most violent countries in the world today are found in parts of Central America, South America and Sub-Sarahan Africa.

Figure 7.1 Homicides per 100,000 people in 2012

Homicides per
100,000 people

<3
3–6
6.1–10
10.1–20
20.1–50
>50

Source: World Health Organization, Global Health Observatory Data Repository, accessed online July 2015.

However, these national-level estimates obscure considerable variation in the burden of such violence within countries and regions. Isolating the 'urban' part of endemic violence is a challenge, but generally speaking rates of violent crime are higher in large cities than small ones or rural areas. In El Salvador, for example, 63 per cent of all homicides took place in cities of at least 50,000 people, which are home to only 51 per cent of the population (UNODC 2013: 28). However, this generalisation sometimes leads to the problematic conclusion that large urban centres *necessarily* generate more violent crime. As Table 7.2 demonstrates, this general conclusion doesn't hold when we compare individual cities. For example, Casablanca is twice the size of Cape Town, which had a homicide rate over 40 times higher in 2009. Similarly, Panama City is home to less than half a million people yet had a homicide rate over three times that of the megacity of São Paulo. Indeed, some argue that there is no obvious

Table 7.2   *Homicide rates and city size in 2009*

Region	Country	City	Homicides per 100,000 people	Population c. 2010
Africa	South Africa	Cape Town	59.9	3,430,992
	Uganda	Kampala	15.3	1,516,210
	Zambia	Lusaka	9.5	2,191,225
	Morocco	Casablanca	1.4	6,861,739
	Ghana	Accra	1.3	2,070,463
Americas	Venezuela	Caracas	122.0	1,942,652
	Panama	Panama City	34.6	430,299
	Brazil	São Paulo	10.8	11,152,344
	Mexico	Mexico City	8.4	8,555,272
	USA	New York	5.6	8,174,959
Asia	Mongolia	Ulan Bator	10.2	1,144,954
	Philippines	Quezon City	5.3	2,761,720
	Thailand	Bangkok	4.0	5,701,394
	India	Mumbai	1.3	18,394,912
	Sri Lanka	Colombo	1.2	561,314
Europe	Estonia	Tallinn	7.3	403,862
	Netherlands	Amsterdam	4.4	1,027,860
	Belgium	Brussels	3.0	157,673
	Germany	Berlin	1.8	3,292,365
	United Kingdom	London	1.6	8,250,205
	Portugal	Lisbon	0.2	552,700

Source: UNODC Homicide Statistics 2013, accessed online July 2015.

and automatic link between urbanism and violence, even in cities naturally prone to conflicts arising from the sheer size and diversity of their populations (Rodgers 2010; Muggah 2012).

Globally, endemic violence is the most pervasive form of violence experienced by urban residents, and 46 out of the 50 most violent cities in the world in 2013 were not in countries experiencing armed conflicts (Muggah 2014: 4). Moreover, when assault, theft, human trafficking and non-lethal interpersonal violence are factored in, millions of people are directly affected by some form of crime or violent incident every year. While there has been much commentary in recent years on the secular decline in violent crime in the rich world (see for example Zimring 2006), the same is not true for many regions of the developing world. In the period 2004–2012, although there have been decreases in Asia and Southern Africa (with partial reversal in the latter), homicide rates have been on the increase in North Africa and Eastern Africa as well as Latin America (especially central America) (UNODC 2013).

In the Latin American context, the fear and insecurity that constitute everyday life for many people undermines community and inter-group relations, leading to the atomisation of urban social life. The following quote from *Don* Sergio, a resident of a low-income neighbourhood in Managua, Nicaragua illustrates the point:

> Nobody does anything for anybody anymore, nobody cares if their neighbour is robbed. Nobody does anything for the common good. There's a lack of trust, you don't know whether somebody will return you your favours, or whether he won't steal your belongings when your back is turned. It's the law of the jungle here.
>
> (Rodgers 2006: 271)

Crime is also highly gendered in its effects. Globally, men are four times more likely to die from interpersonal violence and nearly six times more likely to die from collective violence than women according to WHO data. But women suffer extensively from other crimes such as sexual violence. A shocking 30 per cent of women globally have experienced physical or sexual violence at the hands of an intimate partner in their lifetime. Particular ethnic or racial groups are also especially vulnerable, as Caldeira (2000) illustrates with respect to young black men in her study of São Paulo.

While high levels of violent crime are very serious, the fear and anxiety generated by the perception of crime and endemic violence often outstrips the actual level of danger. Fear and anxiety have social and developmental consequences as well (Bannister and Fyfe 2001: 808). In Mexico City, widespread acts of violence not only impact on the city directly, but also haunt the minds of its residents: insecurity has become 'a "phantom" that wanders the metropolis' (Pansters and Berthier 2007: 36), causing people to restrict their movement, avoid leaving home at night and retreat into private spaces. The fear of crime has increasingly led to private, individual responses in the form of fortress-like gated communities. When viewed historically, this private retreat into walled enclaves over the past quarter of a century is one of the most significant spatial trends of urban development.

Gated communities predominate where inequalities are pronounced and public security is inadequate, and the phenomenon is now very widespread in countries such as Brazil and South Africa, where socioeconomic inequalities are particularly acute (Caldeira 2000; Beall et al. 2002; Lemanski 2004; Lemanski et al. 2008). In some cases, residents have fenced entire neighbourhoods, with gated access being monitored by armed private security guards, making it necessary for metropolitan authorities to legislate against the privatisation of public space. Even some public police stations are protected by private security companies (humansecurity-cities.org 2007: 27).

While walled enclaves and private security arrangements are generally confined to wealthy elites, there are many examples of low-income settlements becoming private, fortified spaces as well. For example, in Johannesburg, some former hostels built to house migrant labourers now operate as low-income private gated communities. Initially formed by ethnic minorities for protection, these 'fortified' hostels often come to cloak shady and illicit activities, leading to them being effectively closed to the police and emergency services at night, which is not in the best interests of many of their residents (Beall et al. 2002). The private claims to public space that underpin the phenomenon of gated communities are given an interesting analytical twist by Lemanski and Oldfield (2009), who compare gated communities in South Africa with so-called 'land invasions' by the urban poor. They argue that both the poor who 'invade' and squat on urban land and the elites and middle classes living in gated communities are pursuing the same goal: a secure home from which they can exercise their right to live and work in the city. Despite the underlying similarity of these claims and the right to security on which they are predicated, the state responds very differently, considering the latter as rational residents while disparaging the former as unreasonable criminals.

When people fortify themselves in closed compounds and exclude themselves from life in the city, a geography of fear emerges based on insider-outsider exclusions. People provide and maintain their own infrastructure and services and become disconnected from their surroundings, eschewing public space and opting out of local governance. Walled homes and fortress settlements are a long way from Jane Jacobs' (1961) notion that busy public places offer natural surveillance and better safety due to more 'eyes on the street', and bring to mind Peter Marcuse's question: 'Do walls in the city provide security – or do they create fear?' (Marcuse 1997: 101). Moreover, fear coupled with a desire to establish a safe refuge from violence also often results in voluntary segregation along ethnic, religious and racial lines, resulting in further mistrust and antagonisms between groups struggling to get by in a complex environment – sometimes with disastrous consequences. In Indian cities in particular, spatial segregation and communal violence between Hindus and Muslims living in such enclaves are widespread, being particularly evident in Mumbai and the Gujarati city of Ahmedabad (Varshney 2002). In 2002, communal violence between Hindus and Muslims in the city of Ahmedabad resulted in the deaths of over 1,000 people (Chandhoke et al. 2007).

When cities gain a reputation for crime and violence their ability to generate wealth is undermined. As we discussed in Chapter 3, cities are dynamic economic spaces, but only if the benefits of agglomeration outweigh the costs. Crime and endemic violence represent significant costs and serve to encourage potential investors and skilled and educated workers – who can take their assets elsewhere – to move away, robbing such cities of the talent needed to compete in the global economy (UN-Habitat 2007). Almost 30 per cent of firms in Latin America, Africa and Asia cite crime as the key concern about the prospects for their business (World Bank 2011). For those who don't have the option to leave, the costs of doing business in an insecure environment limit their scope for growth. The need to spend more revenue on private security and insurance diverts resources away from investing in the future. It is difficult to quantify the effects of crime and violence on economic development, but the available evidence indicates that the effects are significant. In Brazil, for example, it was estimated that some 10 per cent of annual GDP is spent on private security and insurance (Brennan-Galvin 2002), and the collapse of the country's public health system in the 1980s and 1990s has been attributed to the high costs associated with managing the sheer volume of homicides and crime-related injuries (UN-Habitat 2007: 46).

While lawless cities may discourage legitimate investors and consume resources that could otherwise be spent in pursuit of development, they frequently attract illicit operators eager to take advantage of poorly policed and regulated urban spaces. Indeed, the global criminal economy 'is solidly rooted in the urban fabric, providing jobs, income, and social organisation to a criminal culture, which deeply affects the lives of low-income communities and of the city at large' (Castells 2000: 395). Transnational criminal networks gravitate towards cities by virtue of their infrastructure and the cover they provide through the density of the built environment. They often penetrate the urban social fabric and affect the smooth operation of urban governance. These organisations thrive in the cities of low- and middle-income countries, with research suggesting that cities in Africa, Central Asia and Latin America provide particularly fertile ground for their activities (UN-Habitat 2007: 45).

From the local to global level, efforts to tackle crime and endemic violence in cities are complicated by a feedback cycle that renders crime-ridden and violent cities prone to responses that engender further crime and violence (see Figure 7.2). Failed public security, weak urban governance, unequal access to economic opportunity for certain parts of urban society, and the naturalisation of a state of fear and insecurity within a city can become self-reinforcing phenomena (Agostini et al. 2007). Breaking this cycle requires an understanding of the root causes of crime and violence, which include socioeconomic, political and demographic factors, as discussed above.

The situation is often most acute in informal settlements that are virtually abandoned by public security services (Pinheiro 1996; Méndez et al. 1999). In the same way that informal water vendors step in to provide water in un-serviced slums, in the absence of publicly provided security, other social forms – such as vigilante groups and gangs – emerge to fill the void (Koonings and Kruijt 2007). In such contexts, gangs can serve a social function, acting as 'a form of local-level social structuration in the face of socio-political breakdown' (Rodgers 2006: 269). However, these local level institutions can easily be perverted. When such informal institutions fill a law and order void vacated by the state, they are usually left to operate with impunity, often with violent outcomes.

Figure 7.2 *The vicious cycle of public security failure*

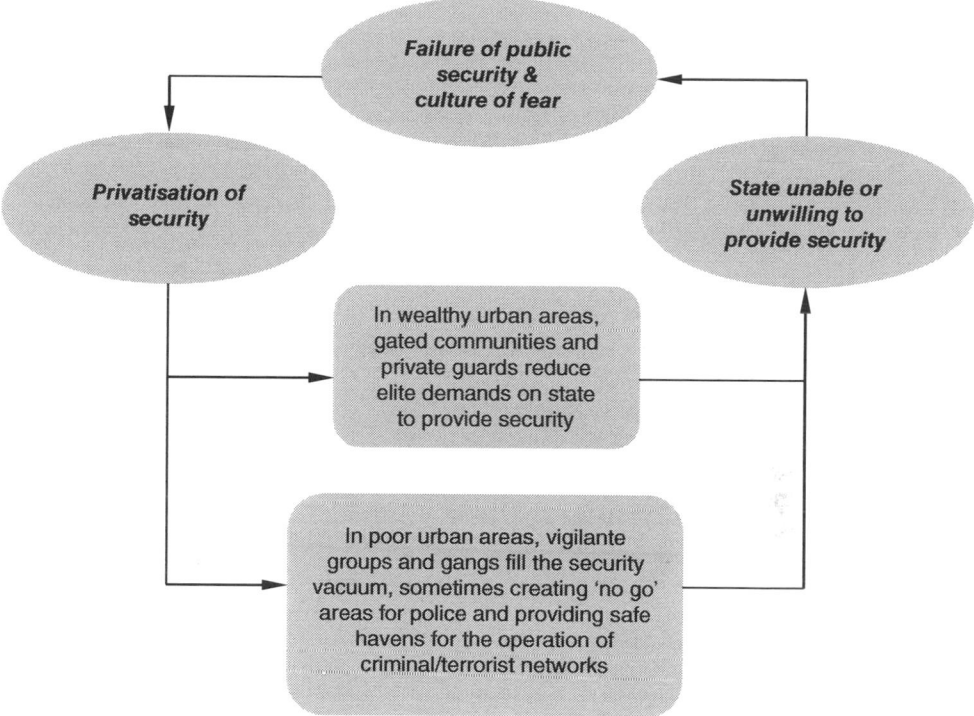

Source: Adapted from humansecurity-cities.org (2007: 26).

It is important to remember that in addition to concentrating risk factors for crime and homicide, cities also play host to protective factors with the potential to limit violence, including higher levels of education, better health facilities and the potential for expanding safety-enhancing infrastructure such as street lighting (UNODC 2013: 27). Some efforts to reduce urban endemic violence have been surprisingly effective in extremely challenging contexts (see Box 7.3). Above all, a sound city-level public security force can help to mitigate crime and violence by managing conflicts in a dispassionate manner. This, however, is a huge challenge. Responsibility for public security does not lie exclusively at the city level, and many metropolitan governments and municipalities do not have their own police force, relying on protective and correctional services being supplied by national governments. These are not always forthcoming. Strengthening the multi-scalar links to improve public security is an important – and frequently overlooked – development priority.

Box 7.3 Unlikely successes in urban violence reduction

Many efforts to reduce violence rely on micro-level initiatives such as alcohol restriction, handgun controls or increased police presence in particular areas. However, these often prove ineffective in the medium to long term, due to dwindling funding or interference by criminal or elite interests – especially when the underpinning institutional context is not conducive to sustained success. For decreases in violence to be sustained over time, policy changes need to be integrated into wider reforms to social and political institutions.

Bogota in Colombia and Recife in Brazil are two cities associated with extremely high levels of endemic violence that have managed to engineer dramatic violence reductions in recent years. Bogota was one of the most dangerous cities in Latin America in the 1980s and early 1990s, with a rate of 80 homicides per 100,000 inhabitants in 1993. By 2006 this had been reduced to below 20 and this has been sustained since, dropping even further to a 30-year low of 13 per 100,000 in 2013. Recife, meanwhile, had one of the highest rates in the world, with 90 homicides per 100,000 in the 1999, though this had declined to under 30 by 2013 in stark contrast to many other cities in North-Eastern Brazil.

In both cases, it was not isolated policy initiatives of the kind described above that generated these dramatic improvements. Important in both was the role of 'critical junctures' related to the emergence of novel political actors on the urban scene in the context of political reforms, which created openings for innovations in public security that were effectively acted upon. In Bogota, the introduction of publicly elected mayors and the election of Antanas Mockus, who took a particular interest in the problem of urban violence, was followed by a focus on creating a new civic culture and reforms to how police were trained and resourced. In Recife, a new state governor was elected on the back of middle class concerns about the effects of violence on their interests, marking a critical break with the pre-existing indifference to violence. New agencies were soon established to collect and manage data on crime, as well as improve police accountability and oversight, with dramatic effects. Also significant were concurrent efforts to improve state–citizen relations generally.

Improving urban security thus does not depend on a single isolated policy or action. While in both cases technocratic and managerial shifts were central to initiating violence reduction, this was built upon underlying political shifts and also required major changes to local civic culture and political institutions to be sustained. Crucially, public security innovations received broad-based buy-in from key stakeholders and civil society in both cities, soon becoming part of the institutional configuration in the city, which heightened potential for sustained violence reduction beyond the presence of the initial promoters.

Source: Hoelscher and Nussio (forthcoming).

War, terrorism and urban reconstruction

In parallel with heightened awareness of urban violence as a development issue has been rapidly mounting interest in the role of cities in contemporary warfare (Graham 2004; 2010; Abrahamsen et al. 2009; Coward 2008; 2009; Beall et al. 2013; Muggah 2014). In the context of interstate war, cities very often form major targets for attack. Enhanced awareness of (and aversion to) casualties in the contemporary era strengthens the preference for aerial bombardment of cities as the primary route to military victory, particularly where Western democracies with advanced weapons capabilities are involved (Beall et al. 2013). Moreover, as resource- and population-intensive sites, cities are not only targets for attack but increasingly sites of resistance as well, especially in developing countries where armed forces may be weak or fragmented and unable to mount effective resistance. Thus in response to onslaughts on cities such as Basra, Baghdad, Kandahar and Kabul, belligerents in countries under attack have increasingly resorted to forms of 'asymmetric warfare' that tend to involve unpredictable acts of urban terror as opposed to more 'conventional' military approaches (Hills 2004).

This 'urbanisation of insurgency' has posed enormous challenges to conventional military practice. The experience that US troops faced in Mogadishu in 1993, and in many cities in the Middle East subsequently, has contributed to the 'new military urbanism' through which the American army in particular has had to reform military tactics and strategies with cities in mind (Rosenau 1997; Graham 2010). The experience of the Iraq War has shown that military hardware is not enough. Alongside high-tech air bombardment, US troops went in with wire cutters, ladders and mirrors to help them see around corners, and had to engage in very basic forms of street fighting at ranges of six and 60 rather than 60,000 feet. With or without technology, urban warfare is difficult because cities are rarely empty and the challenge of distinguishing between combatants and non-combatants is particularly acute (Hills 2004).

In civil wars cities also play a strategic role, though this is often as the perceived 'end-point' of conflict rather than a target of bombardment. Indeed, in a large number of civil conflicts in the developing world, cities have remained relative havens of stability for large parts of the war. For example, Giustozzi (2009) explains how in Afghanistan cities were 'eyes of the storm' for long periods in the country's many civil wars, particularly before the 2001 US-led invasion. Likewise, the city of Quetta on the conflict-prone Afghanistan–Pakistan border has historically witnessed relative peace between its communities even in the context of conflicts in the wider region (Gazdar et al. 2010). The same can also be said of Gulu in Northern Uganda during the brutal civil war there (Branch 2013), or Juba in South Sudan during the second Sudanese civil war from 1983 to 2005 (Martin and Mosel 2011). One reason why cities can remain relative 'safe havens' is that they are deliberately protected from the ravages of conflict due to their role as economic hubs in war economies. Elites themselves also often reside in capital cities, and therefore wield consolidated coercive power to prevent conflict from having too much impact on these places (Beall and Goodfellow 2014).

A consequence of cities acting as spaces of refuge in this way is that it provides further impetus for urban growth, which can often be extremely rapid in conflict situations – and this is precisely when city governments are least well-placed to cope with growth. In the course of Angola's civil war from 1975 to 2002 the capital Luanda grew five times in size (UNPD 2011), while conflict played a clear role in transforming Gulu from a small provincial town in the 1980s to Uganda's second largest city by 2007 (Branch 2013).

One of the key contemporary challenges that has stimulated interest in urban security is a heightened awareness of terrorism since the collapse of New York's Twin Towers on 11 September 2001. This event threw into stark relief the vulnerability of cities everywhere to terrorist attacks, as well as the political impact potential of such 'spectacular violence' (Goldstein 2004). In the years since, this has been confirmed through countless large-scale terrorist attacks in cities from Madrid and London to Mosul, Mumbai and Nairobi. These events can be seen to signify the 'implosion of global and national conflicts into the urban world' (Appadurai 1996: 152–153). Terrorism is not a new phenomenon, but its role on the global stage has clearly increased in the period since 9/11. While the number of terrorist attacks per year was steadily declining in the immediate period after the Cold War, the number of casualties per terrorist incident has steadily increased over the same period. Since the onset of conflict in Syria in 2011, and the rise of Islamic State across parts of the Middle East, the casualty rates have become even worse, with 20,000 people killed in terrorist attacks in 2014 as compared with 2,000 ten years previously (GPI 2015). These deaths are, however, geographically concentrated; 85 per cent occurred in just five countries: Afghanistan, Iraq, Nigeria, Pakistan and Syria (ibid.).

Urban terrorism can be especially shocking when the choice of particular cities seems to bear little direct relation to terrorists' stated political aims, effectively rendering the city itself as 'collateral damage' as well as the civilians affected. Chosen on the basis of their visibility or symbolic significance rather than being strategic sites in their own right, the bombing of busy urban tourist centres such as Sousse in Tunisia or Sharm el Sheikh in Egypt is often aimed not at the towns *per se* but rather to communicate messages across the globe. In a sense, urban terror constitutes an international language; a means of projecting political messages across the global media landscape. However, the impacts of terrorist attacks on cities in low- and middle-income countries can be particularly devastating where limited resources and capacities inhibit recovery efforts. For example, the *al-qa'ida* bombing of a hotel in Mombasa in November 2002 was to the residents of that city the last straw in the region's downward spiral into poverty and banditry (Richards 2002). Moreover, in the context of weak governments and failing economies, such as in Afghanistan, it is even more difficult to undertake reconstruction, both physical and social (Beall and Esser 2005).

It is important to remember that terrorism – defined as violent acts threatened or employed against civilian targets and for political objectives (Barker 2003) – can be perpetrated by states as well as non-state groups. Sometimes this takes the form of 'urbicide', which refers to deliberate efforts to destroy buildings, infrastructure and the social fabric of an urban environment for political purposes (Graham 2004; Coward 2008). In many ways, the deep vulnerability of many urban dwellers is best epitomised by instances of violent population displacement and slum evictions. A particularly extreme example was Operation Murambatsvina in Zimbabwe, which translates literally from the Shona dialect as 'drive out the rubbish' or, more euphemistically, 'restore order' (Potts 2006). Part of the intention was to eradicate illegal (in other words, informal) activities in the capital city of Harare. Within six weeks an estimated 700,000 urban residents lost their homes and livelihoods, with up to 2.4 million people affected in some way. Overnight, self-help housing and informal structures were declared illegitimate as bulldozers and demolition squads run by youth militia led the assault, which resulted in injury and even death, giving rise to the before and after pictures seen in Plates 7.1a and 7.1b. Operation Murambatsvina has been interpreted as a pre-emptive strike against an urban-based political opposition by way of a highly militarised national-level operation within the context

Plate 7.1a *An informal settlement in Zimbabwe before Operation Murambatsvina*

Source: DigitalGlobe.

Plate 7.1b *An informal settlement in Zimbabwe after Operation Murambatsvina*

Source: DigitalGlobe.

of Zimbabwe's severe economic and political crisis (Potts 2006). While extreme, this kind of state action is by no means unique. In Jakarta, for example, state-perpetrated violence against informal settlement-dwellers has been common, rapid and devastating: in one instance from 2003, a city mayor oversaw the eviction of around 6,000 residents from one neighbourhood in just one day (Human Rights Watch 2006).

In the aftermath of war or other forms of extreme collective violence, the financial and technical challenges of repairing the physical fabric of a city can be overwhelming. Basic infrastructure, such as roads, water, sewerage and power supplies are often left in a dysfunctional state, while buildings might be destroyed completely or in part. The violent assault on Baghdad's infrastructure by US forces during the first Gulf War in 1990–1991 nearly reduced what was previously a fairly advanced economy to a 'pre-industrial age' (Banarji 1997: 199), with the second Iraq war further destroying the city's economic foundations until they rivalled those of 'the most degraded nations in the Global South' (Schwartz 2007: 21).

In such contexts, the reconstruction process also offers unprecedented opportunities for profit. Transnational corporations often step in to execute reconstruction contracts worth millions – sometimes billions – of dollars, with donors footing the bill, raising concerns that 'reconstruction' diverts large volumes of aid resources otherwise destined for poverty alleviation and development (Beall et al. 2006). In some situations – such as reconstruction efforts in the wake of conflicts in Afghanistan and East Timor – there is evidence that a focus on big infrastructure projects in the capital city, without due attention to poverty reduction and job creation in either the countryside or city itself, had detrimental effects (Esser 2013; Moxham and Carapic 2013). Reconstruction efforts also sometimes overlook the deeper and less visible legacies of conflict. Hostilities disrupt the rhythms and practices of daily life by destroying the assets of ordinary citizens, preventing children from attending school, encouraging the emigration of those with the means to do so (often the most wealthy and educated) and internal displacement of those seeking refuge. Many cities also come out of conflict severely spatially divided; Beirut, Jerusalem, Mostar and Nicosia are particularly stark examples (Calame and Charlesworth 2011). Cities emerging from conflict are therefore confronted with the challenges of rebuilding a shattered society in physically and socially circumscribed spaces. Hence reconstruction is not merely a question of bricks and mortar; it requires identifying and confronting the social and psychological inheritances of a conflict.

The very features that make cities prone to violence of all kinds – i.e. their size, density, diversity and concentrations of wealth and power – also make cities promising spaces for reconciliation and reconstruction. Where societal divisions are inscribed in the fabric of a city, the dividing lines themselves are often the first sites of reconstruction through deconstruction: the Berlin Wall was brought down in a dramatic rejection of Cold War polarities; South Africans wilfully defied (legislated) residential segregation in cities; and residents of Nicosia regularly crossed the 'green line' separating the Greek and Turkish sides of the city. These dividing lines, which often seem to be 'intractable urban fixtures,' can be transformed into symbolic spaces of reconciliation and unity, performing 'important roles in stitching together torn cities' (Bollens 2001: 187). More generally,

> The city is important in peace building because it is in the streets and neighborhoods of urban agglomerations that there is the negotiation over, and clarification of, abstract concepts such as democracy, fairness and tolerance... Peace building in cities seeks not the well-publicized handshakes of national political elites, but rather the more mundane, but ultimately more meaningful handshakes and smiles of ethnically diverse urban neighbors.
>
> (Bollens 2006: 67)

Urban planners, in particular, can play an important role in the aftermath of crisis by translating seemingly abstract and insoluble conflicts into (literally) concrete, tractable public projects that 'constitute leading implementation edges of new democratic goals and mechanisms' (Bollens 2006: 75). But the process of planning and implementing reconstruction strategies in cities requires more than optimistic planners and a vision of peace. Negotiating initiatives and priorities must involve all stakeholders, from the community level, city level and national level. In Afghanistan, for instance, the reconstruction of Kabul after the 2001 US-led invasion was complicated by poor coordination between various stakeholders, and numerous property disputes arising from war and regime change, which resulted in protracted land tenure disputes and in some cases evictions and land seizures (Beall and Esser 2005).

Whether or not urban reconstruction efforts succeed depends upon the willingness and capacity of the stakeholders involved to seize the window of opportunity that a peace settlement provides. The aftermath of conflict inevitably entails a period of flux, in which new visions for society can be advanced or thwarted, new strategies for inclusive governance can be institutionalised or destabilised. Failure to seize this opportunity can result in a reversion to hostilities and the creation of new conflicts. In other cases, the opportunity has been seized, such as in the Rwandan capital Kigali which was rapidly securitised and has remained remarkably free of violent conflict since the 1994 genocide. However, as illustrated in Box 7.4, this has come at a cost. While efforts to reduce violence in cities after conflict are clearly vital, the manner in which this is undertaken has important implications. If reducing violence among groups in society takes place only through top-down securitisation and the repression of political conflict – at the expense of institutionalising political institutions to work through conflicts – this poses risks for sustainable peace and development.

Box 7.4 Kigali: from epicentre of genocide to 'model city'

Immediately after the 1994 genocide, the Rwandan capital Kigali was virtually empty, the majority of its population having fled or been killed. Within weeks, hundreds of thousands of exiles poured into the city, mostly from the Tutsi ethnic group. These were joined in even larger numbers two years later by returning Hutu refugees, who had fled to the neighbouring Congo immediately after the genocide in fear of reprisals. Rwanda's urban population grew at 18 per cent annually around this time: a rate virtually unprecedented anywhere in the world since at least 1950. Conditions were extremely ripe for urban violence, with genocide survivors who had lost entire families sometimes having to share their houses with genocidaires. Youth unemployment and land disputes were rife. Initially, many people were hounded out of their houses, revenge killings proliferated and informal security squads pacified the slums with bullets.

Against this harrowing backdrop, a new urban vision was conceived in which Kigali was seen as a litmus test for the success of the 'new Rwanda'. The strategy to securitise the city proceeded on several fronts. Without a formal police force yet established, the Rwandan Patriotic Front government brought its military discipline to bear on the city through informal community-based policing. Meanwhile, the government built on Rwanda's long-established social hierarchies to inculcate a culture of upward accountability and 'checking on your neighbour', institutionalised through local government-led reconstruction activities and community works programmes. A sustained clampdown on informal activity in the city as well as tight restrictions on local political activity have also played a role in undercutting potential 'breeding grounds' for collective mobilisation that might fuel further conflict.

The securitisation of Kigali thus proceeded through pervasive surveillance and an all-encompassing emphasis on discipline and order, rather than the hi-tech, barbed-wire military urbanism evident in some other post-conflict cities. Today the city is remarkably safe by regional and even global standards, and Kigali was awarded a UN-HABITAT 'Scroll of Honour' Award in 2008 for 'many innovations in building a model, modern city' accompanied by a substantial reduction in crime. Particularly remarkable is the fact that this was accomplished with such a small formal police force – there were only 800 police in 2006 for a city of almost a million.

Despite these successes, there are problems in pursuing securitisation at the expense of developing effective political institutions. The latter process has been undermined by the banning of local political party activism, constraints on the media and laws limiting the scope of political debate. Kigali's securitisation has been no mean feat, but underpinning it is a certain fragility. This comes brutally to the surface in periodic grenade attacks in the city, which often result in fatalities but sometimes go unreported as the discourse of security is so important to the government. The constraints on urban politics have led some to question whether, despite the reduction in overt violence, conflict has simply been suppressed – and perhaps further violence deferred – rather than being channeled into institutions to make Kigali's success sustainable.

Source: Goodfellow and Smith (2013).

Conclusion

A degree of human security is a prerequisite for development. Without a minimal level of security, people are not free to invest in their own (or their children's) future, and without a minimal level of security of person and property, economies cannot generate wealth. In countries at war with themselves, cities are sometimes the only areas that continue functioning economically. However, when violence is itself urban-based, whatever the cause, the economic and social viability of cities is compromised. Reducing violence in cities is an extremely complex task that cannot be understood without attention to politics. Well-trained, well-equipped police forces are key ingredients to ensuring human security, but strong communities and open forums for dialogue between groups with conflicting interests are equally necessary to prevent these conflicts of interests from devolving into open, armed hostilities. Achieving these conditions requires political, policy and planning processes that are inclusive and responsive.

Unfortunately, many cities are becoming increasingly fortified, militarised spaces, with violence and securitisation reinforcing one another in a vicious cycle. Perceived security threats are being translated into urban planning and governance strategies designed to control, survey and defend urban space. This defensive mentality undermines the prospect for cities to be 'privileged spaces of democratic innovation' (Borja and Castells 1997: 246) by inscribing difference and divisions in the urban fabric. What is needed – now more than ever – is a focus on the ways in which urban political conflicts can be managed through progressive planning and governance strategies that accommodate contestation, cooperation and compromise. It is to these issues that we now turn.

Summary

- Development can be related to conflict and violence in a number of ways, depending on how one defines these terms. For example, some see development as a process aiming to reduce violence, while others see development as an inherently violent process in itself.
- In the post-Cold War era there has been a shift in the definition of security, from a focus on secure nation-states to 'human security'. This refers to a concern with mitigating the various forms of insecurity affecting ordinary citizens, irrespective of national boundaries.
- Violence can be conceptualised as politically, economically or socially motivated. In practice, however, these distinctions are often blurred and analysts have adopted a wide range of different approaches to categorising violence.
- Different forms of violence within and beyond the city are often related in complex ways. For example, civil wars and national level violence can bleed into more localised forms of endemic violence through 'violence chains'.
- In many urban centres, interpersonal violence is the most pervasive source of violence resulting in more deaths globally than war. Violence and crime also contributes to segregation, fear, capital flight and increased expenditure on private security, all of which affect development.
- Cities are significant sites of conflict in interstate warfare and are vulnerable to acts of international terrorism due to their density and symbolic importance.
- There are many reconstruction challenges in contexts where key institutions and the basic fabric of urban life have been destroyed.

In the scramble for profitable post-conflict reconstruction, conventional development goals can often get sidelined.

- While cities can be important sites of reconciliation and peace-building, it is important that planners in post-conflict situations do not revert to top-down securitisation and repression, but rather work through conflicts with all associated stakeholders.

Discussion questions

1. How do different definitions of 'development' affect how we understand the relationship between development, violence and conflict?
2. 'Urban violence is either political, economic or social'. Discuss.
3. What are the various interlinkages between violent urban crime and development?
4. Why are so many states, armed groups and terrorist organisations motivated to commit acts of 'urbicide'?
5. How should urban planners best approach post-conflict reconstruction and what are some of the main challenges they are likely to face?

Further reading

Abrahamsen, R., Hubert, D. and Williams, M. C. (2009) Special Issue of *Security Dialogue* on Urban Insecurities, 40(4–5).

Fox, S. and Beall, J. (2012) 'Mitigating conflict and violence in African cities', *Environment and Planning C: Government and Policy*, 30(6), 968–981.

Beall, J., Goodfellow, T. and Rodgers, D. (eds) (2013), Special Issue of *Urban Studies* on Cities, Conflict and State Fragility in the Developing World, 50(15).

Graham, S. (2010) *Cities under Siege: The New Military Urbanism*. London: Verso Books.

Koonings, K. and Kruijt, D. (2007). *Fractured Cities: Social Exclusion, Urban Violence and Contested Spaces in Latin America*. London: Zed Books.

Moncada, Ed (2013) Special Issue of *Studies in Comparative International Development* on Rethinking Violence and Order in Cities of the Global South, Vol. 48(3), 217–239.

Moser, C. and McIlwaine, C. (2014) Special Issue of *Environment and Urbanization* on Conflict and Violence in Twenty-First Century Cities, 26(2).

Rodgers, D. (2009) 'Slum wars of the 21st century: Gangs, mano dura and the new urban geography of conflict in Central America', *Development and Change*, 40(5), 949–976.

UN-Habitat (2007) *Global Report on Human Settlements 2007: Enhancing Urban Safety Security*. London: Earthscan.

Websites

LSE Crisis States Research Network: Cities and Fragile States component, featuring over thirty Working Papers: www.lse.ac.uk/internationalDevelopment/research/crisisStates/Research/cafs.aspx

Centre for Urban Conflicts Research, University of Cambridge: www.urbanconflicts.arct.cam.ac.uk

Institute for Economics and Peace: www.visionofhumanity.org

UN HABITAT Safer Cities Programme: unhabitat.org/urban-initiatives/initiatives-programmes/safer-cities

UN Office for Drugs and Crime: www.unodc.org

8 Urban governance and politics

- From planning to governance
- Decentralisation and local–central relations
- Participation and deliberative democracy
- Political organisation and urban institutions
- Contentious politics, social movements and protest

Introduction

As Chapter 7 has highlighted, conflict in cities may be inevitable but the degree to which it becomes violent varies enormously. Governance and politics play a crucial role in violence reduction, but limiting violence is not the only reason we need political institutions that can effectively channel and resolve conflicts. Devising institutions and organisations that address conflicts of interest by facilitating group-based bargaining, and that provide platforms for debate, engagement and the development of policies and plans, is fundamental to innovation and development. Governance and politics are thus central processes that underpin any advances (or reversals) in relation to the economic, social and environmental issues discussed thus far. Governing cities is an extremely complex administrative and technical task, and even when problems are resolved at the technical level, proposed solutions may be inequitable and exclusionary in their impacts. This is where politics comes in.

This chapter begins with an overview of how ideas about governing cities have shifted over the past few decades, from an initial emphasis on centralised planning to a more holistic conception of multi-stakeholder governance. We then examine one of the most pervasive governance reform movements of the late twentieth century: decentralisation. This process can have both administrative and political dimensions, resulting in rearrangements of central–local government relations which have important implications for cities.

Following this discussion, we turn to a second major theme in governance reform that emerged in roughly the same period: citizen participation. Partly a response to the overly bureaucratic, planning-focused paradigm that dominated urban government in the immediate post-colonial period, this was often closely linked to decentralisation and the perceived need for citizen voice to be institutionalised. Innovations in urban participatory and deliberative government, along with a range of critiques, will be discussed. The chapter then turns to formal political organisations and institutions and how these interact in the arena of urban politics. Theories of urban politics that originated in the Western world will be discussed, along with the problems associated with transposing these to developing country contexts. In addition to political party and organised interest group activity, this section considers the need to take into account informal organisational and institutional forms and how these challenge received understanding of how politics happens in cities. Following on from this, in the final section we consider forms of politics that unfold outside the institutionalised spheres of participation and organised engagement, considering social movements and urban protest, and how these have come to coalesce around the growing interest in the idea of 'the right to the city'.

From planning to governance

From the 1950s to the 1970s, urban planning represented 'the embodiment of the dream of a brave new world', and to practise planning was to engage in a very highly regarded profession (Taylor 2004b: 4). In this period, urban physical planning was seen as an integral part of the then-esteemed practice of development planning (Gans 1963; Conyers and Hills 1984), which reflected a more general faith in modernist state planning globally at the time, as discussed in Chapter 5. Meanwhile, as we saw in Chapter 2, colonial cities had provided opportunities to experiment with forms of social and spatial organisation that would have been difficult to attempt in the metropole (Home 1997) and the zeal for planning was carried through into the initial post-colonial period. Despite this positive attitude towards planning, city master plans consistently underestimated the pace of urban growth (Mabogunje 1990) and often took many years to produce. Consequently, they were 'often out of date even before they were completed' and struggled to keep pace with post-independence economic and demographic change (Taylor 2004b: 4).

This contributed to growing concern about the appropriateness of 'Western' urban planning norms that were still being imported through the use of consultants from Europe and North America (Mabogunje 1990; Watson 2009a).

More generally, by the end of the 1970s, traditional state planning of any kind came to be seen as a source of corruption that actually impeded economic development (Krueger 1991). The growing disenchantment with state-driven development strategies and the rise of neoliberalism meant that traditional, centralised approaches to planning fell out of favour, giving way to a paradigm of 'urban management' in the 1980s. While the idea of urban management had long existed as an administrative counterpart to physical planning (Wekwete 1997), the 1980s reframing of the term turned it on its head, envisaging a much slimmer role for government in the day-to-day management of cities. The new 'urban management' of international development discourse was instead primarily concerned with rolling back the state from service provision roles and facilitating the entry of civil society and private sector interests (Rakodi 1997; Wekwete 1997), with the government acting as 'enabler'.

These ideas were encapsulated in an administrative paradigm known as new public management (NPM). The term was coined by Christopher Hood (1991) to refer to the organisational changes in public administration designed to align it more closely with the methods and management systems of business. In essence, NPM involves strategies to increase market competition, give greater choice to users (demand-driven approaches) and promote efficiency in service delivery (cost-sharing). In practical terms this has meant 'increasing focus on the customer, user fees or charges, contracting out of service delivery to the private and voluntary sectors, public-private partnerships, and outright privatization' (Batley and Larbi 2004: 49). In time it became clear that the state could not be by-passed completely and so a prerequisite of the NPM was to inject into service delivery a more entrepreneurial style, with government moving from a concern to do towards a concern to ensure that things get done (ibid.). The implications of this for urban service delivery were discussed in Chapter 5.

This was the environment in which the Urban Management Programme (UMP) – a collaborative programme involving the World Bank, UNDP and UN-HABITAT – emerged in 1986 (World Bank and UNHCS 1989; Farvacque and McAuslan 1992). The UMP was presented as 'promoting new paradigms' around partnership and the empowerment of local actors (Mehta 2005). Around the time of its emergence scholars were starting to recognise the crisis into which cities had been plunged in the developing world (Stren and White 1989), partly as a consequence of structural adjustment (Mkandawire 2002). Ironically, however, this recognition of crisis further boosted the momentum behind neoliberal managerial approaches because of the hardening conception of the state as problem rather than solution. The idea was to 'replace long-term physical planning, which had no real impact on city development, with daily action-oriented urban management', and the influence of these ideas was such that 'master planning disappeared progressively from the priorities of developing countries' (Biau 2004: 8). Just as the emphasis shifted in practice, so it did in relation to academic discourses. Under the influence of Bates and Lipton's ideas of 'urban bias' (see Chapter 1), cities ceased to be a major development priority; any literature that did concern itself with cities, however, shifted attention from a concern with regional and urban planning (Friedmann and Alonso 1975; Rondinelli 1975) towards concern with defining the scope of urban management (Lee-Smith and Stren 1991; Stren 1993; Mattingly 1994).

As a consequence of this shift, by the turn of the twenty-first century urban plans in the developing world were 'usually to be found in dilapidated condition, perhaps pinned to the wall in a central government ministry or folded into a large technical report' (Watson 2014: 215). Unfortunately, in the context of disillusionment with the state, planning was cast aside without adequate attention to what exactly had made it so unsatisfactory (Taylor 2004b). There was an ambiguity regarding how planning was viewed even among its critics: on the one hand it was rejected because of states' seeming inability to implement; but on the other, the land use and zoning regulations that accompany planning were often criticised on the grounds that they were implemented too harshly and rigidly, forcing the poor into illegality (De Soto 1989). In this respect, the problem was a dual one: the plans being drawn up demanded a level of bureaucratic capacity to implement that was generally not present in developing countries, but even where implementation did happen the plans themselves were entirely inappropriate to low-income, rapidly expanding and highly informalised cities. All too often, however, in the ideological environment of the 1980s the problem was considered to be state planning *per se* rather than the particular type of planning that had dominated since the colonial period.

Given the exclusionary nature of much planning practice, there was good reason to welcome greater attention to non-state actors, partnership and negotiation. Yet rather than addressing the power relations embodied in planning practice directly, the new urban management ideal, in which the state was conceived as a disinterested 'enabler', tended to obscure power inequalities instead of actively attempting to redress them. If the paradigm of state planning had been too top-down and lacked a sophisticated approach to implementation, the management discourse that replaced it was politically naïve and eschewed strategic engagement with the root causes of urban development problems in favour of a focus on the day-to-day. The recognition in the 1990s that a 'management' approach was ill-equipped to deal with the scale of urban development challenges (Stren 1993) subsequently led to another shift, towards a concern with 'urban governance'.

This shift was stimulated by the extent of social dislocation in cities affected by structural adjustment reforms and a subsequent effort to articulate a less technocratic, more forward-looking and inclusive approach to development in general. In the arena of urban development this took shape in UN-HABITAT's Global Campaign for Urban Governance from 1999. The adoption of a discourse of governance reflects an acceptance of the failures of both the state-directed and market-led development paradigms, and the need for a more nuanced approach that recognises the potential positive synergies between states, markets and the wide range of actors that fall into the broad category of 'civil society'. Governance is therefore seen as being about the collaborative efforts of multiple agents both within and outside the government (Warren, 2014). It is not only in countries affected by structural adjustment that new ideas of urban governance began to emerge from the 1990s onwards, but also in countries transitioning from socialism. Box 8.1 explores how in China, market reforms were accompanied by new approaches to governance that aimed to maintain stability in the face of changing economic circumstances.

Although it flows directly from the idea of urban management (in the sense that the state is decentred as the dominant actor), the concept of 'governance' is especially concerned with the *interactions* between various agents involved in governing (Kooiman 1993). In Stoker's

Box 8.1 Urban governance and 'community-building' in China

Under its socialist planned economy, the urban population in China was organised primarily around the work unit (or *danwei*). These units performed multiple functions in relation to service delivery, spatial planning and social control, as well as the organisation of labour. With China's market reforms in the late twentieth century, and the greater movement of population between rural and urban areas in the context of the relaxation of the household regulation (*hukou*) system, the *danwei* was no longer an appropriate or workable form of local organisation. This led the Chinese government to refocus in the 1990s and 2000s on the concept of community (*shequ*) as the new basic unit of urban governance.

The term *shequ* has multiple meanings: it identifies a demarcated urban space with clear boundaries, but also refers to institutions for neighbourhood governance and implies a particular kind of environment where residents live together as neighbours who mobilise resources collectively to protect their entitlements. The word had largely disappeared from public discourse after the Communist Party banned sociology in the early 1950s. However, as market reform threatened to fundamentally destabilise Chinese urban society, *shequ* was enthusiastically adopted and began to manifest as a concrete institutional model. Fearing a loss of state and party control at the local level, the idea of community presented an opportunity to re-establish the regulation of social conduct and a discourse of 'community-building' was adopted.

Drawing to some degree on ideas about the 'third way' and new forms of governance in the Western world, the institution of community was imagined as a space between the market and the state, which could mitigate the disruptions of the former and reinforce the control of the latter in a cost-effective manner. The city of Shenyang in Northeast China, which suffered substantial redundancy and unemployment as the economy was restructured, was among the first to develop an experimental model of community organisation. The authorities re-scaled neighbourhood committees, approximately doubling the number of households under their jurisdiction, and created new physical spaces and buildings for community activity, emphasising the new centrality of the community in urban life. All members of the community are issued a handbook containing a 'pact' between local community cadres and residents, outlining their mutual duties to one another.

In addition to new community centres, an older two-tier Maoist system of residents' committees at the mass neighbourhood level and 'street offices' at the micro level was resurrected, having played only a marginal role when *danwei* dominated local organisation. Through the 'street offices' in particular, local voluntary community 'activists' – who are usually senior citizens steeped in both Maoist and Confucian principles and have ingrained networks in the community – have played a critical role in implementing government policies and projects, ensuring that the new institutional structure continues to facilitate government and party control.

Sources: Bray (2006), Wan (2015).

terms, governance refers to governing styles in which boundaries between and within public and private sectors become blurred (Stoker 1998: 17). It can thus be thought of as the endlessly changing configurations of societal actors working to solve collective problems (Wagenaar 2014). This, of course, is not a conflict-free process, as theorists of governance often acknowledge; recent work on 'interactive governance', for example, highlights 'the complex process through which a plurality of social and political actors with diverging interests interact in order to formulate, promote, and achieve common objectives' (Torfing et al. 2012: 2).

The governance concept therefore comes closer than the management approach ever did to acknowledging the degree to which cities are contested spaces characterised by struggles for resources and control (Healey 2004; Pieterse 2008). It also provides more scope for facilitating the integration of poor and marginalised groups into decision-making (Beall 1996; Devas 2001; Devas et al. 2004). Nevertheless, the idea of governance continued the trend by which government was reconceptualised as an organisation less concerned with 'doing' than 'ensuring that things are done' by (and in partnership with) other actors (Kaul 1997; Batley and Larbi 2004). In Harvey's (1989) terms, it can also be thought of as paralleling the global shift towards urban 'entrepreneurialism', which emphasises public–private partnership as a means to increase efficiency and promote capitalist urban economic development as an overriding goal. As Doornbos (2001) notes, the concern with governance in developing countries – and particularly normative conceptions of 'good governance' – originated not in academia but among donor organisations, epitomised by the work of Kaufmann and his colleagues at the World Bank whose prioritisation of economic growth is self-evident (Kaufmann et al. 2000).

Part of the problem with the concept of 'good governance' is its broad and flexible meaning, which renders it subject to almost infinite interpretation. UN-HABITAT's conception of good urban governance involved a concern with 'sustainability, subsidiarity, equity, efficiency, transparency and accountability, civic engagement and citizenship, and security', as if assuming that these are interdependent and mutually reinforcing (Pieterse 2008: 66). In fact there are likely to be conflicts and trade-offs between these diverse goals, as well as power relations that need to be addressed if any of them are to be fully realised. The indicators used to measure 'good governance' are so broad and diverse, ranging from improved financial management to questions of how much respect citizens have for state authorities, that one scholar termed the concept a 'policy metaphor': something that can justify support for any reform donors are interested in pushing at a given time (Doornbos 2001). Several scholars now argue that the indicators comprising the World Bank's conception of 'good governance' (see Kaufman et al. 1999) may reflect a desirable end-point, but do not offer a realistic, coherent or effective policy framework to enable developing countries to get to that point (Andrews 2008; Khan 2010). Good governance reforms are widely seen as primarily about promoting free markets; but expecting countries at low levels of development to thrive economically at the same time as fully opening their markets, eliminating corruption and conducting their business with complete transparency flies in the face of most historical experience (Chang 2003; North et al. 2009; Khan 2010).

The discourse of 'good governance' can also obscure the intensely political nature of development policy and processes. Agencies such as the World Bank and UN-Habitat have tended to shy away from explicit consideration or public discussion of politics and power dynamics, despite their centrality to urban change (Stren 2009). The consequent need to place politics back at centre stage in how we understand urban governance is increasingly recognised, and critiques of dominant donor conceptions of 'good urban governance' are gaining momentum (Pieterse 2008; Myers 2011). Research is also proliferating on the ways in which informal practices reshape and sometimes dramatically subvert formal governance institutions (Lund 2006; Goodfellow and Titeca 2012). Before turning explicitly to urban politics, we consider two of the most widely adopted reforms associated with donor agendas for 'improving governance' in developing countries: decentralisation and citizen empowerment through participatory practices.

Decentralisation and local–central relations

Broadly defined, decentralisation is the process of transferring power from national to local government institutions and organisations. The advantages of decentralisation are clear in theory: it is supposed to bring government closer to the governed, thereby improving information and resource flows while contributing to democratic consolidation. Proponents suggest that local governments are inherently more accountable to their constituencies than regional or national ones (World Bank 1997), and that this enhanced accountability will improve public service delivery. It has been said that 'decentralisation fever' gripped the international community from the 1980s onwards (Tendler 1997), and it has certainly been a widely adopted policy strategy: Crook and Manor (2000) estimated that by the mid-1990s, 80 per cent of countries were engaged in some form of decentralisation. In 1994 the Urban Management Programme claimed that 63 out of the 75 developing and transitional countries with populations over five million had embarked on devolving power to local authorities in some way (Dillinger 1994: vii). More recently, it has been argued that 'decentralization is being implemented essentially everywhere' (Faguet 2014: 2). Given the ubiquity of decentralisation as a policy reform, there are a number of important questions about whether it has improved urban governance or realised any of the other goals that it was designed to achieve.

Much of the literature on the effects of decentralisation has been concerned with its impacts on particular public sector outcomes including public service provision, corruption, education or health indicators (see Crook 2003; Treisman 2007). The record is very mixed, reflecting both the depth and diversity of decentralisation strategies and the importance of local context (Crook and Manor 1998). Decentralisation has been shown both to improve and worsen most outcomes, depending on time and place (Faguet 2014: 10). This is partly because the term is used to describe a great variety of changes in local–central government relations.

A widely used typology of decentralisation was developed by Rondinelli (1981), who proposed that it generally takes one of three forms: deconcentration, delegation or devolution. *Deconcentration* refers to the geographical dispersal of central government functions. The most basic and least extensive form of decentralisation, deconcentration is basically a shifting of the central government workload outside the national capital city, allowing a greater degree of responsiveness to regions but without any transfer of authority to lower tiers of government. *Delegation* refers to the transferring of responsibility for decision-making and administration of public functions to other organisations, whether they be local governments or parastatal corporations. These organisations have semi-independent authority and are not entirely controlled by the central government but are nevertheless ultimately accountable to it. Finally, *devolution* refers to the most extensive form of decentralisation, involving an actual transfer of authority to lower tiers of government. Rather than just delegating certain responsibilities, devolution implies a significant degree of autonomy with respect to decision-making, finance and management. In such a scenario local authorities are perceived as a separate governmental level linked to clear, legal geographical boundaries, within which they exercise authority and perform public functions with little or no direct control exercised by central authorities. While Rondinelli's typology is essentially about differing *degrees* of decentralisation, a distinction can also be drawn between different *types*, namely administrative, political and fiscal decentralisation. The latter two often lag behind the former (Dillinger 1994).

Given this diversity of decentralisation strategies and the widely varying contexts in which they have been employed, it should come as no surprise that it is difficult to determine whether or not decentralisation has been a positive or a negative force overall. Even within a single country, the effects of decentralisation can vary. For example, in a comprehensive quantitative study of the effects of decentralisation on local service delivery in Bolivia, Faguet (2004) found an overall positive impact on local government responsiveness to local needs but with significant differences across regions and municipalities. In Mexico, the effects of decentralization in different municipalities varied substantially in accordance with the personal agendas and 'political entrepreneurialism' of local leaders, in terms of their effectiveness in acquiring resources from other levels of government (Grindle 2007). In Uganda too, the performance of district councils was found to differ enormously depending on the nature of informal political linkages between central government and local government personnel (Lambright 2011). The importance of understanding the political economy of any particular context in order to assess the likely success of decentralisation reforms has now been taken on board by the World Bank (Eaton et al. 2011).

After decades of inconclusive research on the effects of decentralisation on service delivery and other outcomes, attention has more recently turned to how decentralisation impacts on the quality of governance itself. This is an important question given that the strongest arguments in favour of decentralisation relate to accountability, responsiveness and citizen voice (Faguet 2014). Despite the various theoretical arguments about how decentralisation *should* improve local accountability and citizen engagement, however, there are many theoretical counter-arguments as well as very mixed empirical evidence. Critics of decentralisation have highlighted that there are no *a priori* reasons why more localised forms of governance should be more accountable than at other levels (Heller 2000; Tendler 1997). Power at the local level can be more concentrated and applied far more ruthlessly than when exercised at the national level. Competing interests clustered around a smaller pool of resources can exclude weaker members of society in the scramble of 'pork-barrel politics' (Beall 2005).

Decentralisation policies can also be used instrumentally by politicians at higher tiers of government to win the support of recalcitrant regional or local leaders, thereby bolstering central government power in peripheral areas where opposition to a ruling party prevails. By accommodating local or regional demands for power sharing through decentralisation strategies, resistance can be neutralised (see for example Green 2010; Lindemann 2011). Decentralisation can be used to 'buy' support in other ways too; it has been argued that promises to groups of key supporters are easier to make and fulfil locally than nationally, due to more close-knit relationships between political patrons and their clients (Conyers 2007: 16, see below for a discussion of patron–client relations). The point has also been made that those who benefitted from pre-existing centralised governance and therefore oppose decentralisation may act as 'spoilers' once decentralisation reforms are implemented, actively undermining local authorities – particularly where valuable natural resources are at stake (Jackson 2005; 2007). The effects of decentralisation are therefore extremely complex and contingent, to the extent that a recent journal issue devoted to decentralisation and governance concluded that 'we can only speculate' about what ultimately differentiates decentralisations that promote development transitions from those that undermine them (Faguet 2014: 11).

These general points aside, there are important questions about how cities fit into decentralisation reforms, and how decentralisation affects urban governance specifically. The implications of decentralisation for cities are relatively under-researched, partly due to rural biases in the way decentralisation has been both designed and discussed in less-developed countries, with a large number of major studies focusing on the relationship between decentralisation and rural development (see for example Conyers 1983; Parker 1995; Crook and Manor 1998; Bardhan 2002). The challenge facing cities under conditions of decentralisation is to establish institutional structures and processes that facilitate coordination between many levels and agencies of government to ensure comprehensive action at the metropolitan scale. Post-Apartheid South Africa, for example, actively sought to address this challenge through the institutionalisation of Integrated Development Plans (IDPs), which are the vehicle through which cities consult their residents about priorities and needs and align their strategic plans with those of neighbouring municipalities and government agencies at different levels. Although subject to some criticism for being overly technocratic, these have led to some substantial improvements in service delivery and represent better co-ordination across different spheres of government than has been seen in many decentralisation programmes elsewhere (Binns and Nel 2002; Parnell et al. 2002; Harrison 2008; Schmidt 2008).

A critical difference between governing urban as opposed to rural areas relates to the nature of local government finance. Urban governments have much greater potential to raise local taxes, but also have a much greater burden in terms of service delivery and infrastructure maintenance. Giving city governments powers to tax and spend is critical if decentralisation is to devolve power to local authorities in any meaningful way. After all, taxation is acknowledged to play a central role in state-building (Tilly 1992, Bräutigam et al. 2008, Moore 2004) so without such powers the local state will be unable to build significant capacity of its own, remaining a mere arm of central authorities.

Urban governments are almost always hopelessly underfunded (Devas 2003; Fjeldstad 2006), with many employing a pot-pourri of over-complicated revenue instruments to address this (Brosio 2000; Smoke 2001). These local taxes are generally combined with fiscal transfers from central government, because in most developing countries even large cities do not generate enough local tax to fund their activities without central government help. The most important sources of cities' own revenue are usually property taxes, business licenses and 'user fees' for urban services (Fjeldstad 2006; Fjeldstad and Heggstad 2011). Property tax holds particular potential for urban governance but is hugely underutilised (Monkam and Moore 2015). This is partly because it is extremely challenging administratively, especially where informal construction, irregular street patterns and absent street names render cadastral systems unworkable (Fjeldstad 2006: 8). Having a well-developed cadre of property valuers who can assess the market value of properties in order to tax them effectively is also lacking in many of the poorest countries (Goodfellow 2014a). Consequently, while property tax in developed countries can account for up to 100 per cent of local revenue and 4 per cent of GDP, one study of a selection of developing countries found that it averaged just 0.6 per cent of GDP (Norregaard 2013). In some countries it amounts to far less – just 0.018 per cent of Rwanda's GDP in 2012–13, for example (Goodfellow 2015b). It is also highly political due to the way it targets the visible assets of the wealthy, so often generates formidable political resistance (ibid.).

Without significant and reliable revenues, urban authorities often struggle to find the resources for investment and maintenance of urban services, or even to pay their permanent staff, leaving little if anything for development projects. The funds that are available for infrastructure development or other activities often come in the form of central government transfers that are 'conditional' and therefore leave little room for local autonomy (Goodfellow 2012; Resnick 2014a). Yet it is not only because of finance and administrative capacity issues that decentralised urban authorities struggle to deliver results. The role of politics – and particularly of political interference from central government – is absolutely crucial, both in terms of obstructing local tax collection and other forms of political interference in policy implementation. Especially under conditions of democratic decentralisation, in which different political parties compete to be elected locally, if opposition parties gain power in urban areas (which they often do) this can result in what has been termed 'vertically-divided authority' (Garman et al. 2001). This is where local authorities are controlled by parties that fundamentally oppose the national authority, and can result in various 'strategies of subversion' through which central government attempts to undermine municipal authorities and weaken them politically (Resnick 2014b).

These problems can be particularly severe in major urban centres or capital cities, where stakes are high and the tussle between overlapping tiers of government is particularly intense. A study comparing Bolivia and Colombia found that smaller, more rural district governments were more responsive to citizen demand than the governments of large cities (Faguet and Sánchez 2008). The explanation for this may lie in the complexity of governing large urban agglomerations and the diversity of powerful interests at play. While some large cities are governed by one overarching urban authority, others are governed by a constellation of municipalities, districts or divisions each concerned only with their patch of the city, making coordination difficult. Furthermore, large settlements tend to be significant sources of economic growth, raising the stakes of local decision-making processes and creating incentives for central interference. An example of the kinds of central interference that can cripple urban governance – and even result in re-centralisation of authority – is provided in Box 8.2.

Box 8.2 Decentralisation and re-centralisation in Kampala

Uganda was lauded in the 1990s for its radical decentralisation programme, which was one of the most far-reaching in Africa, shifting substantial amounts of resources to local governments and institutionalising a five-tier system of Local Councils, from the district level right down to the very local parish level. Administrative and fiscal decentralisation were accompanied by substantial political decentralisation, and in the capital city Kampala this resulted in the opposition Democratic Party taking control of the City Council.

Throughout the first decade of the twenty-first century, the National Resistance Movement (NRM) government – in power nationally since 1986 – became increasingly jealous of the powers it had devolved to Kampala, as it repeatedly failed to win a majority of seats in the City Council or take control of the Mayoralty. Problems of service delivery and infrastructure, which were severe given rapid urban growth, corruption and lack of funding, were rendered worse by bad relations between central government and the City Council. Building on public dissatisfaction with the Council, central government began increasingly to undermine many things the Council tried to do – for example with respect to re-organising urban transport or enforcing land use regulations. This was accompanied by an onslaught in the media (much of which is state controlled) against the Council, consistently branding it corrupt, inefficient and responsible for creating urban 'chaos'.

These developments have been interpreted as a carefully orchestrated campaign on the part of the central government to take back control of the city. In 2010, a law was passed that provided for central government to take over the city, drawing on a vague statement in the 1995 constitution about the right to do so in situations of local government failure. The following year, the new Act was implemented through a radical re-centralisation of city governance. Kampala City Council was officially abolished and replaced with the Kampala Capital City Authority, which was staffed by a team of high-powered bureaucrats appointed directly by the President. The role of the democratically elected council was substantially diminished, and the role of mayor reformed into a largely ceremonial position of 'Lord Mayor' with few substantive political powers.

These developments streamlined city governance by reducing conflict between the national and city levels, as well as generating more resource transfers to the city in the context of higher levels of inter-governmental trust. This has produced some positive changes with respect to infrastructure and some urban services, such as waste collection. However, it amounts to a radical reversal of decentralisation, as well as de-democratisation and de-institutionalisation of citizen voice. It remains to be seen whether the new arrangement is capable of producing not only cosmetic improvements but wider improvements in urban wellbeing. While an extreme case, the re-centralisation of governance in Kampala reflects a broader trend across the continent, where an initial zeal for decentralisation has widely given way to mistrust between city and national authorities and a desire on the part of the latter to re-centralise.

Sources: Goodfellow (2010; 2013a), Gore and Muwanga (2014).

Amid the changing administrative arrangements and elite political struggles that often accompany decentralisation programmes, the role ordinary citizens themselves play in governance can often be overlooked. Yet the question of citizen participation is intimately linked to decentralisation, especially in view of the purported aim of the latter to 'bring government closer to the people'. Indeed, the World Bank and other major donors have often pushed the two agendas of decentralisation and participation in tandem, framing both as critical to achieving 'better governance'.

Participation and deliberative democracy

Debates on participation have a long and complex history, and the term itself can have diverse meanings. Literature on modern 'community participation' in North America and Europe took off in the fields of Political Science and Planning in the 1960s, with a key point of reference being Sherry Arnstein's classic 1969 essay, 'A Ladder of Citizen Participation'. In this Arnstein draws up a typology ranging from full citizen control over local affairs, through various forms of 'tokenistic' participation to forms of 'non-participation' and outright manipulation by government authorities (Arnstein, 1969: 217). In subsequent decades the literature on participation within democratic theory and planning theory has flourished (see for example Healey 1997; Forester 1999, Warren 2002). Meanwhile, a parallel discourse on participatory development has evolved within the field of Development Studies, spurred by the 1970s concern with empowerment (see Chapter 1). By this time disillusionment with expert-led forms of planning was setting in within the field of development policy and practice, as noted previously. Although in the sphere of economics this manifested in the 'neoliberal turn' characterised by deregulation, privatisation and liberalisation, in the field of social development there was a growing interest in participatory, community-led approaches to development. In this discourse the actors facilitating 'community participation' were often development practitioners rather than government authorities, and the participants in question generally rural, in the wake of Lipton's (1977) recasting of the development problem as one of urban elites versus rural poor.

Particularly influential here was the work of Robert Chambers (1983; 1992) who pioneered ideas about how to assist the poorest through eliciting their participation in what came to be known as 'participatory rural appraisal'. This involved activities such as drawing up seasonal calendars, trend and change analysis, well-being and wealth ranking, and 'analytical diagramming' (Chambers 1994: 953), in which the 'objects' of development themselves – the poor – would play a central role. This rapidly evolved into an interest in 'participatory development' more broadly, with the World Bank vocally espousing the concept by the mid-1990s. As the mantra of 'participation' became a new mainstream 'grand narrative' (Kothari 2001: 139), critiques emerged, with some authors questioning the very notion of 'community' on which ideas about participation were based and the tendency to gloss over gendered power differentials (Guijt and Shah 1998). The most scathing critique came in 2001, with a group of scholars arguing that participation constituted a 'new tyranny' (Cooke and Kothari 2001). This case was made on the grounds that the dynamics within communities often lead to decisions that reinforce the interests of the already powerful, and that the development actors facilitating participation can override legitimate local decision-making processes. Even more damning was the accusation that participatory planning often actually means the acquisition, structuring and manipulation of local knowledge (ibid.). One author powerfully summarised the case against participation thus:

> 'Participation' in development activities has been translated into a managerial exercise based on 'toolboxes' of procedures and techniques. It has been turned away from its radical roots; we now talk of problem-solving through participation rather than problematization, critical engagement and class [...] While we emphasize the desirability of empowerment, project approaches remain largely concerned with efficiency. While we recognize the importance of institutions, we focus attention only on the highly visible, formal, local institutions, overlooking the numerous communal activities that occur through daily interactions and socially embedded arrangements.
>
> (Cleaver 2001: 53)

In response, some scholars suggested that the problem was not participation *per se* but that participation had widely become misdirected. Participatory development initiatives were, they argued, increasingly focused on 'imminent development' (technical, short-term interventions to engineer particular development outcomes) at the expense of 'immanent development' (the underlying processes through which societal structures evolve) (Hickey and Mohan 2004: 10). Thus,

> in an echo of the critiques of 'good governance' noted above, the point was made that discourses of participation were screening out the political struggles that necessarily underpin development and change. To be truly transformative, participation needs to enhance people's political capabilities: their ability to learn about political rights and representation, form networks and organisations and – critically – to effectively engage the state.
>
> (Williams 2004)

These critiques amount to a refocusing of participation on the question of citizenship. It is in cities, the spaces with which citizenship has been most strongly associated from the time of ancient Athens through to the post-colonial developing world (Hall 1998; Holston 1999; Mamdani 1996), that this politicised vision of participation has emerged most prominently. Notwithstanding the rural focus of the original participatory development paradigm, a series of democratic experiments in cities have evolved in developing countries since the late twentieth century giving new vigour and hope to those concerned with participation. The most significant have been in Latin America, and especially Brazil, where a new focus on citizenship became prominent in the 1990s in the context of democratisation and the rise to power of the Worker's Party (PT) in many municipalities. Against this backdrop the PT began to implement its innovative proposal of the *orçamento participativo* (participatory budget), most famously in the city of Porto Alegre. Through participatory budgeting (PB), which continues there up to the present day, the city's 1.3 million residents meet and vote twice yearly in district-specific plenaries. Attendees debate potential new priorities and elect resident representatives for three-month terms to a city council that meets at least weekly to formulate a final budget proposal for the mayor (see Souza 2001 for a detailed discussion).

The spread of PB coincided with a 'deliberative turn' in democracy scholarship in the last decade of the twentieth century (Dryzek 2000). As a concrete manifestation of deliberative practice, PB was adopted early in other Brazilian cities such as Recife and Belo Horizonte, and rapidly spread to many other cities not only in Brazil but in the Latin American region and subsequently beyond into Europe and the US, to the extent that it is now operational in around 1,500 cities globally (Baiocchi and Ganuza 2014). The model also spread to some much poorer countries. For example, in 2003 a book club set up by university students in Cameroon organised an event on participatory budgeting, which attracted mayors and local officials from across Africa and resulted in a charter to develop PB on the continent. By 2011, 162 municipalities across 23 African countries had adopted PB. Despite the enduring problem of the small size of municipal budgets, since its introduction in 2009 in Cameroon's capital Yaoundé the process has yielded much-needed public works including wells, sanitation and street lighting in areas previously completely neglected by the state (ARI 2014: 32).

Notwithstanding the successes of PB and its celebration the world over, the dangers of translating the Brazilian model across borders without due attention to local context looms large. In a recent evaluation of its global spread, Baiocchi and Ganuza point out that in its original form PB was just one part of a broader set of institutional reforms taking place in Brazil, many of which were not as visible as PB but were crucially important. Indeed, the Brazilian experiment was born out of a generation of political contestation and struggle by marginalised groups of citizens and a social movement that ultimately transformed into a political party; this institutional background was central to the Brazilian success in making PB transformative. 'To put it bluntly', Baiocchi and Ganuza argue, in the global translations of Participatory Budgeting, the communicative dimension has traveled well, but, with very partial exceptions, the empowerment one has not' (Baiocchi and Ganuza 2014: 32). This concern reflects a sense in recent years that, like participation more generally, deliberation needs to be accompanied by particular institutional innovations to facilitate the empowerment of those most likely to be excluded in the course of deliberative discourse. Some of the ways in which different groups have been excluded in one particular experiment with PB are explored in Box 8.3. The problem is not, of course, limited to PB but extends to participatory governance initiatives more generally.

Box 8.3 The limitations of participatory budgeting in Buenos Aires

In the Argentinian capital Buenos Aires, the model of district-level participatory budgeting (PB) adopted in Porto Alegre was imported wholeheartedly in the early 2000s. Despite the potential that Argentina would seem to offer for this kind of democratic reform at this time, having democratised other aspects of urban governance in the 1990s, PB was actually relatively unsuccessful at promoting the kind of local democratic deliberation it is supposed to facilitate. Building on the idea of a 'sociology of absences', Centner argues that three 'techniques of absencing' were responsible for constraining the deliberative potential of PB as it unfolded in Buenos Aires. These represent efforts to create isolated and controlled spaces that prevented effective participation.

The first 'technique of absence' is what he terms 'secession'. This is where members of local community fora fought to have separate meetings, breaking away from larger-scale community negotiations on the grounds that, for example, they were retired and could not be expected to travel by bus to the main community meeting. What this effectively represented was an effort to create likeminded groups for participation that reduced the parameters of deliberation, undercutting the debates that PB is intended to generate and resolve.

The second technique was 'division', through which the organisation of the PB process at particular geographical scales constrained deliberation by setting up separate meetings for residents of different neighbourhoods in the area, to the extent that they never met at all. This meant, for example, that residents of a poorer neighbourhood would never deliberate with those of a richer area even though both were within the same overall jurisdiction for the PB exercise. This meant there was little chance for mutual representation of interests across geographic and socioeconomic divides.

Finally, and most pernicious of all, was the technique of active exclusion. This manifested either where people were denied access in some way by an authority that deemed them unfit to participate, or where they were only allowed to participate by proxy, ostensibly represented by someone deemed to be wiser and more appropriate. In some cases, for example, only leaders of local NGOs or soup kitchens were encouraged to attend, while in others it was very difficult to find out where the meeting even was unless you were already within the circle of invited participants.

Source: Centner (2012).

One of the most significant ways in which exclusionary norms and practices persist in the face of efforts to widen participation relates to gender. Widely lauded cases of participatory governance in parts of India, for example, have had limited success in overcoming entrenched gender roles to facilitate meaningful participation women (Mohanty 2007; Williams et al. 2015). Even in Brazilian cities, the participation of women is mainly evident at the most local levels and drops significantly in the municipal level components of PB, as well as being largely confined to unmarried women and widows (Novy and Leboult 2005; Pateman 2012). In sum, 'to expect marginalisation to disappear solely through the "correct" performance of a set of reformed government practices is probably a forlorn hope' (Williams and Thampi 2013: 1354).

Concerns about the introduction of participatory institutions without due attention to the roots of power inequalities has fuelled interest in ideas of 'empowered participatory governance' drawing on the relative successes of Porto Alegre and Kerala in India (Fung et al. 2003). While Fung and Wright conclude that these successful participatory experiments share key political and design characteristics, such as devolved deliberative decision-making alongside more centralised support and coordination, it is telling that they also depend on a critical enabling condition: a 'rough equality of power' among participants to begin with (Fung et al. 2003: 24). This cannot be taken for granted. Achieving relative equality depends not only on 'self-conscious institutional design' but on 'historical accidents' that (often unintentionally) change power relations. Rebalancing power also depends, however, on the work of groups such as community organisations, labour unions and advocacy groups, which can sometimes develop the capacity over time to check or limit the power of dominant groups (ibid.: 23). It is therefore important to recognise the pivotal role that institutions and organisations play in catalysing social change (Harriss et al. 2004).

Political organisation and urban institutions

This section draws on the common distinction between *institutions* (conceived as 'rules of the game', such as laws and regulations) and *organisations* (conceived as the 'players' of the game, generally groups of individuals united by a common purpose) (North 1990). Parties, trade unions and lobbying groups are thus all types of political organisation, while the rules through which they organise themselves and relate to other organisations (including the government) – such as electoral systems, constitutions and public–private dialogues – are types of institution. In addition to these formal organisations and institutions, there are also critically important *informal* organisations and institutions that shape urban politics and are often considered to be especially important in developing countries. These include organisations such as 'traditional' authorities, gangs and cartels, as well as institutions such as patron–client relations or systems of chieftaincy. Bearing these distinctions in mind, this section begins by reviewing the evolution of theoretical debates on urban politics in the developed world before exploring ways in which we need to adapt or depart from these in order to understand urban politics in developing countries.

Urban politics – i.e. the way in which power is organised and contested in urban areas – has been the subject of extensive research, but primarily in North American and European contexts. As a sub-discipline of political science, the study of urban politics emerged primarily in the United States in the post-war period, notably through the 'community power' debates of the 1950s and 1960s in which the conflicting theories of 'elitism' and 'pluralism' flourished. The former perspective, epitomised by the work of Hunter (1953) and Mills (1956), depicted political power in cities as being entirely dominated by the organisations of the wealthy, with the vast majority of urban dwellers remaining voiceless and powerless. In contrast, Dahl's hugely influential 1961 book on urban government in New Haven, Connecticut argued that in the twentieth century the locus of urban power was less fixed and much harder to pin down than 'elitists' supposed. The increased capacity for different groups to mobilise themselves and form coalitions, he argued, ensured that no one group could dominate decision-making consistently. Dahl's pluralism was swiftly criticised on the grounds that it was blind to a second 'face' of power: the power exercised by certain interests behind the scenes to keep particular items off the political agenda so they do not even become open to decision-making in pluralist arenas (Bachrach and Baratz 1962).

By the 1970s, the elitism-pluralism debate had grown stale and the study of the urban political arena was decentred within political science (Orr and Johnson 2008), fragmenting into sub-disciplines of urban political sociology and various forms of urban political economy. Around this time, Marxist-influenced urban theory flourished in the work of influential thinkers such as Harvey (1973) and Castells (1972). While these approaches attributed the character of urban development primarily to the impersonal logic of capital, a counter-current emerged through the concept of the 'urban growth machine' (Molotch 1976; Logan and Molotch 1987). This posited that the activism, organisation and lobbying of particular elites pursuing their interests in intensive land development is what shapes urban development, refocusing attention on political action and organisation as key elements of the urban question.

The 'growth machine' idea and the related concept of the urban 'growth coalition' (Cox 1978; Elkin 1987) were developed further through urban regime theory (Stone 1989; 1993; Stoker and Mossberger 1994). This explicitly highlights how coalitions between city governments and organised urban interests are formed and maintained. In challenging the idea that city leaders are merely passive facilitators of urban development and growth driven by structural capitalist forces, urban regime theory's key contribution has been to emphasise how hard people work to assemble and maintain effective coalitions that can shape urban development in their interests. While urban regime theorists do not see democratic urban arenas as open and level playing fields in the way that the pluralists did, neither do they suggest that the dominance of particular elite coalitions is inevitable or easy. Critically, growth machine and urban regime approaches placed questions of *organisation* at the forefront of urban politics; those who are most adept at organising and sustaining their coalitions will prevail in political contestation. Politics, in Stone's words, can thus be seen as 'the art of arranging' (Stone 1989: xii). This skews the playing field against the less powerful:

> [G]overning, as opposed to ad hoc decisions or concessions here and there, rests on a level of politics in which substantial resources, complex capacities to plan and execute, and skills in building cooperation and devising forms of coordination are far beyond the ordinary citizen. Thus, urban regime analysis suggests that institutional repair, community development and community organizing, and reshaping civil society are among the steps needed before one can characterize local politics as open and penetrable.
>
> (Stone 2005: 335).

This emphasis on the need for organised vehicles to articulate and represent the interest of marginalised groups echoes some of the concerns of development studies scholars noted above in relation to participation. Indeed, despite being rooted in the American context, the urban regime concept does hold some relevance for cities of the developing world. Unlike the conceptions of governance and participation favoured by many international donor organisations, urban regime theory highlights the unevenness, conflict and struggle in the urban political process (Stone 1993; Painter 2001). Nevertheless, like most theories of urban politics it bears the clear imprint of having been written in and about a society that was the world's most economically advanced and capitalistic at the time. Translating these theories to the less-developed world poses numerous difficulties.

To begin with, ideas of growth machines and urban regimes assume that metropolitan authorities and city mayors have a high degree of autonomy to rule their own affairs, as is generally true in the United States. In some parts of the developing world this is the case, most particularly in Latin America and notably in Colombia (Gutiérrez et al. 2013; Thibert and Osorio 2014). However, in many (if not most) developing countries there is nothing like this degree of urban autonomy due to historical legacies of colonial governance (Fox 2014), which have contributed to the contemporary fiscal constraints and central government interference discussed previously. This simple fact accounts for many critical differences in urban governance across the world. The current celebration of mayors and the transformative agency of city authorities (e.g. Barber 2013, Chakrabarti 2013) has only limited purchase in many developing countries. A further constraint on metropolitan autonomy is the role of international donors, who often wield significant influence through their frequently sporadic and sometimes inconsistent interventions in the urban sphere (Stren 2014). International agencies can complicate urban regimes by introducing resource streams and promoting development goals that may substantially differ from the priorities of domestic political organisations.

A second problem is that ideas from mainstream urban political science assume developed forms of capitalism, and particular kinds of relationships between the public and private sector. For example, Stone's idea of government having to 'blend' its capacities with other actors in order to govern assumes that a relatively clear division between private actors and the state exists in the first place (Fainstein and Fainstein 1983). This is questionable in many developing country contexts, where economic and political elites have often been closely fused since these states were created at independence, rather than being distinct and autonomous bargaining partners (Beall et al. 2013).

A further problem in thinking about urban politics in developing countries through a high-income country lens is that the organisational landscape in less-developed countries looks very different to that in the United States or Europe. Institutionalised, programmatic political parties are a relative rarity in many parts of the developing world; in sub-Saharan Africa, for example, party systems are frequently highly imbalanced and parties themselves weak, dominated by personalistic tendencies and lacking both resources and clear policy agendas (Randall and Svåsand 2002). Opposition parties, in particular, 'are frequently closely identified with the party founder or current leader rather than with a political programme and lack elaborated organizational underpinnings' (ibid.: 38). At the same time, many opposition parties control major cities in Africa, which can lead to the dual problem of poorly organised urban governance alongside persistent interference from central authorities (Resnick 2014a). It is, however, important not to overstate the point about the lack of party institutionalisation in developing countries given how much variation there is, including within Africa (LeBas 2011).

One of the reasons that political parties in developing country cities are often programmatically and ideologically weak is the prevalence of clientelist political relationships. The concept of clientelism and its counterpart, patronage, have long been central to the study of politics in developing countries (Nelson 1979). Clientelism amounts to the use of instruments of patronage such as jobs, contracts and tax breaks by those in positions of political authority in order to shore up political support. A study of poverty and urban governance conducted across ten cities in Africa, Asia and Latin America (Devas et al. 2004) revealed how community leaders benefited personally from their relationships with government officials through being given public sector jobs, access to government grants and a range of financial incentives in return for managing votes at election time. This practice, known as 'vote-banking' is widespread in the cities of low- and middle-income countries, and undermines the foundations of formal political party contestation as it is conceived in 'Western' democratic norms. Disadvantaged urban dwellers will often show preference for clientelism over conflict as long as they perceive this clientelism to be effective and fair within the prevailing rules of the game (Walton 1998: 477). In other words, poor people often collude with those in power when they feel that such a strategy will most likely ensure their short-term interests, rather than engaging in political activism.

As well as weakly institutionalised parties, many less developed countries have relatively little by way of a formal industrial sector and organised workforce, and correspondingly lack an organised labour movement. This is not to say that trades unions and analogous organisations do not exist in developing countries. In many middle-income nations such as South Africa, India and Brazil there is a rich history of labour organisation, and the informal economy has become an increasingly important part of this. The Self Employed Women's Association (SEWA) in India, for example, began with a few hundred members in 1972 and had around a million by 2008, while StreetNet, an organisation of informal street vendors that began in Durban, South Africa, has subsequently 'gone global', building branches across the developing world and holding an international congress every three years. Nevertheless, the fact remains that in most low-income countries the informal workers who comprise the urban majority are not well represented by international labour movements (Gallin 2001), despite plenty of local collective organising and efforts to reach across borders (see Lindell 2010). Most urban workers consequently lack job security, basic rights and that most basic form of political agency: 'the sense of being able to stand up to the boss' (Gallin 2001: 547).

The relative weakness of these various forms of political organisation clearly impedes the potential for any kind of pluralist bargaining environment to emerge, while also narrowing the range of actors able to claim their stake in any dominant urban regimes. However, viewing urban politics through the conceptual lenses outlined above risks producing an overly negative and ethnocentric picture that prevents cities in most of the world from being considered on their own terms. While there might be an absence of anything that looks like institutionalised pluralism or an urban regime of the kind theorised in developed countries, cities in less-developed countries are clearly sites of intense political bargaining and collective negotiation that provide just as much scope for theorisation and analysis as cities in the rich world. A growing body of research highlights the varying ways in which political bargaining environments affect urban development outcomes in developing countries, with donor-driven democratisation and decentralisation initiatives (as well as 'imported' urban planning models) often combining with other historically rooted local norms and institutions to shape the urban political arena (Myers 2005; Harriss 2009; Njeru 2010; Resnick 2013; Goodfellow 2013a; 2015a).

What this emerging work illustrates is that the institutional landscape in developing country cities is different, rather than less complex or sophisticated, than that which has been theorised so extensively in rich countries. It also suggests that clientelism as a concept cannot capture the full richness of urban political bargaining in developing countries. While the institutional arrangements that facilitate the formalised interaction of interest groups, political parties and the state are often weaker in low-income countries, a complex array of alternative informal institutional interactions abound. Understanding urban politics in these cities therefore often requires attention to institutional and organisational forms such as traditional authorities, chieftaincy, religious organisations, criminal gangs, and mafia organisations who maintain effective control over land allocation and *de facto* property rights. Due to the prevalence of informality there is often a wide gap between the official rules and the norms and practices that actually govern urban social and economic interaction. This can result in attempts by both government officials and citizens to combine and negotiate between different (and often conflicting) rule systems in order to get by. Interest in these phenomena has sparked a growing literature on 'hybrid' political organisations and institutions within Development Studies (Beall et al. 2005: Boege et al. 2008; Williams 2010; Booth and Cammack 2013; Goodfellow and Lindemann 2013; Meagher 2012).

The combination of competing formal and informal institutions can result in considerable friction in cities in developing countries, especially in situations of rapid urban growth where resources are scarce and, as indicated above, formal organisations of democratic engagement relatively undeveloped. In such contexts, politics can become highly contentious, and often violent. Parker (2010) proposes an analytical matrix where forms of urban political activity are placed along axes of ideological breadth and institutional 'thickness', with parties in the top right-hand corner (being both ideologically broad and institutionally thick) and spontaneous violent urban contestation in the opposite position (see Figure 8.1). Between the two – and in many cases periodically linked to both – lie urban 'social movements' and the phenomenon of urban protest.

Figure 8.1 *Types of urban political organisation/activity*

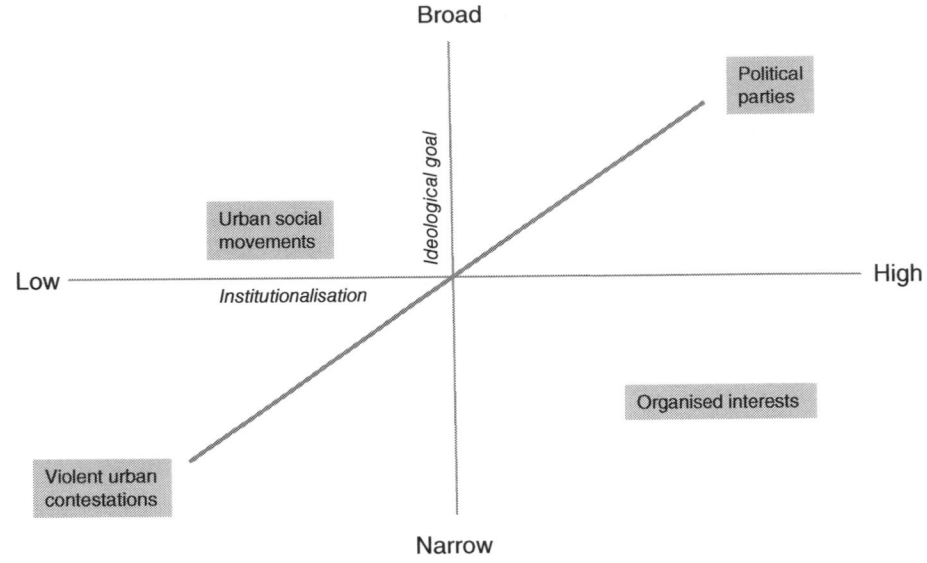

Source: Adapted from Parker 2010: 56.

Contentious politics, social movements and protest

A social movement can be defined as 'a sustained campaign of claim-making, using repeated public performances to advertise this claim, based on organizations, networks, traditions and solidarities that sustain these activities' (Tilly and Tarrow 2007: 8). In seeking to bring about some kind of change, social movements tend to emphasise direct action and protest tactics as well as maintaining a strategic distance from formal party politics. The literature analysing social movements is expansive and diverse, encompassing the analysis of movements as different as the American Civil Rights struggle and contemporary global environmental campaigning. What particularly concerns us here is a sub-category of social movements: namely *urban* social movements. An urban social movement (USM) involves more than a group simply using an urban area as a site of claim-making. For example, although the civil rights movement chose to conduct a March on Washington, there was arguably nothing specifically urban about the movement itself. Urban social movements are thus popular movements that not only play out in cities but 'owe their origin and *raison d'etre* to the city itself' (Parker

2010: 67). True urban social movements are motivated by conflicts and power struggles arising from the distinctively urban aspects of socioeconomic life.

Manuel Castells coined the term 'urban social movement' in his 1972 book *The Urban Question*, which sought to challenge existing approaches to urban social systems and reconstitute an understanding of the urban around the characteristics of what he termed 'collective consumption' in capitalist societies. For Castells, the urban is not defined by capitalist production *per se*, which had no particular spatiality, but by the location of the consumption processes linked to the provision of goods and services necessary to maintain a labour force. USMs constitute expressions of class struggle that manifest over the availability of and access to collective consumption goods. Over time, Castells significantly modified his views, moving further from orthodox Marxism by suggesting that USMs can play autonomous roles in social change, and even facilitate productive linkages between social classes. In his 1983 book *The City and The Grassroots*, he advanced a theory of USMs that was less fixated on class struggle and allowed more room for agency, highlighting the role of USMs in struggles to achieve local self-government or defend cultural identities.

The evolution of Castell's thinking around social movements was accompanied by a broader shift in the social movement discourse away from 'old social movements' concerned with class-based action to 'new social movements' rooted in questions of identity, lifestyle choice and the right to participate. Considering the shifting forms of protest movements, Walton (1998) identifies three types of collective action in response to grievances that dominate at different levels of development and different periods in macroeconomic cycles. The first is *labour action*, which was common in the mid-twentieth century in the developing world and essentially amounts to 'old' social movements whereby workers engage in strikes, demonstrations and other forms of protest against unemployment and policies with adverse implications for their jobs and wages. The second is *collective consumption action*, which emerges in times of high unemployment and where low-income people seek to reduce the cost of consumption of public goods such as urban services and transportation (e.g. by protesting over food and fuel prices), rather than pursue futile efforts to secure wage increases. Such activities have been common since the 1980s. Third is *political and human rights action*, involving mobilisation around issues of social justice, freedom and representation (Walton 1998).

While urban social movements are inherently political in their efforts to challenge existing practices and power structures, they often seek to retain their autonomy from prevailing political agendas and to avoid alignment or affiliation with established political parties. This allows them to negotiate with whoever is in power at different levels of government and build momentum over time (D'Cruz and Satterthwaite 2005: 54–55). The decision to work outside conventional political organisations is usually highly strategic; after all, such movements emerge in the first place because prevailing political arrangements do not provide legitimate avenues for their needs to be addressed (ibid.: 57). Often this strategy is taken to an extreme when social movements deliberately transgress laws and remain illegal in the hope that this will enable them to pressure the government more effectively. Operating on the fringes of legality in contemporary Brazil, urban social movements such as the *União dos Movimentos de Moradia* in São Paulo, Brazil organise illegally to meet their immediate needs as well as pursue wider political agendas (see Box 8.4).

Box 8.4 The *União dos Movimentos de Moradia* in São Paulo

The *União dos Movimentos de Moradia* (Alliance of Housing Movements) or UMM is a large social movement campaigning for low-income housing in São Paulo city. Open to anyone who considers themselves 'Sem Teto' (literally, 'without a roof'), it was inspired by the *Movimento dos Trabalhadores Rurais Sem Terra* (the Rural Landless Workers' Movement), which has been at the forefront of campaigning for land reform since the early 1980s. Their daring occupations of unproductive ranches to set up cooperative farms inspired the UMM's strategy of urban occupations of abandoned buildings in central São Paulo.

This is one of the two main strategies employed by the housing movement, the other being to campaign for housing policy reform through input into consultative and deliberative policy forums at municipal, state and federal level. Although experiments in participatory democracy in Brazilian cities have generated much discussion amongst policymakers and academics, it is the movement's inability to achieve its aims through institutionalised participation alone that has reinforced the need for a parallel strategy on the fringes of legality – that of urban building occupations.

Through its occupations, the movement attempts to draw attention to the chronic spatial segregation of the city, which forces the poor to live in underserviced *favelas* and irregular settlements in the peripheries far from their places of work, whilst an estimated 40,000 buildings remain empty in the centre. This focus on the central districts presents unique challenges for the movement, given that it is a complex environment with multiple stakeholders. It has symbolic value for governments seeking to implement visible 'showcase' infrastructure projects as well as economic value for property developers. However, many buildings are dilapidated and unattractive to both businesses and higher income groups; hence they are left unoccupied.

In 2001 a piece of legislation known as the Statute of the City was passed, in which the 'right to the city' is conceived as a collective right to urban services. The Statute operationalises articles of the progressive 1988 Constitution that enshrine the right to housing. It also reinforces constitutional legislation on the social function of property, and allows for expropriation of abandoned buildings for renovation as low-income housing. The Statute and the Constitution have thus become key weapons used by the UMM to support its claims, while recourse to illegal or semi-legal activity continues to be used when government legislation has not been adequately followed through. In this way the law and rights enshrined within it are used strategically by the movement, who counter accusations that their occupations are criminal by arguing that the government's failure to ensure people's right to housing is itself a crime – and ultimately a more severe one. The use of illegal acts can therefore be considered part of a discourse of 'transgressive citizenship', through which breaking the law is used to draw attention to what the law promises and to raise the political profile of claims to citizenship rights.

Source: Earle (2012).

In reality, it is often difficult to rigidly categorise social movements, which can begin as protests over something very specific but evolve over time into much larger or broader associations and campaigns. For example, the Gezi Park protests that exploded in Istanbul in May 2013 over propositions to build a shopping mall over Istanbul's Taksim Square soon evolved into a much broader resistance movement (Yel and Nas 2013). The line between an association or trade union and a social movement can also be nebulous, with organisations discussed above such as SEWA and StreetNet straddling the divide. Another example is Shack/Slum Dwellers International (SDI), which grew from a small network of NGOs working on Housing Rights in Asia in the 1980s into a major international coalition working closely with UN-HABITAT and Cities Alliance (Patel et al. 2001), essentially forming a global urban social movement. In 2011, SDI's membership reached a million households and by 2014 it had undertaken slum profiling in almost 7,000 settlements in 115 cities (see http://sdinet.org/sdi-timeline).

In the view of Charles Tilly, another major theorist of social movements, to ask what kinds of organisation count as social movements is to miss the point. Tilly argues that

> it is a mistake to think of a social movement as a group of any kind. Instead, the term *social movement* applies most usefully to a sustained *interaction* between a specific set of authorities and various spokespersons for a given challenge to those authorities. The interaction is a coherent, bounded unit in roughly the same sense that a war or political campaign is a unit.
>
> (Tilly 1984: 305)

Building on this, Tilly and colleagues have developed a range of analytical tools for understanding how social movements and other forms of 'contentious interactions' function, focusing on 'political opportunity structures'. Essentially, aggrieved groups will only engage in collective protests where they perceive that opportunities for change exist for at least one of the following reasons: because they have some kind of access to authorities; because repression is in decline; because elites have become divided or because activists have managed to secure the support of certain influential groups (Orum and Dale 2009: 229). Given a favourable opportunity structure, activist groups have recourse to mechanisms and processes such as 'brokerage, certification, mobilization, demobilization and scale shift' (Tilly and Tarrow 2007: 11). These result in various forms of 'contentious performances' and 'repertoires' that may change over time, and can be more or less violent (Tilly and Tarrow 2007). Another important concept in the social movement literature is that of 'framing', whereby groups involved in social movement interactions actively 'assemble' meaning in order to define particular problems and solutions in ways that fit their goals (Rabrenovic 2009).

In many developing countries, contentious performances in the form of urban protest have been on the rise in recent decades. The frequency and intensity of protest does, however, differ markedly between countries. Recent cross-national research indicates that protest frequency is increased in countries with some combination of large urban populations, sluggish economic growth and unstable political regimes that combine characteristics of both dictatorships and democracies (Blanco and Grier 2009; Machado et al. 2011; Urdal and Hoelscher 2012; Fox 2013).

How different political institutions affect opportunity structures to facilitate or constrain urban protest in developing countries has also been explored through qualitative case study research (Harsch 2009; Rodgers 2010; Goodfellow 2013b; 2014b; Engels 2014). A theme that emerges from some of this work is the way in which particular types of contentious performances become 'repeatable and self-sustaining processes' over time (Orum and Dale 2009: 217), as urban dwellers learn from the 'cumulative experience of dissent' (Harsch 2009: 296). Even urban riots themselves can become routine; Goodfellow (2013b) suggests that a series of urban riots in Uganda in the late 2000s evolved into a 'politics of noise', whereby recourse to rioting was the default mode of political participation for urban dwellers disillusioned with formal governance. Unfortunately, unlike the more concerted collective action associated with social movements, this kind of episodic protest and rioting often achieves little aside from very short-term concessions from local authorities.

In many ways, explaining the emergence of social movements and protest is easier than explaining why they are not more widespread. Given the extent of urban poverty and inequality, one must ask why marginalised and excluded populations don't collectively organise more often. A number of authors have tried to explain quiescence among people who suffer conditions of incredible hardship and miscry in cities (Nelson 1979; Scheper-Hughes 1992; Gilbert 1994; Wood 2003; Goodfellow 2013b). For many low-income urban dwellers collective action is confined to activities in pursuit of making ends meet, such as pooling resources, bulk buying of staple foods, savings clubs, income-generating activities and other self-help initiatives – activities that provide important buffers against the kind of vulnerability and risk described in Chapter 4. Without basic resources and security in place it is difficult if not impossible for poor urban households to sustain self-help and mutual assistance, let alone scaled-up public action. Sometimes collective action around social needs can lead to more politicised forms of organisation, but this is invariably localised and often easily suppressed before an effective social movement can emerge. Not surprisingly, then, collective action at scale can be short-lived and fluid. Ephemeral groups emerge, disappear and re-form themselves but rarely remain decisively above the parapet (Beall 2001). All too often the granting of clientelistic favours by those in power in exchange for votes thwarts the potential for the kinds of urban collection action necessary to generate progressive social change (Tendler 2002; Goodfellow and Titeca 2012).

Plate 8.1 *Police confront members of an urban social movement in São Paulo*

Source: Lucy Earle.

Despite this, the recent surge in protest mobilisation in many parts of the world has attracted considerable interest, with a renaissance of Henri Lefebvre's (1968) idea of the 'right to the city' coming to play a key role in how many such protests have been analysed. Lefebvre conceives of the right to the city as not only a right to participation but the right to appropriation: 'to be physically present in the space of the city' and 'to produce urban space so that it meets the needs of inhabitants' (Purcell 2002: 103). As a Marxist, Lefebvre was concerned that the 'exchange value' of urban space had come to dominate at the expense of the 'use value', especially as real estate speculation and construction came to play an increasing role in the capitalism he was observing in Europe in the 1960s and 1970s. As these processes come to be increasingly important in developing countries too, the relevance of his ideas to the developing world has become increasingly apparent. Urban mobilisation by groups demanding access to services, employment or particular urban spaces can be seen as an effort to claim this 'right' to appropriate urban space and participate in city life in ways that meet their needs and desires. Indeed, the political participation that Lefebvre had in mind is not via the 'invited' spaces created by government (Gaventa 2006) but through disruptive processes that aim to remake the city in new ways that challenge existing institutions and dynamics (Kuymulu 2013b: 926).

In his discussion of democracy in Brazil, Holston (2008) introduces the idea of 'insurgent citizenship' to refer to forms of democratic activism, including protest, that have ushered in new ideas of citizenship and innovative forms of metropolitan governance. Indeed, the 'right to the city' idea has been embraced so widely in Brazil that the City Statute passed in 2001 makes explicit reference to it and attempts to enshrine it in law (Brown 2013). Other middle-income countries such as Turkey have also witnessed major protests that might be seen as 'Lefebvrian' in nature. Kuymulu argues that the 2013 Gezi park protests aimed to reclaim ordinary citizens' right to the 'use value' of the city rather than let this be subsumed by capitalist processes that prioritise 'exchange value' and capital accumulation (Kuymulu 2013a: 276). Meanwhile, the idea of 'the right to the city' became so popular in international urban policy circles in the 2000s that it formed the basis for a major UN-HABITAT-UNESCO joint project on urban policies from 2005 to 2008, before being officially adopted as the theme for UN-HABITAT's World Urban Forum in 2010 (Brown 2013).

The enormous popularity of what is actually a rather complex theoretical concept inevitably poses problems. It has been claimed by such a wide range of actors as to risk undermining its value, and as Harvey (2012: xv) points out, '"the right to the city" is an empty signifier. Everything depends on who gets to fill it with meaning'. Kuymulu (2013b) argues that the appropriation of the term by the UN and other organisations has amounted to translating the concept back into 'exchange value' terms, where claiming the right to the city is re-imagined as a market process involving 'new partnerships and business models that could address the bottom of the economic pyramid' (UN-HABITAT 2010: 72). This usage differs vastly from that adopted by social movements focused on the 'use value' of urban space. The current ubiquity of the 'right to the city' idea has thus arguably created a 'conceptual vortex' capable of pulling in completely opposing political projects (Kuymulu 2013b). Whatever the concepts and words used, it is likely that efforts to encourage formal political participation will be paralleled by disruptive protest the more that people become conscious of rights but feel that formal governance arrangements do help them to realise those rights. Both historically and today, contentious urban politics are part of what has driven development and deepened democracy.

Conclusion

Since the mid-twentieth century, changing ideas about how cities should be planned and governed have in large measure paralleled shifting ideas about international development: early faith in state-directed interventions in the form of urban planning gave way to heavily market-oriented conceptions of urban management, and subsequently a conception of good urban governance linked to transparent, accountable and market-enabling institutions. Alongside this broad trajectory has been a growing concern with bringing government closer to communities through decentralisation reforms and increased citizen participation.

These policy prescriptions, however, can only bring about the transformation that many cities so urgently need when the politics is right, and this is the critical element of urban governance that cannot easily be reformed through policy or foreign assistance. Policies designed to improve the condition of the urban poor, and meaningfully expand citizens' voice in urban affairs, are only likely to result in substantial changes when there is a political coalition with a sustained interest in seeing these transformations through to fruition – as for example has been seen to some degree in parts of Colombia, Brazil, South Africa and India. Without this political impetus, plans, regulations, policies and reform programmes usually remain on paper – although they may often be used by politicians and state officials for public relations purposes and implemented selectively for economic or political gain.

It is therefore crucially important to understand how and why favourable political conditions for urban transformation emerge. Unfortunately, the conceptual and theoretical inheritance from the discipline of urban political science offers only very limited insights for cities at lower levels of development. Here, less well-developed capitalist relations, weakly institutionalised conventional forms of political organisation and – crucially – a diverse range of other, context-specific and often poorly understood informal organisations and institutions necessitate different conceptual frameworks for urban politics. After decades in which politics in the developing world (urban and rural) has been commonly characterised with the catch-all concepts of clientelism and patron–client relations, the richness, complexity and variety of urban politics is only just starting to be debated and theorised.

Summary

- There has been a shift in governing cities in the past few decades, from an emphasis on top-down, centralised planning to a more rounded conceptualisation of multi-stakeholder governance.
- The concept of governance arose due to the perceived need to recognise and institutionalise positive synergies and interactions between states, markets and the wide range of actors that fall into the broad category of 'civil society'.
- Despite its prevalence, the idea of 'good governance' is increasingly critiqued due to its multiple and contested meanings and its association with neoliberal policy prescriptions.
- Efforts to improve urban governance have often revolved around decentralisation. This can take many forms and its 'success' depends on a range of factors including financial and administrative capacity and the degree of political interference from central government.

- Closely linked to decentralisation, citizen participation has become a major theme in governance reform in the late twentieth and early twenty-first century. Participatory approaches have had mixed successes and are sometimes critiqued as being overly technical in nature.
- When exploring governance and politics, Euro-American urban theory cannot easily be transported to cities in developing countries. The variations and contextual specificities of political organisations and urban institutions need to be considered.
- Urban social movements are characterised by their diversity, dynamism and fluidity. While there have been numerous examples of successful, transformative collective action, urban protests are often limited by urban dwellers' immediate needs for survival.
- The recent surge in urban protest mobilisation has been accompanied by a renaissance of the concept of the 'right to the city' in international urban policy circles.

Discussion questions

1. How have approaches to governing cities shifted in the past decades? What is the reasoning behind this?
2. Under what conditions is decentralisation likely to be good for urban development?
3. Discuss the strengths and weaknesses of participatory approaches, such as participatory budgeting, in processes of urban planning and development.
4. When thinking about urban organisations and institutions, what are some of the challenges and limitations of transporting Euro-American urban theory to cities in developing countries?
5. With reference to recent urban social movements, discuss the potential for collective action in cities to bring about transformative political change.

Further reading

Devas, Nick, Amis, Philip, Beall, Jo, Grant, Ursula, Mitlin, Diana, Nunan, Fiona and Rakodi, Carole (2004) *Urban Governance, Voice and Poverty in the Developing World.* London: Earthscan.

Doornbos, M. (2001) '"Good governance": The rise and decline of a policy metaphor?', *Journal of Development Studies*, 37(6), 93–108.

Faguet, J. P. (ed.) (2014) Special Issue of *World Development* on Decentralization and Governance, Vol. 53, 2–13.

Goodfellow, T. (2013) 'The institutionalisation of "noise" and "silence" in urban politics: Riots and compliance in Uganda and Rwanda', *Oxford Development Studies*, 41(4), 436–454.

Pieterse, Edgar (2008) *City Futures: Confronting the Crisis of Urban Development.* London: Zed Books.

Resnick, D. (2014a) 'Urban governance and service delivery in African cities: The role of politics and policies', *Development Policy Review*, 32(s1), s3–s17.

Walton, John (1998) 'Urban conflict and social movements in poor countries: Theory and evidence of collective action', *International Journal of Urban and Regional Research*, 22(3), 460–481.

Websites

Cities Alliance, a global coalition of cities and their development partners: www.citiesalliance.org

Global Development Research Center (GDRC), on Urban Governance: gdrc.org/u-gov/index.html

GSDRC, a partnership of research institutes, think-tanks and consultancy organisations with expertise in governance, social development, humanitarian and conflict issues: www.gsdrc.org

Self-Employed Women's Association (SEWA): www.sewa.org

Shack/Slum Dwellers International (SDI): www.sdinet.co.za

9 Shaping city futures

The first 'urban century' is young and there is much work to be done to ensure that it is remembered as one of exceptional progress in the quest for sustainable human development. There is, however, no blueprint for positive action. In the latter half of the twentieth century, the key intellectual debate shaping development policy was essentially around the appropriate role for states and markets respectively in driving development. While this remains a strong current in the field, there has been a noticeable resurgence of interest in the appropriate geographical scale of intervention, and cities are taking centre stage. Indeed, some have even argued that the nation-state has failed to fulfil its promise and that decisive and constructive political action to address the world's biggest challenges is already emerging among city mayors and networks of urban governments (Barber 2013). This could be the way forward in a world of fractious national and international politics: let cities take the lead.

This is a compelling narrative, but it is also problematic in that it paints an overly simplistic picture of the contemporary urban landscape. All cities are embedded in a unique history and locality, even if they all share the possibilities and challenges associated with urbanism. And each is situated within a multi-tiered political context and susceptible to global economic forces that have differential impacts across the global urban system. Shaping city futures is therefore a process that necessarily involves simultaneous engagement at the local, metropolitan, national and global scales.

At the local level, 'bottom-up' initiatives and processes will always be important pillars of urban development. Where successful, initiatives such as participatory budgeting can play important roles in inculcating local civic cultures of planning, increased interest in processes of urban development and an enhanced public understanding of the trade-offs and conflicts that urban transformation entails. It is also important to build strong local institutions and civic organisations that support community action, safe streets, clear lines of communication between citizens and decision-makers, and robust mechanisms for conflict resolution (over land rights, for example). However, there is a limit to the progress that can be achieved at the very local level if the broader challenge of metropolitan or regional governance is not addressed.

While the number and size of cities has exploded over the past century, systems of governing large cities and city-regions have been slow to adapt in rich and poor countries alike. In many places the physical, functional area of a settlement is institutionally fragmented, with multiple local government authorities (e.g. municipalities) and regional authorities (e.g. water boards, transportation agencies and provincial government) overlapping. These complex mosaics of authority overlaid onto constituencies with often competing interests and visions can result in coordination failures that ultimately undermine the potential dynamism of cities. Getting the best from cities and city-regions requires institutionalising mechanisms of effective coordination, planning, accountability and learning across a diverse range of actors and stakeholders.

This metropolitan or city-regional challenge is often compounded by national policy frameworks that do not provide an enabling environment for effective, inclusive urban governance. Too many national development policies have been insufficiently attentive to the spatial dynamics of development, assuming that structural transformation, innovation or the pursuit of market-driven growth bring with them their own coherent spatial organisation. This, as we have seen, is not necessarily the case: cities grow, transform and advance or hinder development for reasons not factored into economic development plans, whether because of the deep-seated drivers discussed in Chapter 2 or more proximate causes such as conflict and environmental change, as discussed in Chapters 6 and 7. City futures thus require national strategies for responding to and – where appropriate – helping to guide urbanisation and urban growth trajectories. This does not just mean urbanisation policies of the kind that have been used to develop growth poles in the past, but national urban policies that outline what states aspire to do in their cities, and how resources and institutions might best be arranged within them (Parnell and Simon 2014). Simply put, 'cities are far too important to be left entirely to the devices of local government' (ibid.: 238). This does not mean the recentralisation of government but rather the creation of national frameworks that support urban governments while liberating them, along the lines of what has been taking place in recent decades in Colombia and Brazil. The critical challenges facing cities in the developing world this century – from job creation and infrastructure to land tenure and climate change adaptation and mitigation – require sustained, concerted engagement from the centre.

However, even the best national urban policy frameworks are unlikely to prove sufficient in themselves; activities to shape city futures also need to be mainstreamed into the global development agenda. After all, the overwhelming majority of global population growth in coming decades will be in towns and cities in developing regions. To be effective, international engagement with the urban challenge needs to be predicated on a sound understanding of processes of development and cannot be considered aside from the problem of poverty; a seemingly obvious point, but not one that is always at the forefront of the minds of urban planners, designers or policymakers. Simply put, in the twenty-first century city futures cannot be separated conceptually or concretely from the achievement of global development objectives. Addressing this connection at the global level requires that international development agencies – from the World Bank and IMF to bilateral donors and global NGOs – take seriously the challenges and opportunities proliferating in cities as fundamental to the process of development. Yet it also requires that those with expertise in urban planning, governance, environmental management and infrastructure – wherever they may be based in the world – turn some of their attention towards cities that are most in need, the overwhelming majority of which are in less-developed countries.

These processes are now, thankfully, underway. The attention being paid in the international development community to the global urban challenge is much greater than it was just seven years ago when the first edition of this book was published. The inclusion of a holistic 'urban' goal among the Sustainable Development Goals being debated and drafted at the time of writing is a powerful indication of this, as is the recent and notable increase in funding for global urban research. Yet these advances are nowhere near enough: many donors and NGOs struggle to devise urban programmes, unfamiliar with the complex nature of urban poverty, the built environment and urban politics. The international urban policy and planning community, meanwhile, too often export models from rich countries without attention to local context and to what it means to plan cities in developing countries with very different institutional environments.

This latter point highlights an important assumption underpinning this book: that the idea of development still retains validity, and it is therefore still meaningful to distinguish between the 'developed' and 'developing' world. While more of a continuum than a hard line, there are indeed significant divides between global regions not only in terms of income, institutions and wellbeing but academic, policy and planning discourses that render the distinction useful. We are, however, aware of the pitfalls of dividing these 'worlds' (or the global 'North' and 'South') too rigidly. Now more than ever, this divide is contested, as the world's economic core tilts towards countries still generally considered 'developing', and high-income countries still struggle to recover from the shockwaves of the global financial crisis that began in 2008.

The sense of a breaking down of the developed/developing divide also comes at a time when the urban–rural divide is increasingly questioned, with ideas of 'planetary urbanisation' – the idea that the 'urban' now permeates all of society, posing a fundamental challenge to Urban Studies itself – gaining ground (Brenner and Schmid 2011; Catterall 2014). If the idea that somehow 'we are all urban now' might seem a victory for those who for decades have fought to get urban issues on the development agenda, there is ample reason for caution. For one thing, the tenets of the urban on which ideas of 'planetary urbanisation' are based are still very much rooted in rich country experiences, with limited applicability in countries where remote rural areas and cities are frequently still worlds apart and urban institutions often differ fundamentally. It is critical therefore to retain a focus on cities and development, not just on the shiny 'future cities' with 'smart' infrastructure that compete in global city league tables (Watson 2014). Second, it cannot be assumed that the fact urban development has landed on the UN and donor agenda means there will be any kind of consensus on how to address the urban challenge. Indeed, the negotiations leading up to the inclusion of an urban SDG suggest fundamental divides in what it means to put cities on the development agenda (Parnell 2016).

This highlights a critical point: there is no escaping the intensely political nature of urban development at every level. For all the improved management systems, information sharing and technological innovation thrown at cities in the interests of 'smart' and ecological urban development, sustained and inclusive development fundamentally requires governance systems that actively manage conflict, rather than suppressing it through violence or clientelism. Such systems are composites of institutions and organisations that channel contestation through active engagement, provide mechanisms for realising rights, support and amplify innovative solutions to everyday challenges, and facilitate the emergence of shared visions for city life and society at large. And translating these visions into material and social reality requires systems for forward planning. In sum, after several decades in which attention has mostly been devoted to governance-as-management and the creation of space for market-based innovation, politics and planning need to reclaim centre stage in shaping city futures.

There are promising recent examples of successful innovations in urban governance that combine politics and planning in ways that may hold lessons for rich and poor countries alike, most notably in highly urbanised middle-income countries such as Colombia and Brazil. Importantly, these successes have been predicated on some old-fashioned party and union politics as well as the emergence of a consensus among elites that change was needed. In other words, broad-based, organised and sustained collective action to reshape the vision of society was key to successful reform – a perennial theme in both urban history and development studies.

References

Abrahamsen, R., Hubert, D. and Williams, M. C. (2009) Special issue on urban insecurities Introduction. *Security Dialogue*, 40(4–5), 399–418.

Acemoglu, D. and Robinson, J. A. (2012) *Why Nations Fail: The Origins of Power, Prosperity and Poverty*. London: Profile Books.

Ades, Alberto F. and Glaeser, Edward L. (1995) 'Trade and circuses: Explaining urban giants', *The Quarterly Journal of Economics*, 110(1), 195–227.

Adger, N., Hughes, T. P., Folke, C., Carpenter, S. R. and Rockstrom, J. (2005) 'Socio-ecolocigal resilience to coastal disasters', *Science*, 309(5737), 1036–1039.

Aditya, A. and Acharyya, R. (2013) 'Export diversification, composition, and economic growth: Evidence from cross-country analysis', *The Journal of International Trade & Economic Development*, 22(7), 959–992.

Adri, Neelopal (2014) 'Climate induced rural–urban migration in Bangladesh: Experience of migrants in Dhaka City', *International Centre for Climate Change and Development Newsletter*, August. Available at: http://www.icccad.net/blog/climate-induced-rural-urban-migration-in-bangladesh-experience-of-migrants-in-dhaka-city, accessed 13 November 2014.

Agostini, Giulia, Chianese, Francesca, French, William and Sandhu, Amita (2007) 'Understanding the processes of urban violence: An analytical framework'. Report prepared for the Cities and Conflict Theme of the Crisis States Research Centre, London School of Economics.

Albert, I. O. (2007) 'Between the state and transporter unions,' in Laurent Fourchard, (ed.), *Gouverner les villes d'Afrique: état, gouvernement local et acteurs privés*. Paris: Karthala, pp. 125–139.

Alkire, S. and Santos, M. (2014) 'Poverty in the developing world: Robustness and scope of the Multidimensional Poverty Index', *World Development*, 59, 251–274.

Alkire, S., Conconi, A. and Seth, S. (2014) 'Multidimensional Poverty Index 2014: Brief Methodological Note and Results'. OPHI Methodological Notes.

Allmendinger, Philip (2002) *Planning Theory*. London: Palgrave MacMillan.

Alusi, A., Eccles, R. G., Edmondson, A. C. and Zuzul, T. (2011) 'Sustainable cities: Oxymoron or the shape of the future?', *Harvard Business School Organizational Behavior Unit Working Paper*, 11–062.

Anderson, David and Rathbone, Richard (2000) 'Urban Africa: Histories in the making', in David Anderson and Richard Rathbone, (eds), *Africa's Urban Past*. Oxford: James Currey and Heinemann, pp. 1–18.

Andrews, M. (2008) 'The good governance agenda: Beyond indicators without theory', *Oxford Development Studies*, 36(4), 379–417.

Anguelovski, I. and Carmin, J. (2011) 'Something borrowed, everything new: Innovation and institutionalization in urban climate governance', *Current Opinion in Environmental Sustainability*, 3(3), 169–175.

Anjaria, J. (2006) 'Urban calamities: A view from Mumbai', *Space and Culture*, 9(1), 80–82.

Annez, P., Buckley, R. and Kalarickal, J. (2010) 'African urbanization as flight? Some policy implications of geography', *Urban Forum*, 21, 221–234.

Ansari, J. (2004) 'Time for a new approach in India', *Habitat Debate*, 10(4), 15.

Anthopoulos, L. G. and Vakali, A. (2012) 'Urban planning and smart cities: Interrelations and reciprocities,' in Federico Álvarez, (ed.), *The Future Internet*. Berlin Heidelberg: Springer, pp. 178–189.

Appadurai, Arjun (1996) *Modernity at Large: Cultural Dimensions of Globalization*. Minneapolis: University of Minnesota Press.

—— (2006) *Fear of Small Numbers: An Essay on the Geography of Anger*. Durham, North Carolina: Duke University Press.

ARI (2014) *The Booklovers, the Mayors and the Citizens: Participatory Budgeting in Yaoundé Cameroon*. London: Africa Research Institute.

Arndt, H. W. (1987) *Economic Development: The History of an Idea*. Chicago: Chicago University Press.

Arnott, Richard J. and Gersovitz, Mark (1986) 'Social welfare underpinnings of urban bias and unemployment', *The Economic Journal*, 96(382), 413–424.

Arnstein, S. R. (1969) 'A ladder of citizen participation', *Journal of the American Institute of Planners*, 35(4), 216–224.

Audefroy, J. F. (2011) 'Haiti: Post-earthquake lessons learned from traditional construction', *Environment and Urbanization*, 23(2), 447–462.

Aylett, A. (2013) 'The socio-institutional dynamics of urban climate governance: A comparative analysis of innovation and change in Durban (KZN, South Africa) and Portland (OR, USA)', *Urban Studies*, 50(7), 1386–1402.

BBC News (2008) 'Riots prompt Ivory Coast tax cuts', 2 April, available at news.bbc.co.uk/2/hi/africa/7325733.stm, accessed 4 February 2009.

Baabereyir, A., Jewitt, S. and O'Hara, S. (2012) 'Dumping on the poor: The ecological distribution of Accra's solid-waste burden', *Environment and Planning-Part A*, 44(2), 297–314.

Bachrach, P. and Baratz, M. S. (1962) 'Two faces of power', *American Political Science Review*, 56(04), 947–952.

Baffour-Awuah, K. G., Hammond, F. N., Lamond, J. E. and Booth, C. (2014) 'Benefits of urban land use planning in Ghana', *Geoforum*, 51, 37–46.

Bahadur, Aditya and Tanner, Thomas (2014) 'Transformational resilience thinking: Putting people, power and politics at the heart of urban climate resilience', *Environment and Urbanization*, 26, 200–214.

Baiocchi, G. and Ganuza, E. (2014) 'Participatory budgeting as if emancipation mattered', *Politics & Society*, 42(1), 29–50.

Bairoch, Paul (1988) *Cities and Economic Development: From the Dawn of History to the Present*. Chicago: University of Chicago Press.

Banarji, Gautam (1997) *The Impact of Modern Warfare: The Case of Iraq. A City for All, Valuing Difference and Working with Diversity*. London: Zed Books, pp. 194–199.

Banerjee, A. V. and Duflo, E. (2011) *Poor Economics: A Radical Rethinking of the Way to Fight Global Poverty*. New York: PublicAffairs.

Bannister, Jon and Fyfe, Nick (2001) 'Introduction: Fear and the city', *Urban Studies*, 38 (5–6), 807–813.

Baran, P. (1957) *The Political Economy of Growth*. New York: Monthly Review Press.

Barber, B. R. (2013) *If Mayors Ruled the World: Dysfunctional nations, rising cities*. Cambridge: Yale University Press.

Bardhan, P. (2002) 'Decentralization of governance and development', *Journal of Economic Perspectives*, 16(4), 185–205.

Barker, J. (2003) *The No-Nonsense Guide to Terrorism*. London: Verso in association with New Internationalist.

Barnett, Jon and Adger, W. Neil (2007) 'Climate change, human security and violent conflict', *Political Geography*, 26, 639–655.

Barrios, S., Bertinelli, L. and Strobl, E. (2006) 'Climatic change and rural–urban migration: The case of sub-Saharan Africa', *Journal of Urban Economics*, 60(3), 357–371.

Bartlett, Sheridan (2003) 'Water, sanitation and urban children: The need to go beyond "improved" provision', *Environment and Urbanization*, 15(2), October, 56–58.

Barton, Jonathan R. (2013) 'Climate change adaptive capacity in Santiago de Chile: Creating a governance regime for sustainability planning', *International Journal of Urban and Regional Research*, 37(6), 1916–33.

Baruah, B. (2010) 'Energy services for the urban poor: NGO participation in slum electrification in India', *Environment and Planning. C, Government & Policy*, 28(6), 1011.

Basant, Rakesh and Chandra, Pankaj (2007) 'Role of educational and R&D institutions in city clusters: An exploratory study of Bangalore and Pune regions in India', *World Development*, 35(6), 1037–1055.

Bates, R. (1981) *Markets and States in Tropical Africa*. Berkeley: University of California Press.

—— (1983) *Essays on the Political Economy of Rural Africa*. Cambridge: Cambridge University Press.

—— (1988) *Toward a Political Economy of Development*. Berkeley: University of California Press.

Bates, R. H. (2001) *Prosperity and Violence*. London and New York: W.W. Norton.

Batley, Richard (1996) 'Public-private relationships and performance in service provision', *Urban Studies*, 33(4–5), 723–751.

Batley, Richard and Larbi, George (2004) *The Changing Role of Government, The Reform of Public Services in Developing Countries*. Basingstoke and New York: Palgrave Macmillan.

Beall, Jo (1995) 'Social security and social networks among the urban poor in Pakistan', *Habitat International*, 19(4), 427–455.

—— (1996) *Urban Governance: Why Gender Matters*. New York: United Nations Development Programme.

—— (1997) 'Thoughts on poverty from a South Asian rubbish dump: Gender, inequality and waste', *IDS Bulletin*, 28(3), July, 73–90.

—— (2001) 'Valuing social resources or capitalising on them? The limits to pro-poor urban governance in nine cities of the South', *International Planning Studies*, 6(4), 357–375.

—— (2002) 'Living in the present, investing in the future – Household security among the poor', in Carole Rakodi with Tony Lloyd-Jones, (eds), *Urban Livelihoods, A People-Centred Approach to Reducing Poverty*. London: Earthscan Publications Limited, pp. 71–87.

—— (2004) 'Surviving in the city: Livelihoods and linkages of the urban poor', in Nick Devas with Philip Amis, Jo Beall, Ursula Grant, Diana Mitlin, Fiona Nunan and Carole Rakodi, (eds), *Urban Governance, Voice and Poverty in the Developing World*. London: Earthscan Publications Ltd., pp. 53–67.

—— (2005) *Funding Local Governance: Small Grants for Democracy and Development*. London: ITDG Publishing.

—— (2007) 'Cities, terrorism and urban wars of the 21st century', Crisis States Programme Series Two, Working Paper No. 9, London: Crisis States Development Research Centre, Development Studies Institute, London School of Economics, February.

Beall, Jo and Esser, D. (2005) *Shaping Urban Futures: Challenges to Governing and Managing Afghan Cities*. Kabul: Afghanistan Research and Evaluation Unit, March.

Beall, Jo and Goodfellow, Tom (2014) 'Conflict and post-war transition in African cities', in S. Parnell and E. Pieterse, (eds), *Africa's Urban Revolution: Policy Pressures*. London: Zed Books, pp. 18–34.

Beall, Jo, Crankshaw, Owen and Parnell, Susan (2002) *Uniting a Divided City: Governance and Social Exclusion in Johannesburg*. London: Earthscan Publications.

Beall, J., Goodfellow, T. and Putzel, J. (2006) 'Introductory article: On the discourse of terrorism, security and development', *Journal of International Development*, 18(1), January, 51–68.

Beall, J., Goodfellow, T. and Rodgers, D. (2013) 'Cities and conflict in fragile states in the developing world', *Urban Studies*, 50(15), 3065–3083.

Beall, J., Mkhize, S. and Vawda, S. (2005) 'Emergent democracy and "resurgent" tradition: Institutions, chieftaincy and transition in KwaZulu-Natal', *Journal of Southern African Studies*, 31(4), 755–771.

Beall, J., Parnell, S. and Albertyn, C. (2015) 'Elite compacts in Africa: The role of area-based management in the new governmentality of the Durban city-region', *International Journal of Urban and Regional Research*, 39(2), 390–406.

Bebbington, Anthony (1999) 'Capitals and capabilities: A framework for analyzing peasant viability, rural livelihoods and poverty', *World Development*, 27(12), 2021–2044.

Becker, C. M. and Morrison, A. R. (1995) 'The growth of African cities: Theory and estimates', in A. Mafeje and S. Radwan, (eds), *Economic and Demographic Change in Africa*. Oxford: Clarendon Press, pp. 109–142.

Becker, C. M., Hamer, A. M. and Morrison, A. R. (1998) *Beyond Urban Bias in Africa: Urbanization in an Era of Structural Adjustment*. Portsmouth, NH: Heinemann.

Berdegué, J. A., Carriazo, F., Jara, B., Modrego, F. and Soloaga, I. (2015) 'Cities, territories, and inclusive growth: Unraveling urban–rural linkages in Chile, Colombia, and Mexico', *World Development*, 73, 56–71.

Bernstein, Henry (2000) 'Colonialism, capitalism, development', in Tim Allen and Alan Thomas, (eds), *Poverty and Development in the 21st Century*. Oxford: Oxford University Press, pp. 241–270.

Bertaud, A. and Malpezzi, S. (2001) 'Measuring the costs and benefits of urban land use regulation: A simple model with an application to Malaysia', *Journal of Housing Economics*, 10(3), 393–418.

Bhan, G. (2009) '"This is no longer the city I once knew". Evictions, the urban poor and the right to the city in millennial Delhi'. *Environment and Urbanization*, 21(1), 127–142.

Biau, Daniel (2004) 'Making city planning affordable to all countries', *Habitat Debate*, 10(4), 7–11.

Biermann, Frank and Boas, Ingrid (2010) 'Preparing for a warmer world: Towards a global governance system to protect climate refugees', *Global Environmental Politics*, 10(1), 60–88.

Binns, Tony and Etienne Nel (2002) 'Devolving development: Integrated development planning and developmental local government in post-apartheid South Africa', *Regional Studies*, 36(8), 921–945.

Birch, Eugénie L. (2008) *The Urban and Regional Planning Reader*. London and New York: Routledge.

Blaikie, P., Cannon, T., Davis, I. and Wisner, B. (1994) *At Risk: Natural Hazards, People's Vulnerability and Disasters*. Abingdon: Routledge.

Blanco, H., Alberti, M., Olshansky, R., Chang, S., Wheeler, S. M., Randolph, J., et al. (2009) 'Shaken, shrinking, hot, impoverished and informal: Emerging research agendas in planning'. *Progress in Planning*, 72(4), 195–250.

Blanco, L. and Grier, R. (2009) 'Long live democracy: The determinants of political instability in Latin America', *The Journal of Development Studies*, 45(1), 76–95.

Blattman, C. and Miguel, E. (2010) 'Civil war'. *Journal of Economic Literature*, 48(1), 3–57.

Boege, V., Brown, A., Clements, K. and Nolan, A. (2008) *On Hybrid Political Orders and Emerging States: State Formation in the Context of 'Fragility'*. Berlin: Berghof Research Center for Constructive Conflict Management.

Bollens, Scott (2001) 'City and soul – Sarajevo, Johannesburg, Jerusalem, Nicosia', *City*, 5(2), 169–187.

—— (2006) 'Urban planning and peace building', *Progress in Planning*, 66, 67–139.

Bond, Patrick (2000) *Cities of Gold, Townships of Coal: Essays on South Africa's New Urban Crisis*. Trenton, New Jersey: Africa World Press.

Booth, D. and Cammack, D. (2013) *Governance for Development in Africa: Solving Collective Action Problems*. London: Zed Books.

Borja, Jordi and Castells, Manuel (1997) *Local & Global: The Management of Cities in the Information Age*. London: Earthscan Publications Limited.

Boserup, E. (1970) *Women's Role in Economic Development*. London: Earthscan Publications.

Bouquet, E. (2009) 'State-led land reform and local institutional change: Land titles, land markets and tenure security in Mexican communities', *World Development*, 37(8), 1390–1399.

Bouwer, L. M. (2011) 'Have disaster losses increased due to anthropogenic climate change?', *Bulletin of the American Meteorological Society*, 92, 39–46.

Branch, A. (2013) 'Gulu in war… and peace? The town as camp in northern Uganda', *Urban Studies*, 50(15), 3152–3167.

Brand, P. and Dávila, J. D. (2011) 'Mobility innovation at the urban margins: Medellín's Metrocables', *City*, 15(6), 647–661.

Braudel, Fernand (1984) *Civilization and Capitalism, 15th–18th Century*. Vol. 3, *The Perspectives of the World*. New York: Harper & Row.

Bräutigam, D., Fjeldstad, O-H. and Moore, M. (eds) (2008) *Taxation and State Building in Developing Countries*. Cambridge: Cambridge University Press.

Bray, D. (2006) 'Building "community": New strategies of governance in urban China', *Economy and Society*, 35(4), 530–549.

Brecht, Henrike, Deighmann, Uwe and Wang, Hyoung Gun (2013) 'A Global Urban Risk Index', World Bank Policy Research Working Paper No. 6506. Washington DC: The World Bank.

Bredenoord, J., Van Lindert, P. and Smets, P. (2014) *Affordable Housing in the Urban Global South: Seeking Sustainable Solutions*. Abingdon: Routledge.

Brennan-Galvin, Ellen (2002) 'Crime and violence in an urbanizing world,' *Journal of International Affairs*, 56(1), 123–145.

Brenner, N. and Schmid, C. (2011) 'Planetary urbanisation', in Matthew Gandy, (ed.), *Urban Constellations*. Berlin: Jovis, pp. 10–13.

Brett, E. A. (2009) *Reconstructing Development Theory*. New York: Palgrave Macmillan.

Briggs, John (2011) 'The land formalisation process and the peri-urban zone of Dar es Salaam, Tanzania', *Planning Theory and Practice*, 12(1), 131–137.

Briones, M. R. (2011) 'Crossers at crossing: Narratives of work and aspirations of transgender informal workers in Los Baños, Laguna', *Philippine Quarterly of Culture and Society*, 39(1), 1–26.

Bromley, R. (1978) 'Introduction–The urban informal sector: why is it worth discussing?', *World Development*, 6(9), 1033–1039.

Brosio, G. (2000) *Decentralization in Africa*. Washington DC: International Monetary Fund.

Brown, A. (2013) 'The right to the city: Road to Rio 2010', *International Journal of Urban and Regional Research*, 37(3), 957–971.

Brown, Alison and Lloyd-Jones, Tony (2002) 'Spatial planning, access and infrastructure', in Carole Rakodi, (ed.), *Urban Livelihoods: A People-Centred Approach to Reducing Poverty*. London: Earthscan.

Brown, Oli, Hammill, Anne and McLeman, Robert (2007) 'Climate change as the "new" security threat: Implications for Africa', *International Affairs*, (83)6, 1141–1154.

Brülhart, M. and Sbergami, F. (2009) 'Agglomeration and growth: Cross-country evidence', *Journal of Urban Economics*, 65(1), 48–63.

Budds, J. and McGranahan, G. (2003) 'Are the debates on water privatization missing the point? Experiences from Africa, Asia and Latin America', *Environment and Urbanization*, 15(2), 87–114.

Bulkeley, H. (2010) 'Cities and the governing of climate change', *Annual Review of Environment and Resources*, 35(2), 229–253.

——— (2013) *Cities and Climate Change*. London: Routledge.

Bulkeley, H. and Betsill, M. (2005) 'Rethinking sustainable cities: Multilevel governance and the "urban" politics of climate change'. *Environmental politics*, 14(1), 42–63.

Bulkeley, H. and Castán Broto, V. (2013) 'Government by experiment? Global cities and the governing of climate change', *Transactions of the Institute of British Geographers*, 38(3), 361–375.

Bulkeley, H., Broto, V. C. and Edwards, G. (2012) 'Bringing climate change to the city: Towards low carbon urbanism?', *Local Environment*, 17(5), 545–551.

Bunnell, T. and Ann Miller, M. (2011) 'Jakarta in post-Suharto Indonesia: Decentralisation, neo-liberalism and global city aspiration', *Space and Polity*, 15(1), 35–48.

Burgess, R. (1992) 'Helping some to help themselves: Third world housing policies and development strategies', in Kosta Mathey, (ed.), *Beyond Self-Help Housing*. London/New York: Mansell Publishing, pp. 75 91.

Burton, I., Kates, R. W. and White, G. F. (1993) *The Environment as Hazard*. New York: Guilford Press.

Butterworth, Douglas and Chance, John K. (1981) *Latin American Urbanization*. Cambridge: Cambridge University Press.

Buur, L. (2008) 'Democracy and its discontents: Vigilantism, sovereignty and human rights in South Africa', *Review of African Political Economy*, 35(118), 571–584.

Byerlee, D. (1974) 'Rural–urban migration in Africa: Theory, policy and research implications', *International Migration Review*, 8(4), 543–566.

Calame, J. and Charlesworth, E. (2011) *Divided Cities: Belfast, Beirut, Jerusalem, Mostar, and Nicosia*. Philadelphia: University of Pennsylvania Press.

Caldeira, Theresa (2000) *City of Walls: Crime, Segregation and Citizenship in São Paulo*. Berkeley: University of California Press.

Campbell-Lendrum, Diarmid and Corvalan, Carlos (2007) 'Climate change and developing-country cities: Implications for environmental health and equity', *Journal of Urban Health: Bulletin of the New York Academy of Medicine*, 84(1), i109–i117.

Cardoso, F. H. (1972) 'Dependency and development in Latin America', *New Left Review*, 74, 83–95.

Carney, Diana (1998) 'Implementing the sustainable livelihoods approach', in Diana Carney, (ed.), *Sustainable Rural Livelihoods; What Contribution can we Make?* London: Department for International Development, pp. 3–23.

Carroll, T. (2012) 'Illusions of unity: The paradox between mega-sporting events and nation building', *Exchange: The Journal of Public Diplomacy*, 3(1), 2.

Carter, T. R. (1996) 'Assessing climate change adaptations: The IPCC guidelines', in Joel B. Smith, Neeloo Bhatti, Gennady V. Menzhulin, et al. (eds), *Adapting to Climate Change*. New York: Springer, pp. 27–43.

Casanova-Dorotan, F. G. (2010) 'Informal Economy Budget Analysis in Philippines and Quezon City'. WIEGO Working Paper No. 12. Women in Informal Employment: Globalizing and Organizing.

Castells, M. (1972) *The Urban Question*. Cambridge, MA: MIT Press.

—— (1983) *The City and the Grassroots*. Berkeley: University of California Press.

—— (2000) 'Urban sociology in the twenty-first century', in Ida Susser, (ed.), *The Castells Reader on Cities and Social Theory*. Blackwell, Oxford, pp. 390–406.

Castells, Manuel and Alejandro Portes (1989) 'World underneath: The origins, dynamics and effects of the informal economy', in Alejandro Portes, Manuel Castells and Lauren A. Benton, (eds), *The Informal Economy: Studies in Advanced and Less Developed Countries*. Baltimore: John Hopkins University Press, pp. 11–40.

Castro, José Esteban (2005) 'Water-borne diseases', in Tim Forsyth, (ed.), *Encyclopedia of International Development*. London and New York: Routledge, pp. 751–752.

Catterall, B. (2014) 'Towards the Great Transformation': (11) Where/what is culture in '"Planetary Urbanisation"? Towards a new paradigm', *City*, 18(3), 368–379.

Centner, R. (2012) 'Techniques of absence in participatory budgeting: Space, difference and governmentality across Buenos Aires', *Bulletin of Latin American Research*, 31(2), 142–159.

Cervero, Robert and Golub, Aaron (2007) 'Informal transport: A global perspective', *Transport Policy*, 14 (6), 445–457.

Chakrabarti, V. (2013) *A Country of Cities: A Manifesto for an Urban America*. New York: Metropolis Books.

Chakravorty, S. (2000) 'From colonial city to globalizing city? The far-from-complete spatial transformation of Calcutta', in P. Marcuse and R. van Kempen, (eds), *Globalizing Cities: A New Spatial Order?* London and Cambridge: Blackwell, pp. 56–77.

Chambers, R. (1983) *Rural Development: Putting the Last First*. Harlow, Essex: Longmans.

—— (1992) 'Rural appraisal: Rapid, relaxed and participatory', IDS Discussion Paper 311, Institute of Development Studies, Sussex, UK.

—— (1994) 'The origins and practice of participatory rural appraisal', *World Development*, 22(7), 953–969.

—— (2004) 'Ideas for development: Reflecting forwards', IDS Working paper 238, Brighton: IDS.

Chambers, Robert and Gordon Conway (1992) 'Sustainable rural livelihoods: Practical concepts for the 21st century', IDS Discussion Paper 296, Brighton: Institute of Development Studies.

Chandhoke, N., Priyadarshi, P., Tyagi, S. and Khanna, N. (2007) 'The displaced of Ahmedabad', *Economic and Political Weekly*, 42(43), 10–14.

Chang, Ha-Joon (2003) *Globalisation, Economic Development and the Role of the State*. London: Zed Books.

Charlton, S. (2013) 'State ambitions and peoples' practices: An exploration of RDP housing in Johannesburg' (Doctoral dissertation, University of Sheffield).

Chen, Martha Alter (2006) 'Rethinking the informal economy: Linkages with the formal economy and the formal regulatory environment', in Basudeb Guha-Khasnobis, Ravi Kanbur and Elinor Ostrom, (eds), *Linking the Formal and Informal Economy: Concepts and Policies*. Oxford: Oxford University Press, pp. 75–92.

—— (2012) 'The informal economy: Definitions, theories and policies. Women in informal economy globalizing and organizing', WIEGO Working Paper, 1–26.

Childe, G. V. (1950) 'The urban revolution', *Town Planning Review*, 21(1), 3–17.

Christian Aid (2007) *Human Tide: The Real Migration Crisis*. Christian Aid Report.

Christian, T. J. (2012) 'Trade-offs between commuting time and health-related activities', *Journal of Urban Health*, 89(5), 746–757.

Christiansen, L. and Todo, Y. (2013) 'Poverty reduction during the rural urban transformation: The role of the missing middle', Policy Research Working Paper 6445. The World Bank.

Cleaver, F. (2001) 'Institutions, agency and the limitations of participatory approaches to development', in B. Cooke and U. Kothari, (eds), *Participation: The New Tyranny?* London: Zed Books, pp. 36–55.

Cohen, B. (2004) 'Urban growth in developing countries: A review of current trends and a caution regarding existing forecasts', *World Development*, 32(1), 23–51.

Collier, P. (2007) *The Bottom Billion: Why the Poorest Countries are Failing and What Can Be Done About It*. Oxford: Oxford University Press.

Collier, P. and Hoeffler, A. (2004) 'Greed and grievance in civil war', *Oxford Economic Papers*, 56(4), 563–595.

Connelly, Steve (2007) 'Mapping sustainable development as a contested concept', *Local Environment: The International Journal of Justice and Sustainability*, 12(3), 259–278.

Conyers, D. (1983) 'Decentralization: The latest fashion in development administration?', *Public Administration and Development*, 3, 97–109.

—— (2007) 'Decentralisation and service delivery: Lessons from sub-Saharan Africa', *IDS Bulletin*, 38(1), Institute of Development Studies, Sussex, UK.

Conyers, D. and Hills, P. (1984) *An Introduction to Development Planning in the Third World*. Chichester: Wiley.

Cooke, B. and Kothari, U. (2001) *Participation: The New Tyranny?* London: Zed Books.

Cooper, Fredrick (2002) *Africa since 1940: The Past of the Present*. New York: Cambridge University Press.

Cooper-Knock, S. J. (2014) 'Policing in intimate crowds: Moving beyond 'the mob' in South Africa', *African Affairs* 113(453), 563–582.

Cornia, Giovanni Andrea, Jolly, Richard and Stewart, Frances (eds) (1987) *Adjustment with a Human Face*, Volume 1. Oxford: Oxford University Press.

Cornwall, A. and Nyamu-Musembi, C. (2004) 'Putting the "rights-based approach" to development into perspective', *Third World Quarterly*, 25(8), 1415–1437.

Coutard, O. (2008) 'Placing splintering urbanism: Introduction', *Geoforum*, 39(6), 1815–1820.

Coward, M. (2008) *Urbicide: The Politics of Urban Destruction* (Vol. 66). Abingdon: Routledge.

—— (2009) 'Network-centric violence, critical infrastructure and the urbanization of security', *Security Dialogue*, 40(4–5), 399–418.

Cowen, M. and Shenton, R. (1996) *Doctrines of Development*. London and New York: Routledge.

Cox, K. (1978) 'Local interests and urban political processes in market societies', in K. R. Cox, (ed.), *Urbanization and Conflict in Market Societies*. Chicago: Maarofa Press, pp. 94–108.

Cramer, C. (2006) *Civil War is Not a Stupid Thing. Accounting for Violence in Developing Countries*. London: Hurst & Company.

Crankshaw, Owen, Gilbert, Alan and Morris, Alan (2000) 'Backyard Soweto', *International Journal of Urban and Regional Research*, 24(4), 841–857.

Crawford, C. and Bell, S. (2012) 'Analysing the relationship between urban livelihoods and water infrastructure in three settlements in Cusco, Peru', *Urban Studies*, 49(5), 1045–1064.

Creutzig, F., Baiocchi, G., Bierkandt, R., Pichler, P. P. and Seto, K. C. (2015) 'Global typology of urban energy use and potentials for an urbanization mitigation wedge', *Proceedings of the National Academy of Sciences*, 112(20), 6283–6288.

Critelli, F. M. (2010) 'Women's rights = Human rights: Pakistani women against gender violence', *J. Soc. & Soc. Welfare*, 37, 135.

Crook, Richard C. (2003) 'Decentralisation and poverty reduction in Africa: The politics of local–central relations', *Public Administration and Development*, 23(1): 77–88.

Crook, R. C. and Manor, J. (1998) *Democracy and Decentralisation in South Asia and West Africa: Participation, Accountability and Performance*. Cambridge: Cambridge University Press.

Crook, Richard and Manor, James (2000) 'Democratic decentralization', Operations Evaluation Department Working Paper. The World Bank, Washington DC.

Cugurullo, F. (2013) 'How to build a sandcastle: An analysis of the genesis and development of Masdar City', *Journal of Urban Technology*, 20(1), 23–37.

D'Cruz, Celine and Satterthwaite, David (2005) 'Building homes, changing official approaches: The work of urban poor organizations and their federations and their contributions to meeting the millennium development goals in urban areas', International Institute for Environment and Development, Poverty Reduction in Urban Areas Series, working paper 16. May 2005.

Davis, J. C. and J. V. Henderson (2003) 'Evidence on the political economy of the urbanization process', *Journal of Urban Economics*, 53, 98–125.

Davis, K. (1965) 'The urbanization of the human population', *Scientific American*, 213 (September), 40–53.

Davis, M. (2006) *Planet of Slums*. London: Verso.

De Boeck, F. and Plissart, M. F. (2005) *Kinshasa: Tales of the Invisible City*. Leuven: Leuven University Press.

de Soto, Hernando (1989) *The Other Path: The Economic Answer to Terrorism*. New York: Basic Books.

—— (2000) *The Mystery of Capital*. New York: Basic Books.

Denzin, N. K. (2009) 'The elephant in the living room: Or extending the conversation about the politics of evidence', *Qualitative Research*, 9(2), 139–160.

Devas, Nick (2001) 'Who runs cities? The relationship between urban governance, service delivery and poverty', University of Birmingham, Theme Paper 4, Urban Governance, Poverty and Partnerships programme.

—— (2003) 'Can city governments in the south deliver for the poor?: A municipal finance perspective', *International Development Planning Review*, 25(1), 1–29.

Devas, Nick, Amis, Philip, Beall, Jo, Grant, Ursula, Mitlin, Diana, Nunan, Fiona and Rakodi, Carole (2004) *Urban Governance, Voice and Poverty in the Developing World*. London: Earthscan Publications Ltd.

Dillinger, William (1994) *Decentralization and its Implications for Urban Service Delivery*. Urban Management Programme: UNDP/UNCHS/World Bank.

Dodman, David (2009) 'Blaming cities for climate change? An analysis of urban greenhouse gas emissions inventories', *Environment and Urbanization*, 21(1), 185–201.

Dodman, D. and Satterthwaite, D. (2008) 'Institutional capacity, climate change adaptation and the urban poor', *IDS Bulletin*, 39(4), 67–74.

Donkor, Kwabena (2002) 'Structural adjustment and mass poverty in Ghana', in Peter Townsend and David Gordon, (eds), *World Poverty, New Policies to Defeat an Old Enemy*. Bristol: The Policy Press, pp. 197–232.

Doornbos, M. (2001) '"Good governance": The rise and decline of a policy metaphor?', *Journal of Development Studies*, 37(6), 93–108.

Drakakis-Smith, David (2000) *Third World Cities* (second edition). London: Routledge.

Drechsel, Pay, Scott, Christopher A., Raschid-Sally, Liqa, Redwood, Mark and Bahri, Akica. (2010) *Wastewater Irrigation and Health: Assessing and Mitigating Risk in Low-Income Countries*. London: Earthscan with the International Development Research Centre and the International Water Management Institute.

Dryzek, J. S. (2000) *Deliberative Democracy and Beyond: Liberals, Critics, Contestations*. Oxford: Oxford University Press.

Duffield, M. R. (2001) *Global Governance and the New Wars: The Merging of Development and Ssecurity* (Vol. 87). London: Zed books.

Duflo, E., Galiani, S. and Mobarak, M. (2012) 'Improving access to urban services for the poor: Open issues and a framework for a future research agenda', J-PAL Urban Services Review Paper. Cambridge, MA: Abdul Latif Jameel Poverty Action Lab.

Dugard, J. (2001) *From Low-intensity War to Mafia War: Taxi Violence in South Africa, 1987–2000* (Vol. 4). Johannesburg, South Africa: Centre for the Study of Violence and Reconciliation.

Dupont, V. D. (2011) 'The dream of Delhi as a global city', *International Journal of Urban and Regional Research*, 35(3), 533–554.

Durand-Lasserve, A. and Royston, L. (ed.) (2002) *Holding their Ground: Secure Land Tenure for the Urban Poor in Developing Countries*. London: Earthscan.

Duranton, G. (2014) 'Growing through cities in developing countries', *The World Bank Research Observer*, 30(1), 39–73.

Duranton, Gilles and Puga, Diego (2001) 'Nursery cities: Urban diversity, process innovation, and the life cycle of products', *American Economic Review*, 91(5), 1454–1477.

—— (2004) 'Micro-foundations of urban agglomeration economies', *Handbook of regional and urban economics*, 4, 2063–2117.

Dyson, Tim (2001) 'A partial theory of world development: The neglected role of the demographic transition in the shaping of modern society', *International Journal of Population Geography*, 7, 1–24.

Earle, L. (2012) 'From insurgent to transgressive citizenship: Housing, social movements and the politics of rights in São Paulo', *Journal of Latin American Studies*, 44(01), 97–126.

—— (2014) 'Stepping out of the twilight? Assessing the governance implications of land titling and regularization programmes', *International Journal of Urban and Regional Research*, 38(2), 628–645.

Easterly, W. (2001) 'The lost decades: Developing countries' stagnation in spite of policy reform, 1980–1998', *Journal of Economic Growth*, 6, 135–157.

—— (2006) *The White Man's Burden: Why the West's Efforts to Aid the Rest Have Done So Much Ill and So Little Good*. Oxford: Oxford University Press.

Eaton, Kent, Kaiser, Kai and Smoke, Paul J. (2011) *The Political Economy of Decentralization Reforms: Implications for Aid Effectiveness*. Washington DC: The World Bank.

Edensor, T. and Jayne, M. (2012) *Urban Theory Beyond the West: A World of Cities*. London: Routledge.

Ejigu, A. G. (2012) 'Socio-spatial tensions and interactions: An ethnography of the condominium housing of Addis Ababa, Ethiopia', in Mélanie Robertson, (ed.), *Sustainable Cities: Local Solutions in the Global South*. 97–112

Elkin, S. I. (1987) *City and Regime in the American Republic*. Chicago: University of Chicago Press.

Elson, D. (ed.) (1991) *Male Bias in the Development Process*. Manchester: Manchester University Press.

Engels, B. (2014) 'Contentious politics of scale: The global food price crisis and local protest in Burkina Faso', *Social Movement Studies*, 14(2), 180–194.

Engels, Friedrich (1845) 'The Condition of the Working Class in England', Marx/Engels Internet archive. http://www.marxists.org/archive/marx/works/1845/condition-working-class/index.htm, accessed 4 February 2009.

Erdman, G. and Engel, U. (2007) 'Neopatrimonialism reconsidered: Critical review and elaboration of an elusive concept', *Commonwealth and Comparative Politics* 45(1), 95–119.

Escobar, A. (1992) 'Reflections on "development": grassroots approaches and alternative politics in the Third World', *Futures*, 24(5), 411–436.

—— (1995) *Encountering Development: The Making and Unmaking of the Third World*. Princeton, New Jersey: Princeton University Press.

—— (2004) 'Development, violence and the new imperial order', *Development*, 47(1), 15–21.

Esser, D. (2013) 'The political economy of post-invasion Kabul, Afghanistan: Urban restructuring beyond the North–South divide', *Urban Studies*, 50(15), 3084–3098.

Estache, A. and Wren-Lewis, L. (2009) 'Toward a theory of regulation for developing countries: Following Jean-Jacques Laffont's lead', *Journal of Economic Literature*, 47(3), 729–770.

Evans, Barbara (2005) *Securing Sanitation: The Compelling Case to Address the Crisis*. Stockholm: Stockholm International Water Institute (SIWI).

Everett, M. (2001) 'Evictions and human rights: Land disputes in Bogota, Colombia', *Habitat International*, 25, 453–471.

Faguet, Jean-Paul (2003) 'Decentralization and local government in Bolivia: An overview from the bottom up', Crisis States Programme Series One, Working Paper No. 29, London: Crisis States Research Centre, Development Studies Institute, London School of Economics, May.

Faguet, Jean-Paul (2004) 'Does decentralization increase government responsiveness to local needs? Evidence from Bolivia', *Journal of Public Economics*, 88(3–4), 867–893.

—— (2014) 'Decentralization and governance', *World Development*, 53, 2–13.

Faguet, J. P. and Sanchez, F. (2008) 'Decentralization's effects on educational outcomes in Bolivia and Colombia', *World Development*, 36(7), 1294–1316.

Fainstein, N. I. and Fainstein, S. S. (1983) *Regime Strategies, Communal Resistance, and Economic Forces*. New York: Longman.

Fanon, F. (1965) *The Wretched of the Earth* (Vol. 390). New York: Grove Press.

FAO (2002) 'Land tenure and rural development', FAO Land Tenure Series 3. Rome: Food and Agricultural Organization, United Nations.

Farvacque, Catherine and McAuslan, Patrick (1992) *Reforming Urban Land Policies and Institutions in Developing Countries*. Washington DC: Urban Management Programme.

Fay, Marianne and Opal, Charlotte (2000) 'Urbanization without Growth: A not-so-uncommon phenomenon', World Bank Policy Research Working Paper No. 2412, Washington DC: The World Bank.

Fearon, J. D. and Laitin, D. D. (2003) 'Ethnicity, insurgency, and civil war', *American Political Science Review*, 97(01), 75–90.

Fehling, M., Nelson, B. D. and Venkatapuram, S. (2013) 'Limitations of the millennium development goals: A literature review', *Global Public Health*, 8(10), 1109–1122.

Feige, Edgar L. (1990) 'Defining and estimating underground and informal economies: The new institutional economics approach', *World Development*, 18(7), 989–1002.

Ferguson, James (1990) *The Anti-Politics Machine: 'Development',
Depoliticization and Bureaucratic Power in Lesotho*. Cambridge:
Cambridge University Press.

Fischel, William A. (2004) 'An economic history of zoning and a cure for its
exclusionary effects', *Urban Studies*, 41(2), 317–340.

Fjeldstad, O-H. (2006) *Local Revenue Mobilization in Urban Settings in Africa*,
Bergen Chr. Michelsen Institute, WP 2006: 15.

Fjeldstad, O-H. and Heggstad, K. (2011) 'Local government revenue
mobilisation in Anglophone Africa Brighton Institute for Development
Studies', Working Paper EP1.

Folke, C. (2006) 'Resilience: The emergence of a perspective for social–
ecological systems analyses', *Global Environmental Change*, 16(3),
253–267.

Forester, J. (1999) *The Deliberative Practitioner: Encouraging Participatory
Planning Processes*. Cambridge, MA: MIT Press.

Fox, S. (2012) 'Urbanization as a global historical process: Theory and evidence
from sub-Saharan Africa', *Population and Development Review*, 38(2),
285–310.

—— (2013) 'The political economy of urbanisation and development in sub-
Saharan Africa', PhD thesis. The London School of Economics and
Political Science.

—— (2014) 'The political economy of slums: Theory and evidence from sub-
Saharan Africa', *World Development*, 54, 191–203.

Fox, S. and Beall, J. (2012) 'Mitigating conflict and violence in African cities',
Environment and Planning C: Government and Policy, 30(6), 968–981.

Fox, S. and Hoelscher, K. (2012) 'Political order, development and social
violence', *Journal of Peace Research*, 49(3), 431–444.

Frank, A. G. (1967) *Capitalism and Underdevelopment in Latin America:
Historical Studies of Chile and Brazil*, New York: Monthly Review Press.

Frankenhoff, C. A. (1967) 'Elements of an economic model for slums in a
developing economy', *Economic Development and Cultural Change*, 16(1),
27–36.

Freund, Bill (2007) *The African City*. Cambridge: Cambridge University Press.

Friedmann, J. (1967) 'Regional planning and nation-building: An agenda for
international research', *Economic Development and Cultural Change*,
16(1), 119–129.

—— (1986) 'The World City Hypothesis', *Development and Change*, 17,
69–84.

—— (2005) *China's Urban Transition*. Minneapolis: University of Minnesota
Press.

—— (2007) 'The wealth of cities: Towards an assets-based development of
newly urbanizing regions', *Development and Change*, 38(6), 987–998.

Friedmann, J. and Alonso, W. (eds) (1975) *Regional Policy: Readings in Theory
and Applications*. Cambridge, MA: MIT Press.

Friedmann, John and Wolff, G. (1982) 'World city formation: An agenda for
research and action', *International Journal of Urban and Regional
Research*, 15(1), 269–283.

Fung, A., Wright, E. O. and Abers, R. (2003) *Deepening Democracy: Institutional Innovations in Empowered Participatory Governance* (Vol. 4). London: Verso.

Gakenheimer, R. (1999) 'Urban mobility in the developing world', *Transportation Research Part A: Policy and Practice*, 33(7), 671–689.

Galiani, S. and Schargrodsky, E. (2010) 'Property rights for the poor: Effects of land titling', *Journal of Public Economics*, 94(9), 700–729.

Gallin, D. (2001) 'Propositions on trade unions and informal employment in times of globalisation', *Antipode*, 33(3), 531–549.

Galor, Oded (2005) 'The demographic transition and the emergence of sustained economic growth', *Journal of the European Economic Association*, 3(2–3), 494–504.

Galtung, J. (1969) 'Violence, peace, and peace research', *Journal of Peace Research*, 6(3), 167–191.

Gandy, M. (2004) 'Rethinking urban metabolism: Water, space and the modern city', *City*, 8(3), 363–379.

Gans, Herbert J. (1963) 'Social and physical planning for the elimination of urban poverty', *Washington University Law Quarterly*, 2, 2–18.

Garman, C., Haggard, S. and Willis, E. J. (2001) 'Fiscal decentralisation: A political theory with Latin American cases', *World Politics*, 53(2), 205–36.

Gaventa, J. (2006) 'Finding the spaces for change: A power analysis', *IDS Bulletin*, 27(6), 23–33.

GaWC (2012) 'The World According to GaWC 2012'. Available online: http://www.lboro.ac.uk/gawc/world2012.html, accessed 11 July 2015.

Gazdar, H. and Mallah, H. B. (2013) 'Informality and political violence in Karachi', *Urban Studies*, 50(15), 3099–3115.

Gazdar, H., Kaker, S. A. and Khan, I. (2010) 'Buffer zone, colonial enclave or urban hub? Quetta: Between four regions and two wars', Working Paper No. 69, Series 2, Crisis States Research Centre, London School of Economics and Political Science.

Gilbert, Alan (1994) *The Latin American City*. Nottingham: Russell Press.
—— (2002) 'On the mystery of capital and the myths of Hernando de Soto: What difference does legal title make?', *International Development Planning Review*, 24(1), 1–19.
—— (2007) 'The return of the slum: Does language matter?', *International Journal of Urban and Regional Research*, 31(4), 697–713.

Gilbert, Alan and Crankshaw, Owen (1999) 'Comparing South African and Latin American experience: Migration and housing mobility', *Urban Studies*, 36(1), 2375–2400.

Giustozzi, Antonio (2009) 'The eye of the storm: Cities in the vortex of Afghanistan's civil wars', Crisis States Research Centre working papers series 2, 62. Crisis States Research Centre, London School of Economics and Political Science, London, UK.

Glaeser, Edward (2011) *Triumph of the City*. London: Pan Macmillan.

Glaeser, Edward and Kahn, Matthew (2009) 'The greenness of cities: Carbon dioxide emissions and urban development', *Journal of Urban Economics*, 67(3), 404–418.

Glaeser, Edward L. and Sacerdote, Bruce (1999) 'Why Is there more crime in cities?', *The Journal of Political Economy*, 1076), Part 2: Symposium on the Economic Analysis of Social Behavior in Honor of Gary S. Becker, S225–S258.

Glaeser, E. L., Kallal, H. D., Scheinkman, J. A. and A. Schleifer (1992) 'Growth in cities', *Journal of Political Economy*, 100, 1126–1152.

Godard, X. (2011) 'Poverty and urban mobility: Diagnosis toward a new understanding', in H. T. Dimitriou and R. Gakenheimer, (eds), *Urban Transport in the Developing World: A Handbook of Policy and Practice.* Cheltenham: Edward Elgar, pp. 232–261.

—— (2013) 'Comparisons of urban transport sustainability: Lessons from West and North Africa', *Research in Transportation Economics*, 40(1), 96–103.

Godschalk, D. R. (2003) 'Urban hazard mitigation: Creating resilient cities', *Natural Hazards Review*, 4(3), 136–143.

Goldstein, D. (2004) *The Spectacular City: Violence and Performance in Urban Bolivia.* Durham: Duke University Press Books.

Gómez-Ibáñez, J. A. and Meyer, J. R. (1993) *Going Private: The International Experience with Transport Privatization.* Washington DC: The Brookings Institute.

Goodfellow, T. (2010) '"The bastard child of nobody"? Anti-planning and the institutional crisis in contemporary Kampala', Crisis States Research Centre Working Paper 67.2, London School of Economic and Political Science.

—— (2012) 'State effectiveness and the politics of urban development in East Africa: A puzzle of two cities, 2000–2010' (Doctoral dissertation, The London School of Economics and Political Science).

—— (2013a) 'Planning and development regulation amid rapid urban growth: Explaining divergent trajectories in Africa', *Geoforum*, 48, 83–93.

—— (2013b) 'The institutionalisation of "noise" and "silence" in urban politics: Riots and compliance in Uganda and Rwanda', *Oxford Development Studies*, 41(4), 436–454.

—— (2013c) 'Urban planning in Africa and the politics of implementation: Contrasting patterns of state intervention in Kampala and Kigali', in V. Arlt, E. Macamo and B. Obrist, (eds), *Living the City in Africa: Processes of Invention and Intervention.* Berlin: Lit Verlag, pp. 45–62.

—— (2014a) 'Rwanda's political settlement and the urban transition: Expropriation, construction and taxation in Kigali', *Journal of Eastern African Studies*, 8(2), 311–329.

—— (2014b) 'Legal manoeuvres and violence: Law making, protest and semi-authoritarianism in Uganda', *Development and Change*, 45(4), 753–776.

—— (2015a) 'Taming the "rogue" sector: Studying state effectiveness in Africa through informal transport politics', *Comparative Politics*, 47(2), 127–147.

—— (2015b) 'Taxing the urban boom: Property taxation and land leasing in Kigali and Addis Ababa', ICTD Working Paper 38. Brighton: Institute for Development Studies.

Goodfellow, T. and Lindemann, S. (2013) 'The clash of institutions: Traditional authority, conflict and the failure of "hybridity" in Buganda', *Commonwealth & Comparative Politics*, 51(1), 3–26.

Goodfellow, T. and Smith, A. (2013) 'From urban catastrophe to "model" city? Politics, security and development in post-conflict Kigali', *Urban Studies*, 50(15), 3185–3202.

Goodfellow, T. and Titeca, K. (2012) 'Presidential intervention and the changing "politics of survival" in Kampala's informal economy', *Cities*, 29(4), 264–270.

Gore, C. D. and Muwanga, N. K. (2014) 'Decentralization is dead, long live decentralization! Capital city reform and political rights in Kampala, Uganda', *International Journal of Urban and Regional Research*, 38(6), 2201–2216.

GPI (2015) *Global Peace Index (2015)*. Sydney: Institute for Economics and Peace.

Graham, Stephen (2003) 'Lessons in urbicide', *New Left Review*, 19 (January–February), 63–78.

—— (ed.) (2004) *Cities, War, and Terrorism: Towards an Urban Geopolitics*. Oxford: Blackwell Publishing Ltd.

—— (2010) *Cities under Siege: The New Military Urbanism*. London: Verso Books.

Graham, Stephen and Marvin, Simon (2001) *Splintering Urbanism: Networked Infrastructures, Technological Mobilities and the Urban Condition*. London: Routledge.

Green, E. (2010) 'Patronage, district creation, and reform in Uganda', *Studies in Comparative International Development*, 45(1), 83–103.

Grindle, Merilee S. (2007) *Decentralization, Democratization, and the Promise of Good Governance*. Princeton, NJ: Princeton University Press.

Gugler, Josef (ed.) (2004) *World Cities beyond the West: Globalization, Development and Inequality*. Cambridge: Cambridge University Press.

Guijt, I. and Shah, M. K. (eds) (1998) *The Myth of Community: Gender Issues in Participatory Development*. London: Intermediate technology publications.

Gulyani, S. and Talukdar, D. (2008) 'Slum real estate: The low-quality high-price puzzle in Nairobi's slum rental market and its implications for theory and practice', *World Development*, 36(10), 1916–1937.

—— (2010) 'Inside informality: The links between poverty, microenterprises, and living conditions in Nairobi's slums', *World Development*, 38, 1710–1726.

Gutiérrez, F., Pinto, M., Arenas, J. C., Guzmán, T. and Gutiérrez, M. (2013) 'The importance of political coalitions in the successful reduction of violence in Colombian cities', *Urban Studies*, 50(15), 3134–3151.

Habitat, U. N. (2005) *Financing Urban Shelter: Global Report on Human Settlements 2005*. London: Earthscan and UN Habitat.

Haddad, Lawrence, Ruel, Marie T. and Garrett, James L. (1999) 'Are urban poverty and undernutrition growing? Some newly assembled evidence', *World Development*, 27(11), 1891–1904.

Hagedorn, J. (2007) *Gangs in the Global City: Alternatives to Traditional Criminology*. Urbana and Chicago: University of Illinois Press.

Hägerstrand, Torsten (1978) 'An equitable urban structure', in L.S. Bourne and J.W. Simmons, (eds), *Systems of Cities: Readings on Structure, Growth and Policy*. New York: Oxford University Press, pp. 556–561.

Hajat, S., Armstrong, B. G., Gouveia, N. and Wilkinson, P. (2005) 'Mortality displacement of heat-related deaths: A comparison of Delhi, São Paulo and London', *Epidemiology*, 16, 613–620.

Hall, Peter (1966) *The World Cities*. London: Weidenfeld and Nicolson.

—— (1998) *Cities in Civilization*. London: Weidenfeld and Nicolson.

—— (2002) *Cities of Tomorrow* (Third Edition). Oxford: Blackwell.

Handzic, K. (2010) 'Is legalized land tenure necessary in slum upgrading? Learning from Rio's land tenure policies in the Favela Bairro Program', *Habitat International*, 34(1), 11–17.

Hanson, Gordon H. (2001) 'U.S.–Mexico integration and regional economies: Evidence from border-city Pairs', *Journal of Urban Economics*, 50(2), 259–287.

Harbom, L. and Wallensteen, P. (2009) 'Armed conflict, 1946–2008', *Journal of Peace Research*, 46(4), 477–487.

Hardoy, Jorge and Satterthwaite, David (1989) *Squatter Citizen, Life in the Urban Third World*. London: Earthscan Publications Limited.

Hardoy, Jorge, Mitlin, Diana and Satterthwaite, David (2001) *Environmental Problems in an Urbanizing World: Finding Solutions in Cities in Africa, Asia and Latin America*. London: Earthscan Publications.

Harms, E. (2012) 'Beauty as control in the new Saigon: Eviction, new urban zones, and atomized dissent in a Southeast Asian city', *American Ethnologist*, 39(4), 735–750.

Harris, J. R. and Todaro, M. (1970) 'Migration, unemployment and development: A two-sector analysis', *American Economic Review*, 60, March, 126–142.

Harrison, Philip (2008) 'The origins and outcomes of South Africa's integrated development plans', in Mirjam van Donk, Mark Swilling, Edgar Pieterse and Susan Parnell, (eds), *Consolidating Developmental Local Government, Lessons from the South African Experience*. Cape Town: UCT Press, pp. 321–337.

Harriss, Barbara (1978) 'Quasi-formal employment structures and behaviour in the unorganized urban economy, and the reverse: Some evidence from South India', *World Development*, 6 (9–10), 1077–1086.

Harriss, J. (2009) 'Compromised democracy: Observations on popular democratic representation from urban India', in O. Törnquist, N. Webster and K. Stokke, (eds), *Rethinking Popular Representation*. New York: Palgrave Macmillan, pp. 161–178.

Harriss, J., Stokke, K. and Törnquist, O. (2004) *Politicising Democracy: The New Local Politics of Democratization*. Houndmills: Palgrave Macmillan.

Harsch, E. (2009) 'Urban protest in Burkina Faso', *African Affairs*, 108(431), 263–288.

Hart, Keith (1973) 'Informal income opportunities and urban employment in Ghana', *The Journal of Modern African Studies*, 11, 61–89.

Harvey, D. (1973) *Social Justice and the City* (Vol. 1). Oxford: Blackwell Publishers Ltd.

—— (1989) 'From managerialism to entrepreneurialism: The transformation in urban governance in late capitalism', *Geografiska Annaler. Series B. Human Geography*, 71(1), 3–17.

—— (2012) *Rebel Cities: From the Right to the City to the Urban Revolution*. London: Verso Books.

He, S. (2010) 'New-build gentrification in Central Shanghai: Demographic changes and socioeconomic implications', *Population, Space and Place*, 16(5), 345–361.

Healey, Patsy (1997) *Collaborative Planning: Shaping Places in Fragmented Societies*. Vancouver: University of British Columbia Press.

—— (2004) 'Creativity and urban governance', *Policy Studies*, 25(2), 87–102.

Heid, B., Larch, M. and Riaño, A. (2013) 'The rise of the Maquiladoras: A mixed blessing', *Review of Development Economics*, 17(2), 252–267.

Heinrichs, Dirk, Krellenberg, Kertsin and Fragkias, Michail (2013) 'Urban responses to climate change: Theories and governance practice in cities of the global South', *International Journal of Urban and Regional Research*, 37(6), 1865–1878.

Heisler, Helmuth (1971) 'The creation of a stabilized urban society: A turning point in the development of Northern Rhodesia/Zambia', *African Affairs*, 70(279), 125–145.

Heller, Patrick (2000) 'Degrees of democracy: Some comparative lessons from India', *World Politics*, 52(4), 484–519.

Henderson, J. V., Kuncoro, A. and Turner, M. (1995) 'Industrial development in cities', *Journal of Political Economy*, 103, 1067–1085.

Henderson, J. V., Storeygard, A. and Weil, D. (2012) 'Measuring economic growth from outer space', *American Economic Review*, 102(2), 994–1028.

Henderson, Vernon (2003) 'The urbanization process and economic growth: The so-what question', *Journal of Economic Growth*, 8, 47–71.

Herbst, J. (2004) 'Let them fail: State failure in theory and practice: Implications for policy', in Robert I. Rotberg, (ed.), *When States Fail: Causes and Consequences*. Princeton, NJ: Princeton University Press, pp. 302–318.

Herzer, D. and Nowak-Lehmann, D. F. (2006) 'What does export diversification do for growth? An econometric analysis', *Applied Economics*, 38(15), 1825–1838.

Heynen, N. C., Kaika, M. and Swyngedouw, E. (2006) *In the Nature of Cities: Urban Political Ecology and the Politics of Urban Metabolism* (Vol. 3). Abingdon: Taylor & Francis.

Hickey, S. and Mohan, G. (2004) *Participation: From Tyranny to Transformation?* London: Zed Books.

Hidalgo, D., Pereira, L., Estupiñán, N. and Jiménez, P. L. (2013) 'TransMilenio BRT system in Bogota, high performance and positive impact–Main results of an ex-post evaluation', *Research in Transportation Economics*, 39(1), 133–138.

Hills, Alice (2004) 'Continuity and discontinuity: The grammar of urban military operations', in Stephen Graham, (ed.), *Cities, War and Terrorism: Towards an Urban Geopolitics*. Oxford: Blackwell, pp. 231–246.

Hinderink, Jan and Titus, Milan (2002) 'Small towns and regional development: Major findings and policy implications from comparative research', *Urban Studies*, 39(3), 379–391.

Hirschman, A. O. (1958) *Strategy of Economic Development*. New Haven, CT: Yale University Press.

Hobsbawm, Eric (2005) 'Cities and Insurrection', *Global Urban Development*, 1(1), 1–8.

Hodson, M. and Marvin, S. (2009) '"Urban ecological security": A new urban paradigm?', *International Journal of Urban and Regional Research*, 33(1), 193–215.

Hoelscher, K. (2015) 'Politics and social violence in developing democracies: Theory and evidence from Brazil'. *Political Geography*, 44, 29–39.

Hoelscher, K. and Nussio, E. (2015) 'Understanding unlikely successes in urban violence reduction', *Urban Studies*, published online before print, June 10. doi: 10.1177/0042098015589892

Holston, J. (1999) *Cities and Citizenship*. Durham and London: Duke University Press.

—— (2008) *Insurgent Citizenship: Disjunctions of Democracy and Modernity in Brazil*. Princeton, NJ: Princeton University Press.

Home, R. (1997) *Of Planting and Planning: The Making of British Colonial Cities*. London: E & FN Spon.

Home, R. K. (1990) 'Town planning and garden cities in the British colonial empire 1910–1940', *Planning Perspectives*, 5(1), 23–37.

Hood, Christopher (1991) 'A public management for all seasons?', *Public Administration*, 69, 3–19.

Hopkins, Lewis D. (2001) *Urban Development: The Logic of Making Plans*. Washington DC: Island Press.

Huillery, E. (2009) 'History matters: The long-term impact of colonial public investments in French West Africa', *American Economic Journal: Applied Economics*, 1(2), 176–215.

Hulme, David and McKay, Andrew (2005) 'Identifying and measuring chronic poverty: Beyond monetary measures', Chronic Poverty Research Centre Working Paper #30. Pages 1–16.

Human Rights Watch (2006) *Condemned Communities: Forced Evictions in Jakarta*. New York: Human Rights Watch.

Humansecurity-cities.org (2007) *Human Security for an Urban Century, Local Challenges, Global Perspectives*. Ottawa: Foreign Affairs and International Trade Canada, March.

Hunter, F. (1953) *Community Power Structure: A Study of Decision Makers*. Chapel Hill: University of North Carolina Press.

Huxley, M. (2006) 'Spatial rationalities: Order, environment, evolution and government', *Social and Cultural Geography*, 7(5), 771–787.

IETA (2015) Tokyo: An emissions Trading case study, http://www.edf.org/sites/default/files/tokyo-case-study-may2015.pdf, accessed 6 July 2015.

ILO (2002) *Women and Men in the Informal Economy: A Statistical Picture*. Geneva: International Labour Office.

—— (2011) *Statistical Update on Employment in the Informal Economy*. Geneva: International Labour Organisation Department of Statistics.

—— (2013a) *The Informal Economy and Decent Work: A Policy Resource Guide, Supporting Transitions to Formality*. Geneva: International Labour Organisation.

—— (2013b) *Women and Men in the Informal Economy: A Statistical Picture* (second edition). Geneva: International Labour Office.

—— (2014) *Global Employment Trends 2014: Risk of a Jobless Recovery?* Geneva: International Labour Office.

InfoChange News (2005) 'The price the poor pay in Mumbai', InfoChange News and Features, October. Available at: http://infochangeindia.org/agenda/the-politics-of-water/the-price-the-poor-pay-in-mumbai-pune.html, accessed 6 October 2015.

IPCC (2014) *Climate Change 2014: Impacts, Adaptation, and Vulnerability* (Vol. 1). International Panel on Climate Change. Cambridge: Cambridge University Press.

Jabeen, H. (2014) 'Adapting the built environment: The role of gender in shaping vulnerability and resilience to climate extremes in Dhaka', *Environment and Urbanization*, 26(1), 147–165.

Jackson, P. (2005) 'Chiefs, money and politicians: Rebuilding local government in post-war Sierra Leone', *Public Administration and Development*, 25, 49–58.

—— (2007) 'Reshuffling an old deck of cards? The politics of decentralisation in Sierra Leone', *African Affairs*, 106, 95–111.

Jacobs, Jane (1961) *The Death and Life of Great American Cities*. London: Penguin Books.

—— (1969) *The Economy of Cities*. New York: Random House.

—— (1984) *Cities and the Wealth of Nations*. New York: Vintage Books.

Jäger, J., Frühmann, J., Günberger, S. and Vag, A. (2009) 'Environmental Change and Forced Migration Scenarios Project Synthesis Report. May 14 2009.

Jamal, V. and Weeks, J. (1988) 'The vanishing rural–urban gap in sub-Saharan Africa', *International Labour Review*, 127(3), 271–292.

Jellinek, Lea (2000) 'Jakarta, Indonesia, Kampung culture or consumer culture?' in Nicholas Low, Brendan Gleeson, Ingemar Elander and Rolf Lidskog, (eds), *Consuming Cities, The Urban Environment in the Global Economy after the Rio Declaration*. London: Routledge, pp. 265–280.

Jenkins, P., Smith, H. and Wang, Y. P. (2006) *Planning and housing in the rapidly urbanising world*. Abingdon: Routledge.

Jones, G. A. and Corbridge, S. (2010) 'The continuing debate about urban bias: the thesis, its critics, its influence and its implications for poverty-reduction strategies', *Progress in Development Studies*, 10(1), 1–18.

Jørgensen, Steen Lau and Van Domelen, Julie (2001) 'Helping the poor manage risk better: The role of social funds', in N. Lustig, (ed.), *Shielding the Poor, Social Protection in the Developing World*. Washington DC: The Brookings Institution Press and Inter-American Development Bank, pp. 91–107.

Joss, S. (2010) 'Eco-cities: A global survey 2009'. The Sustainable City VI: Urban Regeneration and Sustainability, WIT Press, Southampton, pp. 239–250.

Joy, Claire and Hardstaff, Peter (2005) *Dirty Aid, Dirty Water, The UK Government's Push to Privatise Water and Sanitation in Poor Countries*. London, World Development Movement, February.

Junhua, L. (1997) 'Beijing's old and dilapidated housing renewal', *Cities*, 14(2), pp. 59–69.

Kabeer, N. (1994) *Reversed Realities, Gender Hierarchies in Development Thought*. London: Verso Press.

Kagawa, Ayako and Tuksra, Jan (2002) 'The process of land tenure formalisation in Peru', in Payne, G., (ed.), *Land Rights and Innovation: Improving Tenure Security for the Urban Poor*. London: ITDG Publications.

Kaiser, Edward J., Godshalk, David R. and Chapin, Jr., F. Stuart (1995) 'The land planning arena', in Eugénie L. Birch, (ed.), *The Urban and Regional Planning Reader*. London and New York: Routledge, pp. 207–215.

Kaldor, M. (1999) *New and Old Wars: Organized Violence in a Global Era*. Cambridge: Polity.

—— (2007) *Human Security*. Cambridge: Polity.

Kalvyas, S. N., Shapiro, I. and Masoud, T. E. (eds) (2008) *Order, Conflict, and Violence*. Cambridge: Cambridge University Press.

Kanji, Nazneen (2002) 'Social funds in sub-Saharan Africa: How effective for poverty reduction?', in Peter Townsend and David Gordon, (eds), *World Poverty, New Policies to Defeat an Old Enemy*. Bristol: The Policy Press, pp. 233–250.

Kapagama, Pascal and Waterhouse, Rachel (2009) 'Portrait of Kinshasa: A city on the edge', Crisis States Research Centre Working Paper No: 53 (series 2).

Karaman, O. (2013) 'Urban renewal in Istanbul: Reconfigured spaces, robotic lives', *International Journal of Urban and Regional Research*, 37(2), 715–733.

Kaufmann, D., Kraay, A., and Zoido-Lobatón, P. (2000) 'Governance matters'. *Finance & Development*, 37(2), pp. 10–13.

Kaul, Mohan (1997) 'The new public administration: Management innovations in government', *Public Administration and Development*, 17(1), 13–26.

Kennedy, C., Pincetl, S. and Bunje, P. (2011) 'The study of urban metabolism and its applications to urban planning and design', *Environmental Pollution*, 159(8), 1965–1973.

Khan, Mushtaq (2010) 'Political settlements and the governance of growth-enhancing institutions of Soas', Working Paper. London: School of Oriental and African Studies.

Killick, T. (2004) 'Politics, evidence and the new aid agenda', *Development Policy Review*, 22(1), 5–29.

King, Anthony D. (1976) *Colonial Urban Development: Culture, social Power and Environment*. London: Routledge and Kegan Paul.

—— (1990) *Urbanism, Colonialism and the World-Economy*. London: Routledge.

Knox, Paul and Taylor, Peter (eds) (1995) *World Cities in a World System*. Cambridge: Cambridge University Press.

Kooiman, J. (1993) *Modern Governance: New Government: Society Interactions*. London: Sage.

Koonings, K. and Kruijt, D. (2007) *Fractured Cities: Social Exclusion, Urban Violence and Contested Spaces in Latin America*. London: Zed Books.

Kothari, U. (2001) 'Power, knowledge and social control in participatory development in Bill Cooke and Uma Kothari' (eds), *Participation: The New Tyranny?* London: Zed Books, pp. 139–52.

Krueger, Anne (1991) 'Government failures in development'. Cambridge, MA: National Bureau of Economic Research, Working Paper No. 33.

Kumar, Ajay and Barrett, Fanny (2008) *Stuck in Traffic: Urban Transport in Africa*. Washington DC: The World Bank.

Kumar, Chandra Bhushan (2013) 'Climate change and Asian cities: So near yet so far', *Urban Studies*, 50(7), 1456–1468.

Kumar, Sunil (2001) 'Embedded tenures: Private renting and housing policy in India', *Housing Studies*, 16(4), 425–442.

—— (2002) 'Round pegs and square holes: Mismatches between poverty and housing policy in urban India', in Peter Townsend and David Gordon, (eds),

World Poverty, New Policies to Defeat an Old Enemy. Bristol: The Policy Press, pp. 271–295.

Kuymulu, M. B. (2013a) 'Reclaiming the right to the city: Reflections on the urban uprisings in Turkey', *City*, 17(3), 274–278.

—— (2013b) 'The vortex of rights: "Right to the city" at a crossroads', *International Journal of Urban and Regional Research*, 37(3), 923–940.

Laczko, F. and Aghazarm, C. (eds) (2009) *Migration, Environment and Climate Change: Assessing the Evidence*. Geneva: International Organization for Migration.

Lal, Deepak (1985) *The Poverty of 'Development Economics'*. Cambridge, MA: Harvard University Press.

Lambright, G. M. (2011) *Decentralization in Uganda: Explaining Successes and Failures in Local Governance*. Boulder: Lynne Reiner.

Leander, A. (2003) 'Wars and the un-making of states: Taking Tilly seriously in the contemporary world', in Stefano Guzzini and Dietrich Jung, (eds), *Contemporary Security Analysis and Copenhagen Peace Research*. London and New York: Routledge, pp. 69–80.

LeBas, A. (2011) *From Protest to Parties: Party-building and Democratization in Africa*. Oxford: Oxford University Press.

—— (2013) 'Violence and urban order in Nairobi, Kenya and Lagos, Nigeria', *Studies in Comparative International Development*, 48(3), 240–262.

Lee-Smith, Diana and Stren, Richard E. (1991) 'New perspectives on African urban management', *Environment and Urbanization*, 3(1), 23–36.

Lefebvre, H. (1968) *Le droit à la ville*. Anthropos, Paris.

Legg, S. and McFarlane, C. (2008) 'Ordinary urban spaces: Between postcolonialism and development', *Environment and Planning. A*, 40(1), 6.

Leitman, Josef (1994) 'The World Bank and the Brown Agenda: Evolution of a revolution', *Third World Planning Review*, 16(2), 117.

Lele, S. M. (1991) 'Sustainable development: A critical review', *World Development*, 19(6), 607–621.

Lemanski, C. (2004) 'A new apartheid? The spatial implications of fear of crime in Cape Town, South Africa', *Environment and Urbanization*, 16(2), 101–112.

—— (2011) 'Moving up the ladder or stuck on the bottom rung? Homeownership as a Solution to Poverty in Urban South Africa', *International Journal of Urban and Regional Research*, 35(1), 57–77.

Lemanski, C. and Oldfield, S. (2009) 'The parallel claims of gated communities and land invasions in a Southern city: Polarised state responses', *Environment and Planning. A*, 41(3), 634.

Lemanski, C., Landman, K. and Durington, M. (2008, June) 'Divergent and similar experiences of "gating" in South Africa: Johannesburg, Durban and Cape Town', *Urban Forum*, 19(2), 133–158.

Lewis, A. W. (1954) 'Economic development with unlimited supplies of labor', *Manchester School of Economic and Social Studies*, 22(2), 139–91.

Leys, C. (1996) *The Rise and Fall of Development Theory*. Oxford: James Currey.

Li, B. (2013) 'Governing urban climate change adaptation in China', *Environment and Urbanization*, 25(2), 413–427.

Li, S-M and Song, Yi (2009) 'Redevelopment, displacement, housing conditions, and residential satisfaction: a study of Shanghai', *Environment and Planning. A*, 41(5), 1090–1108.

Li, T. M. (2007) *The Will to Improve: Governmentality, Development, and the Practice of Politics*. Durham and London: Duke University Press.

Li, Zhigang and Wu, Fulong (2006) 'Socioeconomic transformations in Shanghai (1990–2000): Policy impacts in global-national-local contexts', *Cities*, 23(4), 250–268.

Lin, J. Y. (2011) 'New structural economics: A framework for rethinking development', *The World Bank Research Observer*, 1–29.

Lindau, L. A., Hidalgo, D. and Facchini, D. (2010) 'Curitiba, the cradle of bus rapid transit', *Built Environment*, 36(3), 274–282.

Lindell, I. (2010) *Africa's Informal Workers: Collective Agency, Alliances and Transnational Organizing in Urban Africa*. London: Zed Books.

Lindemann, Stefan (2011) 'Just another change of guard? Broad-based politics and Civil War in Museveni's Uganda', *African Affairs*, 110(440), 387–416.

Linsky, A. S. (1965) 'Some generalizations concerning primate cities 1', *Annals of the Association of American Geographers*, 55(3), 506–510.

Liotta, Peter H. (2007) 'Human security and cities in the Greater Near East', in Humansecurity-cities.org Human Security for an Urban Century, Local Challenges, Global Perspectives, Vancouver: Canadian Consortium on Human Security.

Lipton, M. (1977) *Why Poor People Stay Poor: Urban Bias in World Development*. London: Maurice temple Smith; Cambridge, MA: Harvard University Press.

Livi-Bacci, Massimo (2001) *A Concise History of World Population* (3rd edition). Oxford: Blackwell Publishers.

Logan, J. and Molotch, H. (1987) *Urban Fortunes. The Political Economy of Place*. Berkeley, University of California.

Lohani, Bindu N. (2010) 'Asia 2010–2020: The Decade of Sustainable Transport in Asia and the Role of the Asia Development Bank'. Presented at the Asia 2010–2020: The Decade of Sustainable Transport in Asia and the Role of the Asian Development Bank, Bangkok, 23–25 August.

Lombard, M. (2012) 'Land tenure and urban conflict: A review of the literature'. GURC Working Paper Series: University of Manchester.

Lourenço-Lindell, I. (2002) *Walking the Tight Rope: Informal Livelihoods and Social Networks in a West African City*. Stockholm: Acta Universitatis Stockholmiensis.

Lund, C. (2006) 'Twilight institutions: Public authority and local politics in Africa', *Development and Change*, 37(4): 685–705.

Lynch, Kenneth (2005) *Rural–Urban Interaction in the Developing World*. New York: Routledge.

Lyons, M. (2009) 'Building back better: The large-scale impact of small-scale approaches to reconstruction', *World Development*, 37(2), 385–398.

Mabogunje, Akin L. (1990) 'Urban planning and the post-colonial state in Africa: A research overview', *African Studies Review*, 33(2), 121–203.

McAuslan, P. (1985) *Urban Land and Shelter for the Poor*. London: International Institute for Environment and Development.

McFarlane, C. (2008) 'Urban shadows: Materiality, the "Southern city" and urban theory', *Geography Compass*, 2(2), 340–358.

McFarlane, C. and Rutherford, J. (2008) 'Political infrastructures: Governing and experiencing the fabric of the city', *International Journal of Urban and Regional Research*, 32(2), 363–374.

MacGaffey, Janet and Banzenguissa-Ganga, Remy (2000) *Congo-Paris: Transnational Traders on the Margins of the Law*. Bloomington, IN: Indiana University Press.

McGranahan, G. (2013) 'Evolving health risks: Housing, water and sanitation, and climate change', in Elliott D. Sclar, Nicole Volavka-Close and Peter Brown, (eds), *The Urban Transformation: Health, Shelter and Climate Change*. Abingdon: Earthscan, pp. 15–41.

—— (2015) 'Realizing the right to sanitation in deprived urban communities: Meeting the challenges of collective action, coproduction, affordability, and housing tenure', *World Development*, 68, 242–253.

McGranahan, G., Balk, D. and Anderson, B. (2007) 'The rising tide: Assessing the risks of climate change and human settlements in low elevation coastal zones', *Environment and Urbanization*, 19(1), 17–37.

McGranahan, G., Mitlin, D., Satterthwaite, D., Tacoli, C. and Turok, I. (2009) *Africa's Urban Transition and the Role of Regional Collaboration*. London: International Institute for Environment and Development.

Machado, F., Scartascini, C. and Tommasi, M. (2011) 'Political institutions and street protests in Latin America', *Journal of Conflict Resolution*, 55(3), 340–365.

McIlwaine, C. (2014) 'Everyday urban violence and transnational displacement of Colombian urban migrants to London, UK', *Environment and Urbanization*, 26(2), 417–426.

McIntosh, Roderick J. and McIntosh, Susan Keech (1981) 'The Inland Niger Delta before the Empire of Mali: Evidence from Jenne-Jenno', *Journal of African History*, 22, 1–22.

Maloney, William F. (1999) 'Does informality imply segmentation of labour markets? Evidence from sectoral transitions in Mexico,, *World Bank Economic Review*, 13(2), 275–302.

Mamdani, Mahmood (1996) *Citizen and Subject: Contemporary Africa and the Legacy of Late Colonialism*. Princeton, New Jersey: Princeton University Press.

Mangin, William (1967) 'Latin American squatter settlements: A problem and a solution', *Latin American Research Review*, 2(3), 65–98.

Maplethorpe (2013) 'Climate Change Vulnerability Index 2013 – Most at risk cities. Available at: http://www.preventionweb.net/files/29649_maplecroftccvisubnationalmap.pdf, accessed 11 November 2014.

Marcuse, Peter (1997) 'Walls of fear and walls of support', in N. Elllin, (ed.), *Architecture of Fear*. New York: Princeton Architectural Press, pp. 101–114.

Marshall, Alfred (1920) *Principles of Economics: An Introductory Volume*. London: Macmillan.

Martin, E. and Mosel, I. (2011) *City Limits: Urbanisation and Vulnerability in Sudan: Juba Case Study*, London: Overseas Development Institute, pp. 53–79.

Martin, Richard (1983) 'Upgrading', in J. Skinner Reinhard and M. J. Rodell, (eds), *People, Poverty and Shelter: Problems of Self-Help Housing*. London: Methuen.

Marvin, S. (1992) 'Urban policy and infrastructure networks', *Local Economy*, 7(3), 225–247.

Marx, K. and Engels, F. (1848) *The Communist Manifesto*. London: Penguin Classics.

Mattingly, Michael (1994) 'Meaning of urban management', *Cities*, 11(3), 201–205.

Maxwell, D. G. (1995) 'Alternative food security strategy: A household analysis of urban agriculture in Kampala', *World Development*, 23(10), 1669–1681.

Mazumdar, D. (1987) 'Rural–urban migration in developing countries', in E. S. Mills, (ed.), *Handbook of Regional and Urban Economics: Volume II*. Amsterdam: Elsevier, pp. 1097–1128.

Meadows, D. H., Meadows, D. L., Randers, J. and Behrens, W. W. (1972) *The Limits to Growth*. New York: Universe Books.

Meagher, David and Farhana, Tithe (2012) 'Climate change drives rural–urban migration to Dhaka's slums', *Global South Development Magazine*, July 28. Available at: http://gsdmagazine.org/2012/07/28/climate-changes-drives-rural-urban-migration-to-dhakas-slums, accessed 12 November 2014.

Meagher, K. (2012) 'The strength of weak states? Non-state security forces and hybrid governance in Africa', *Development and Change*, 43(5), 1073–1101.

Meagher, Kate (1995) 'Crisis, informalization and the urban informal sector in sub-Saharan Africa', *Development and Change*, 26(2), 259–284.

—— (2003) 'A Back Door to Globalisation? Structural Adjustment, Globalisation & Transborder Trade in West Africa', *Review of African Political Economy*, 95, 57–75.

Mehta, Dinesh (2005) 'Our common past: The contribution of the urban management programme', *Habitat Debate*, 11(4), 6–7.

Méndez, Juan, O'Donnell, Guillermo and Sérgio Pinheiro, Paul (eds) (1999) *The (Un)rule of Law and the Underprivileged in Latin America*. Notre Dame, IN: University of Notre Dame Press.

Meth, P. (2003) 'Rethinking the "domus" in domestic violence: Homelessness, space and domestic violence in South Africa', *Geoforum*, 34(3), 317–327.

—— (2009) 'Marginalised men's emotions: Politics and place', *Geoforum*, 40(5), 853–863.

—— (2014) 'Cities, crime and development' in V. Desai and R. Potter (eds), *The Companion to Development Studies* (3rd Edition). Oxford: Routledge, pp. 324–328.

Milgram, B. L. (2012) 'Reconfiguring margins: Secondhand clothing and street vending in the Philippines', *Textile*, 10(2), 200–221.

Mills, C. Wright (1956) *The Power Elite*. Oxford: Oxford University Press.

Mitlin, D. (2011) 'Shelter finance in the age of neo-liberalism', *Urban Studies*, 48(6), 1217–1233.

Mitlin, D. and Satterthwaite, D. (2013) *Urban Poverty in the Global South: Scale and Nature*. Abingdon: Routledge.

Mkandawire, Thandika (2002) 'The terrible toll of post-colonial "rebel movements" in Africa: Towards an explanation of the violence against the peasantry', *The Journal of Modern African Studies*, 40(02), 181–215.

Mobarak, A. M. (2005) 'Democracy, volatility, and economic development', *Review of Economics and Statistics*, 87(2), 348–361.

Mohanty, R. (2007) 'Gendered subjects, the state and participatory spaces: The politics of domesticating participation in rural India', in A. Cornwall and V. S. Coelho, (eds), *Spaces for Change? The Politics of Citizen Participation in Democratic Arenas*. New York and London: Zed Books, pp. 76–94.

Molotch, H. (1976) 'The city as a growth machine: Toward a political economy of place', *American Journal of Sociology*, 82(2), 309–332.

Momsen, J. H. (2004) *Gender and Development*. London: Routledge.

Moncada, E. (2013) 'The politics of urban violence: Challenges for development in the global South', *Studies in Comparative International Development*, 48(3), 217–239.

Monkam, N. and Moore, M. (2015) *How Property Tax would Benefit Africa*. London: Africa Research Institute.

Moomaw, R. L. and Alwosabi, M. A. (2004) 'An empirical analysis of competing explanations of urban primacy evidence from Asia and the Americas', *The Annals of Regional Science*, 38(1), 149–171.

Moore, M. (2004) 'Revenues, state formation, and the quality of governance in developing countries', *International Political Science Review*, 25(3), 297–319.

Morris, J. and Winchester, S. (1983) *Stones of Empire: The Buildings of the Raj*. Oxford: Oxford University Press.

Moser, C. O. (1993) *Gender Planning and Development: Theory, Practice and Training*. London and New York: Routledge.

—— (1996) 'Confronting crisis: A comparative study of household responses in four poor urban communities', Environmentally Sustainable Development Studies and Monograph Series No. 8, Washington DC: The World Bank.

—— (1998) 'The asset vulnerability framework: Reassessing urban poverty reduction strategies', *World Development*, 26(1), 1–19.

—— (2004) *Urban Violence and Insecurity: An Introductory Roadmap*. London: IIED International Institute for Environment and Development.

—— (2007) 'Asset accumulation policy and poverty reduction', in Caroline Moser, ed., *Reducing Global Poverty, The Case for Asset Accumulation*. Washington DC: The Brookings Institution, pp. 83–103.

Moser, C. O. and McIlwaine, C. (2006) 'Latin American urban violence as a development concern: Towards a framework for violence reduction', *World Development*, 34(1), 89–112.

—— (2014) 'New frontiers in twenty-first century urban conflict and violence', *Environment and Urbanization*, 26(2), 331–344.

Moser, C. O. and Peake, L. (eds) (1987) *Women, Human Settlements and Housing*. New York: Tavistock Press.

Moser, C. O. and Rodgers, D. (2005) *Change, Violence and Insecurity in Non-conflict Situations*. London: Overseas Development Institute.

—— (2012) 'Understanding the tipping point of urban conflict: Global policy report'. Urban Tipping Point Project Working Paper No, 7.

Moxham, B. and Carapic, J. (2013) 'Unravelling Dili: The crisis of city and state in Timor-Leste', *Urban studies*, 50(15), 3116–3133.

Moyo, D. (2009) *Dead Aid: Why Aid is Not Working and How There is Another Way for Africa*. London: Allen Lane.

Muggah, R. (2012) *Researching the Urban Dilemma: Urbanization, Poverty and Violence*. Ottawa: International Development Research Centre.

—— (2014) 'Deconstructing the fragile city: Exploring insecurity, violence and resilience', *Environment and Urbanization*, 26(2), 345–358.

Mumford, L. (1937) 'What is a city', Architectural Record, reprinted in Richard Le Gates and Fredric Stout, (eds), *The City Reader* (third edition), London: Routledge, pp. 93–96.

—— (1961) *The City in History*. London: Martin Secker and Warburg Ltd.

Myers, Garth Andrew (2003) *Verandahs of Power: Colonialism and Space in Urban Africa*. Syracuse: Syracuse University Press.

—— (2005) *Disposable Cities: Garbage, Governance and Sustainable Development in Urban Africa*. Aldershot: Ashgate.

—— (2011) *African Cities: Alternative Visions of Theory and Practice*. London: Zed books.

Myers, Norman (2002) 'Environmental Refugees: A Growing Phenomenon of the 21st Century', *Philosophical Transactions: Biological Sciences*, 357(1420), 609–613.

—— (2005) 'Environmental refugees: An emergent security issue', paper presented at the 13th Economic Forum, Prague, 23–27 May 2005.

Myrdal, G. (1957) *Economic Theory and Underdeveloped Regions*. New York: Harper and Row.

Nakamura, S. (2014) 'Impact of slum formalisation on self-help housing construction: A case of slum notification in India', *Urban Studies*, 51(16), 3420–3444.

Nantulya, Vinand M. (2002) 'The neglected epidemic: Road traffic injuries in developing countries', *British Medical Journal*, 324(7346), 1139–1141.

Narayan, Deepa, Chambers, Robert, Shah, Meera and Petesch, Patti (2000b) *Voices of the Poor, Crying Out for Change*. Oxford and New York: Oxford University Press for the World Bank.

Narayan, Deepa, with Patel, Raj, Schafft, Kai, Rademacher, Anne, and Koch-Schulte, Sarah (2000a) *Voices of the Poor, Can Anyone Hear Us?* Oxford and New York: Oxford University Press for the World Bank.

Nathan, M. and Overman, H. (2013) 'Agglomeration, clusters, and industrial policy', *Oxford Review of Economic Policy*, 29(2), 383–404.

Nelson, Joan M. (1979) *Access to Power: Politics and the Urban Poor in Developing Nations*. Princeton: Princeton University Press.

Neuman, Michael (1998) 'Does planning need the plan?', *Journal of the American Planning Association*, 64(2), 208–220.

Neumayer, E. and Barthel, F. (2010) 'Normalizing economic loss from natural disasters: A global analysis', *Global Environmental Change*, 21, 13–24.

Newman, E. (2009) 'Conflict research and the "decline" of civil war', *Civil Wars*, 11(3), 255–278.

Niederhafner, S. (2013) 'The governance modes of the Tokyo Metropolitan Government Emissions Trading Scheme', Technical report, Hitotsubashi University, Japan.

Niño, P., Jr., Coronado, R., Fullerton, T., Jr. and Walke, A. (2015) 'Cross-border homicide impacts on economic activity in El Paso', *Empirical Economics*, 1–17, doi: 10.1007/s00181-015-0924-0.

Njeru, J. (2010) '"Defying" democratization and environmental protection in Kenya: The case of Karura Forest reserve in Nairobi', *Political Geography*, 29(6), 333–342.

Njoh, A. (2004) 'The experience and legacy of French colonial urbanmplanning in sub-Saharan Africa', *Planning Perspectives*, 19(4), 435–454.

Norregaard, J. (2013) *Taxing Immovable Property: Revenue Potential and Implementation Challenges*. Washington DC: International Monetary Fund.

North, Douglass C. (1990) *Institutions, Institutional Change, and Economic Performance*. Cambridge: Cambridge University Press.

—— (1995) 'The new institutional economics and third world development', in John Harriss, Janet Hunter and Colin Lewis, (eds), *The New Institutional Economics and Third World Development*. London: Routledge, pp. 17–26.

North, D. C., Wallis, J. J. and Weingast, B. R. (2009) *Violence and Social Orders: A Conceptual Framework for Interpreting Recorded Human History*. Cambridge: Cambridge University Press.

Novy, A. and Leuboult, B. (2005) 'Participatory budgeting in Porto Alegre: Social innovation and the dialectical relationship of state and civil society', *Urban Studies*, 42(11), 2023–2036.

O'Connor, Anthony (1983) *The African City*. London: Hutchinson.

Okpala, Don (2008) *Regional Overview of the Status of Urban Planning and Planning Practice in Anglophone (Sub-Saharan) African Countries*. Nairobi: UN-HABITAT.

Ondakie, A. B., Serbeh-Yiadom, K. C. and Asfaw, M. (2015) 'The condominium scheme as a strategy for housing slum dwellers: The case of Gofa Mebrat Hail Condominium, Addis Ababa, Ethiopia', *Developing Country Studies*, 5(9), 159–165.

Orleans Read, S., Friend, R., Canh Toan, V., Thinphanga, P., Sutarto, R. and Singh, D. (2013) '"Shared learning" for building urban climate resilience: Experiences from Asian cities', *Environment and Urbanization*, 26, 393–412.

Orr, Marion and Johnson, Valerie C. (2008) *Power in the City: Clarence Stone and the Politics of Inequality*. Lawrence, Kansas: University Press of Kansas.

Orum, A. M. and Dale, J. G. (2009) *Introduction to Political Sociology: Power and Participation in the Modern World*. New York: Oxford University Press.

Owusu, F. (2007) 'Conceptualizing livelihood strategies in African cities: Planning and development implications of multiple livelihood strategies', *Journal of Planning Education and Research*, 26(4), 450–465.

Oxfam (2007) 'Adapting to climate change: What's needed in poor countries, and who should pay', Oxfam Briefing Paper 104, Banbury: Oxfam.

Oyeniyi, Bukola (2007) 'Road Transport Workers' Union: The paradox of negotiating socio-Economic and political space in Nigeria'. Paper prepared for the conference on 'Informalising Economies and New Organising Strategies in Africa', 20–22 April, Uppsala: Nordic Africa Institute.

Pacione, Michael (2005) *Urban Geography: A Global Perspective* (Second Edition). Oxon: Routledge.

Painter, J. (2001) 'Regulation, regime and practice in urban politics', in B. Jessop, (ed.), *Regulation Theory and the Crisis of Capitalism: Developments and Extensions*. Cheltenham: Edward Elgar, pp. 122–143.

Palma, G. (1981) 'Dependency and development: A critical overview', in Dudley Seers, (ed.), *Dependency Theory: A Critical Assessment*. London: Frances Pinter, pp. 20–78.

Pansters, Wil and Hector Castillo Berthier (2007) 'Mexico City', in K. Koonings and D. Kruijt, (eds), *Fractured Cities: Social Exclusion, Urban Violence and Contested Spaces in Latin America*. London: Zed Books, pp. 36–56.

Parfitt, T. (2013) 'Modalities of violence in development: Structural or contingent, mythic or divine?', *Third World Quarterly*, 34(7), 1175–1192.

Parikh, P., Fu, K., Parikh, H., McRobie, A. and George, G. (2015) 'Infrastructure provision, gender, and poverty in Indian slums', *World Development*, 66, 468–486.

Parker, Andrew N. (1995) 'Decentralization: The way forward for rural development?'. Washington DC: The World Bank, World Bank Policy Research Working Paper No. 1475.

Parker, Geoffrey (2004) *Sovereign City: The City-state Through History*. London: Reaktion Books Ltd.

Parker, S. (2010) *Cities, Politics & Power*. London: Routledge.

Parnell, S. (2015) 'Fostering transformative climate adaptation and mitigation in the African city: Opportunities and constraints of urban planning', in Pauleit, S., Coly, A., Fohlmeister, S. et al., *Urban Vulnerability and Climate Change in Africa*. London: Springer International Publishing, pp. 349–367.

—— (2016) 'Defining a global urban agenda', *World Development*, 78, 529–540.

Parnell, S. and Simon, D. (2014) 'National urbanisation and urban strategies: Necessary but absent policy instruments', in S. Parnell and E. Pieterse, (eds), *Africa's Urban Revolution*. London: Zed Books, pp. 237–256.

Parnell, S. and Walawege, R. (2011) 'Sub-Saharan African urbanisation and global environmental change', *Global Environmental Change*, 21, S12–S20.

Parnell, Susan, Simon, David and Vogel, Coleen (2007) 'Global environmental change: Conceptualising the growing challenges for cities in poor countries', *Area*, 39(3), 357–369.

Parnell, Susan, Pieterse, Edgar, Swilling, Mark and Wooldridge, Dominique (2002) *Democratising Local Government: The South African Experience*. Cape Town: University of Cape Town Press.

Parr, John B. (1999a) 'Growth-pole strategies in regional and economic planning: A retrospective view (Part 1: Origins and advocacy)', *Urban Studies*, 36(7), 1195–1215.

—— (1999b) 'Growth-pole strategies in regional and economic planning: A retrospective view (Part 2: Implementation and outcome)', *Urban Studies*, 36(8), 1247–1268.

Patel, Raj (2008) *Stuffed and Starved: The Hidden Battle for the World Food System*. New York: Melville House.

Patel, S. Burra, S. and D'Cruz, C. (2001) 'Slum/Shack DwellersInternational (SDI)-foundations to treetops', *Environment and Urbanization*, 13(2), 45–59.

Pateman, C. (2012) 'Participatory budgeting revisited', *Perspectives on Politics*, 10(1), 7–19.

Patibandla, Murali and Petersen, Bent (2002) 'Role of transnational corporations in the evolution of a high-tech industry: The case of India's software industry', *World Development*, 30(9), 1561–1577.

Paul, Bimal Kanti (2009) 'Why relatively fewer people died? The case of Bangladesh's Cyclone Sidr', *Natural Hazards*, 50(2), 289–304.

Payan, T. (2014) 'Ciudad Juárez: A perfect storm on the US–Mexico border', *Journal of Borderlands Studies*, 29(4), 435–447.

Payne, Geoffrey (1997) *Urban Land Tenure and Property Rights in Developing Countries: A Review*. London: ODA/Intermediate Technology Publications.

—— (2001) 'Urban land tenure policy options: Titles or rights?', *Habitat International*, 25, 415–429.

Payne, G., Durand-Lasserve, A. and Rakodi, C. (2009) 'The limits of land titling and home ownership', *Environment and Urbanization*, 21(2), 443–462.

Peattie, Lisa (1987) 'An idea in good currency and how it grew: The informal sector', *World Development*, 15(7), 851–860.

Pelling, M. (2003) *Natural Disaster and Development in a Globalizing World*. London: Routledge.

—— (2010) *Adaptation to Climate Change: From Resilience to Transformation*. London: Routledge.

Pelling, M. and Wisner, B. (2009) *Disaster Risk Reduction: Cases from Urban Africa*. London: Earthscan.

Perlman, J. E. and UN-HABITAT (2014) *Urban Informality: Marginal or Mainstream?* New York: UN-Habitat.

Perry, Guillermo E., Maloney, William F., Arias, Omar S., Fajnzylber, Pablo, Mason, Andrew D. and Saavedra-Chanduvi, Jaime (2007) *Informality: Exit and Exclusion*. Washington DC: The World Bank.

Pflieger, G. and Matthieussent, S. (2008) 'Water and power in Santiago de Chile: Sociospatial segregation through network integration', *Geoforum*, 39(6), 1907–1921.

Pieterse, Edgar (2008) *City Futures: Confronting the Crisis of Urban Development*. London: Zed Books.

Pilkington, Ed (2008) 'Eaten up', *Guardian*, 29 July.

Pincetl, S., Bunje, P. and Holmes, T. (2012) 'An expanded urban metabolism method: Toward a systems approach for assessing urban energy processes and causes', *Landscape and Urban Planning*, 107(3), 193–202.

Pinheiro, Paul Sérgio (1996) 'Democracy without citizenship', *NACLA Report on the Americas*, 30(2), 17–23.

Porter, L. (2010) *Unlearning the Colonial Cultures of Planning*. Farnham: Ashgate Publishing, Ltd.

Porter, Michael (2000) 'Location, competition, and economic development: Local clusters in a global economy', *Economic Development Quarterly*, 14(1), 15–34.

Portes, Alejandro and Haller, William (2005) 'The informal economy', in N. J. Smelser and R. Swedberg, (eds), *The Handbook of Economic Sociology*. New Jersey: Princeton University Press, pp. 403–425.

Potter, R., Conway, D., Evans, R. and Lloyd-Evans, S. (2012) *Key Concepts in Development Geography*. Sage, London.

Potts, D. (2006) '"Restoring order"? Operation Murambatsvina and the urban crisis in Zimbabwe', *Journal of Southern African Studies*, 32(2), 273–291.

—— (2007) 'The State and the informal sector in sub-Saharan African economies: Revisiting debates on dualism', Crisis States Research Centre Working Paper No. 18 (series two), London: Crisis States Research Centre, Development Studies Institute, London School of Economics, October.

—— (2010) *Circular Migration in Zimbabwe & Contemporary Sub-Saharan Africa*. Rochester, NY: Boydell & Brewer.

—— (2011) 'Shanties, slums, breeze blocks and bricks: (Mis)understandings about informal housing demolitions in Zimbabwe', *City*, 15(6), 709–721.

Pryer, Jane (2003) *Poverty and Vulnerability in Dhaka Slums*. Aldershot: Ashgate.

Puga, D. (2010) 'The magnitude and causes of agglomeration economies', *Journal of Regional Science*, 50(1), 203–219.

Pugh, Cedric (1995) 'The role of the World Bank in housing', in Brian Aldrich and Ranvinder Sandhu, (eds), *Housing the Urban Poor, Policy and Practice in Developing Countries*. London: Zed Books, pp. 34–92.

—— (2001) 'The theory and practice of housing sector development for developing countries, 1950–99', *Housing Studies*, 16(4), 399–423.

Purcell, M. (2002) 'Excavating Lefebvre: The right to the city and its urban politics of the inhabitant', *GeoJournal*, 58(2), 99–108.

Pye, L. W. (1969) 'The political implications of urbanization and the development process', in G. Breese, (ed.), *The City in Newly Developing Countries: Readings on Urbanism and Urbanization*. Englewood Cliffs, NJ: Prentice Hall, pp. 401–406.

Quadir, F. (2013) 'Rising donors and the new narrative of "South–South" cooperation: What prospects for changing the landscape of development assistance programmes?', *Third World Quarterly*, 34(2), 321–338.

Rabbani, F., Qureshi, F. and Rizvi, N. (2008) 'Perspectives on domestic violence: Case study from Karachi, Pakistan', *Eastern Mediterranean Health Journal*, 14(2), pp. 415–426.

Rabrenovic, G. (2009) 'Urban social movements', *Theories of Urban Politics*, 2, 239–254.

Rakodi, Carole (1997) *The Urban Challenge in Africa*. Tokyo and New York: United Nations University Press.

—— (1999) 'A capital assets framework for analysing household livelihood strategies', *Development Policy Review*, 17(3), September, pp. 315–342.

—— (2001b) 'Urban governance and poverty: Addressing needs, asserting claims: An editorial introduction', *International Planning Studies*, 6(4), 343–356.

Rakodi, Carole with Lloyd-Jones, Tony (eds) (2002) *Urban Livelihoods, A People-Centred Approach to Reducing Poverty*. London: Earthscan Publications Limited.

Randall, V. and Svåsand, L. (2002) 'Party institutionalization in new democracies', *Party Politics*, 8(1), 5–29.

Räty, R. and Carlsson-Kanyama, A. (2010) 'Energy consumption by gender in some European countries', *Energy Policy*, 38(1), 646–649.

Ravallion, M., Chen, S. and Sangraula, P. (2007) 'New evidence on the urbanization of global poverty', *Population and Development Review*, 33(4), 667–701.

Reader, John (2004) *Cities*. London: Vintage.

Redclift, M. (1987) *Sustainable Development: Exploring the Contradictions*. London: Methuen and Company.

Redwood, M. (ed.) (2012) *Agriculture in Urban Planning: Generating Livelihoods and Food Security*. London: Routledge.

Redwood, M. and Wakely, P. (2012) 'Land tenure and upgrading informal settlements in Colombo, Sri Lanka', *International Journal of Urban Sustainable Development*, 4(2), 166–185.

Rees, W. E. (1992) 'Ecological footprints and appropriated carrying capacity: What urban economics leaves out', *Environment and Urbanization*, 4(2), 121–130.

Ren, X. (2013) *Urban China*. Cambridge: Polity Press.

Resnick, D. (2012) 'Opposition parties and the urban poor in African democracies', *Comparative Political Studies*, 45(11), 1351–1378.

—— (2013) *Urban Poverty and Party Populism in African Democracies*. Yale: Cambridge University Press.

—— (2014a) 'Urban governance and service delivery in African cities: The role of politics and policies', *Development Policy Review*, 32(s1), s3–s17.

—— (2014b) 'Strategies of subversion in vertically-divided contexts: Decentralisation and urban service delivery in Senegal', *Development Policy Review*, 32, (S1): s61–s80.

Reuveny, Rafael (2007) 'Climate change-induced migration and violence conflict', *Political Geography*, 26: 656–673.

Rex, T. R. (2014) *Trade between the United States and México, with a Focus on the Border Area*. The United States-México Series of Background Reports. W. P. Carey School of Business, Arizona State University.

Richards, Susan (2002) 'More trouble in paradise', Open Democracy. Available at http://www.opendemocracy.net/conflict-witnessconflict/article_848.jsp, accessed 6 October 2015.

Rigon, A. (2014) 'Building local governance: Participation and elite capture in slum-upgrading in Kenya', *Development and Change*, 45(2), 257–283.

Rizzo, Matteo (2002) 'Being taken for a ride: Privatisation of the Dar es Salaam transport system 1983–1998', *Journal of Modern African Studies*, 40 (1), 133–157.

Robinson, J. (2002) 'Global and world cities: A view from off the map', *International Journal of Urban and Regional Research*, 26(3), 531–554.

Robinson, Jennifer (2006) *Ordinary Cities*. London: Routledge.

Rodgers, D. (2006) 'Living in the shadow of death: Gangs, violence and social order in urban Nicaragua, 1996–2002', *Journal of Latin American Studies*, 38(02), 267–292.

—— (2009) 'Slum wars of the 21st century: Gangs, mano dura and the new urban geography of conflict in Central America', *Development and Change*, 40(5), 949–976.

—— (2010) 'Urban violence is not (necessarily) a way of life: Towards a political economy of conflict in cities', in J. Beall, R. Kanbur and B.

Guha-Khasnobis, (eds), *Urbanization and Development: Multidisciplinary Perspectives*. Oxford: Oxford University Press, pp. 235–250.

Rodgers, Dennis (2004) '"Disembedding" the city: Crime, insecurity and spatial organization in Managua, Nicaragua', *Environment and Urbanization*, 16(2), October, 113–124.

Rodgers, D. and O'Neill, B. (2012) 'Infrastructural violence: Introduction to the special issue', *Ethnography*, 13(4), 401–412.

Rodrik, D. (ed.) (2003) *In Search of Prosperity: Analytic Narratives on Economic Growth*. Princeton, NJ: Princeton University Press.

Rodrik, D., Subramanian, A. and Trebbi, F. (2004) 'Institutions rule: The primacy of institutions over geography and integration in economic development', *Journal of Economic Growth*, 9(2), 131–165.

Rolnik, R. (2013) 'Ten years of the City Statute in Brazil: From the struggle for urban reform to the World Cup cities', *International Journal of Urban Sustainable Development*, 5(1), 54–64.

Rondinelli, Dennis A. (1975) *Urban and Regional Development Planning: Policy and Administration*. Ithaca: Cornell University Press.

—— (1981) 'Government decentralization in comparative theory and practice in developing countries', *International Review of Administrative Science*, 47 (2), 133–147.

—— (1994) 'Urbanization policy and economic growth in sub-Saharan Africa: The private sector's role in urban development', in James D. Tarver, (ed.), *Urbanization in Africa: A Handbook*. Westport: Greenwood Press, pp. 362–387.

Rosenau, W. (1997) 'Every room is a new battle: The lessons of modern urban warfare', *Studies in Conflict and Terrorism*, 20(4), 371–394.

Rosenthal, Stuart and Strange, William (2004) 'Evidence on the nature and sources of agglomeration economies', in J. V. Henderson and J. Thisse, (eds), *Handbook of Regional and Urban Economics*, 4, 2119–2171.

Rostow, W. W. (1960) *The Stages of Economic Growth: A Non-communist Manifesto*. Cambridge: Cambridge University Press.

Rotberg, R. I. (2002) 'Failed states in a world of terror', *Foreign Affairs-New York*, 81(4), 127–141.

Roy, A. (2006) 'Praxis in the time of empire', *Planning Theory*, 5(1), 7–29.

—— (2009) 'Why India cannot plan its cities: Informality, insurgence and the idiom of urbanization', *Planning Theory*, 8(1), 76–87.

Roy, A. and AlSayyad, N. (eds) (2004) *Urban Informality: Transnational Perspectives from the Middle East, Latin America, and South Asia*. New York: Lexington Books.

Rudolph, S. and Kawakatsu, T. (2012) 'Tokyo's Greenhouse Gas Emissions Trading Scheme: A Model for Sustainable Megacity Carbon Markets?', Joint Discussion Paper, Series on Economics, Available at: https://www.uni-marburg.de/fb02/makro/forschung/magkspapers/25-2012_rudolph.pdf, accessed 6 October 2015.

Rutherford, J. and Coutard, O. (2014) 'Urban energy transitions: Places, processes and politics of socio-technical change', *Urban Studies*, 51(7), 1353–1377.

Rydin, Y. (2013) *The Future of Planning*. Bristol: Policy Press.

Sambanis, N. (2004) 'What is civil war? Conceptual and empirical complexities of an operational definition', *Journal of Conflict Resolution*, 48(6), 814–858.

Sasaki, Satoshi, Suzuki, Hiroshi, Igarashi, Kumiko, Tambatamba, Bushimbwa and Mulenga, Philip (2008) 'Spatial analysis of risk factor of cholera outbreak for 2003–2004 in a peri-urban area of Lusaka, Zambia', *American Journal of Tropical Medicine and Hygiene*, 79(3), 414–421.

Sassen, Saskia (1991) *The Global City: New York, London, and Tokyo.* Princeton, NJ: Princeton University Press.

—— (2002) 'Locating cities on global circuits', *Environment and Urbanization*, 14(1), 13–30.

—— (2012) *Cities in a World Economy* (fourth edition). Thousand Oaks, California: Pine Forge Press.

Satterthwaite, David (2006a) 'Small urban centres and large villages: The habitat for much of the world's low-income population', in Cecilia Tacoli, (ed.), *The Earthscan Reader in Rural–Urban Linkages*. London: Earthscan, pp. 15–38.

—— (2006b) 'Towards a real-world understanding of less ecologically damaging patterns of urban development', *Environment and Urbanization*, 18(2), 1–6.

—— (2008a) 'Cities' contribution to global warming: Notes on the allocation of greenhouse gas emissions', *Environment and Urbanization*, 20(2), 539–549.

—— (2013) 'The political underpinnings of cities' accumulated resilience to climate change', *Environment and Urbanization*, 25(2), 381–391.

—— (2007) 'The transition to a predominantly urban world and its underpinnings'. Human Settlements Discussion Paper – Urban Change 4, London: International Institute for Environment and Development.

Satterthwaite, D. and D. Dodman (2013) 'Towards resilience and transformation for cities within a finite planet', *Environment and Urbanization*, 25(2), 291–298.

Satterthwaite, David, Huq, Saleemul, Pelling, Mark, Reid, Hannah and Romero Lankao, Patricia (2007) 'Adapting to climate change in urban areas, the possibilities and constraints in low- and middle-income nations', IIED Human Settlements Discussion Paper Series, Theme: Climate Change and Cities No. 1, London: International Institute for Environment and Development, October. Available at: http://pubs.iied.org/pdfs/10549IIED. pdf?, accessed 6 October 2015.

Schatzberg, M. G. (1979) 'Islands of privilege: Small cities in Africa and the dynamics of class formation', *Urban Anthropology*, 8(2), 173–190.

Scheper-Hughes, Nancy (1992) *Death without Weeping: The Violence of Everyday Life in Brazil*. Berkeley and Los Angeles: University of California Press.

Scheper-Hughes, N. and Bourgois, P. I. (eds) (2004) *Violence in War and Peace: An Anthology*. Oxford: Blackwell Pub.

Schindler, Seth (2013) 'Producing and contesting the formal/informal divide: Regulating street hawking in Delhi, India', *Urban Studies*, 51(12), 2596–2612.

Schindler, S. and Kishore, B. (2015) 'Why Delhi cannot plan its "new towns": The case of solid waste management in Noida', *Geoforum*, 60, 33–42.

Schmidt, David (2008) 'From spheres to tiers: Conceptions of local government in South Africa in the period 1994–2006' in Mirjam van Donk, Mark Swilling, Edgar Pieterse and Susan Parnell, (eds), *Consolidating Developmental Local Government, Lessons from the South African Experience*. Cape Town: UCT Press, pp. 109–152.

Schwartz, M. (2007) 'Neo-liberalism on crack: Cities under siege in Iraq', *City*, 11(1), 21–69.

Scoones, Ian (1998) 'Sustainable rural livelihoods: A framework for analysis', IDS Working Paper 72, Brighton: Institute of Development Studies at the University of Sussex.

Scott, Allen J. (2001) 'Introduction', in Allen J. Scott, (ed.), *Global City-Regions: Trends, Theory, Policy*. Oxford: Oxford University Press, pp. 1–8.

Scott, Allen J. and Storper, Michael (2003) 'Regions, globalization, development', *Regional Studies*, 37(6–7), 579–593.

Scott, Allen, Agnew, John, Soja, Edward and Storper, Michael (2001) 'Global city-regions', in Allen J. Scott, (ed.), *Global City-Regions: Trends, Theory, Policy*. Oxford: Oxford University Press, pp. 11–30.

Segbers, Klaus (ed.) (2007) *The Making of Global City Regions: Johannesburg, Mumbai/Bombay, São Paulo, and Shanghai*. Baltimore: Johns Hopkins University Press.

Sen, Amartya (1981) *Poverty and Famines: An Essay on Entitlement and Deprivation*. Oxford: Clarendon Press.

—— (1999) *Development as Freedom*. New York: Anchor Book (Random House).

Sharan, A. (2006) 'In the city, out of place: Environment and modernity, Delhi 1860s to 1960s', *Economic and Political Weekly*, 41(47), 4905–4911.

Shen, L. Y., Ochoa, J. J., Shah, M. N. and Zhang, X. (2011) 'The application of urban sustainability indicators: A comparison between various practices', *Habitat International*, 35(1), 17–29.

Sims, D. (2011) *Understanding Cairo: The logic of a city out of control*. Oxford: Oxford University Press.

Sjaastad, E. and Cousins, B. (2009) 'Formalisation of land rights in the South: An overview', *Land Use Policy*, 26(1), 1–9.

Sjoberg, G. (1960) *The Preindustrial City*. New York: The Free Press.

Smit, B. and Wandel, J. (2006) 'Adaptation, adaptive capacity and vulnerability', *Global Environmental Change*, 16, 282–292.

Smit, B., Burton, I., Klein, R. and Wandel, J. (2000) 'An anatomy of adaptation to climate change and variability', *Climate Change*, 45: 223–251.

Smith, A. (1970; first published 1776) *Enquiry into the Nature and Causes of the Wealth of Nations*. Harmondsworth: Penguin.

Smith, A. and Stirling, A. (2010) 'The politics of social-ecological resilience and sustainable socio-technical transitions', *Ecology and Society*, 15(1), 11.

Smith, Carol A. (1995) 'Types of city-size distributions: A comparative analysis', in Ad van der Woude, Akira Hayami and Jan de Vries, (eds), *Urbanization in History: A Process of Dynamic Interactions*. Oxford: Clarendon Press, pp. 20–42.

Smoke, P. (2001) 'Fiscal decentralization in developing countries: A review of current concepts and practice', Democracy, Governance and Human Rights

Programme Paper No. 2, United Nations Research Institute for Social Development.

Soja, Edward W. (2000) *Postmetropolis: Critical Studies of Cities and Regions*. Oxford: Blackwell.

Solecki, William, Rosenzweig, Cynthia, Hammer, Stephen and Mehrotra, Shagun (2013) 'The urbanization of climate change: Responding to a new global challenge', in Elliott D. Sclar, Nicole Volavka-Close and Peter Brown, (eds), *The Urban Transformation: Health, Shelter and Climate Change*. Abingdon: Earthscan, pp. 197–220.

Southall, A. (ed.) (1979) *Small Urban Centres in Rural Development in Africa*. Madison: University of Wisconsin, African Studies Program.

—— (1988) 'Small towns in Africa revisited', *African Studies Review*, 31(3), 100–113.

Souza, C. (2001) 'Participatory budgeting in Brazilian cities: Limits and possibilities in building democratic institutions', *Environment and Urbanization*, 13(1), 159–184.

Sovani, N. V. (1964) 'The analysis of "over-urbanization"', *Economic Development and Cultural Change*, 12(2), 113–122.

Sridhar, K. S. (2010) 'Impact of land use regulations: Evidence from India's cities', *Urban Studies*, 47(7), 1541–1569.

Stanislawski, Dan (1947) 'Early Spanish town planning in the New World', *Geographical Review*, 37(1) (January), 94–105.

Staudt, Kathleen (2001) 'Informality knows no borders? Perspectives from El Paso-Juárez', *SAIS Review*, 21(1), 123–130.

Stein, A. and Moser, C. (2014) 'Asset planning for climate change adaptation: Lessons from Cartagena, Colombia', *Environment and Urbanization*, 26(1), 166–183.

Stewart, Frances (1985) *Basic Needs in Developing Countries*. Baltimore, JD: Johns Hopkins University Press.

Stoker, G. (1998) 'Governance as theory: Five propositions', *International Social Science Journal*, 50(155), 17–28.

Stoker, G. and Mossberger, K. (1994) 'Urban regime theory in comparative perspective', *Environment and Planning C: Government and Policy*, 12(2), 195–212.

Stone, Clarence (1989) *Regime Politics: Governing Atlanta*. Lawrence, Kansas: University Press of Kansas.

—— (2005) 'Looking back to look forward: Reflections on urban regime analysis', *Urban Affairs Review*, 40, 309.

—— (1993) 'Urban regimes and the capacity to govern', *Journal of Urban Affairs*, 15, 1–28.

Storper, Michael and Venables, Anthony J. (2004) 'Buzz: Face-to-face contact and the urban economy', *Journal of Economic Geography*, 4, 351–370.

Streeten, Paul, with Burki, S., Haq, M., Hicks, N. and Stewart, F. (1981) *First Things First: Meeting Basic Human Needs in Developing Countries*. Oxford and New York: Oxford University Press.

Stren, Richard (1993) '"Urban management" in development assistance: An elusive concept', *Cities*, 10(2), 125–138.

—— (2009) 'Diversity: A challenge to urban governance'. In APSA 2009 Toronto Meeting Paper.

—— (2014) 'Urban service delivery in Africa and the role of international assistance', *Development Policy Review*, 32(S1), s19–s37.

Stren, R. E. and White, R. R. (eds) (1989) *African Cities in Crisis: Managing Rapid Urban Growth*. Boulder, Colorado: Westview Press.

Sullivan, Edward J. and Michel, Matthew J. (2003) 'Ramapo Plus Thirty: The changing role of the plan in land use regulation', *The Urban Lawyer*, 35(1), 75–111.

Sutton, P. C. and Costanza, R. (2002) 'Global estimates of market and non-market values derived from nighttime satellite imagery, land cover, and ecosystem service valuation', *Ecological Economics*, 3, 509–527.

Sutton, P. C., Elvidge, C. D. and Ghosh, T. (2007) 'Estimation of gross domestic product at sub-national scales using nighttime satellite imagery', *International Journal of Ecological Economics and Statistics*, 8, 5–21.

Suzuki, H., Dastur, A., Moffat, S., Yabuki, N. and Maruyama, H. (2010) *Eco2 Cities: Ecological Cities As Economic Cities*. Washington DC: The World Bank.

Swanstrom, T. (2008) *Regional Resilience: A Critical Examination of the Ecological Framework*. Berkeley: Institute of Urban & Regional Development.

Tacoli, Cecilia (1998) 'Rural–urban interactions: A guide to the literature', *Environment and Urbanization*, 10(1), 147–166.

—— (2009) 'Crisis or adaptation? Migration and climate change in a context of high mobility', *Environment and Urbanization*, 21, 513–525.

Tannerfeldt, Göran and Ljung, Pers (2006) *More Urban Less Poor: An Introduction to Urban Development and Management*. London: Earthscan Publications.

Taylor, B. D. and Botea, R. (2008) 'Tilly tally: War-making and state-making in the contemporary third world', *International Studies Review*, 10(1), 27–56.

Taylor, N. (1998) *Urban Planning Theory since 1945*. London: Sage.

Taylor, Peter (2004a) *World City Network: A Global Urban Analysis*. London: Routledge.

—— (2004b) 'Planning for a better future', *Habitat Debate*, 10(4), 4–5.

Tendler, Judith (1997) *Good Government in the Tropics*. Baltimore/London: John Hopkins University Press.

—— (2000) 'Why are social funds so popular?', in Shahid Yusef, Weiping Wu and Simon Everett, (eds), *Local Dynamics in the Era of Globalization, Companion Volume of the World Development Report 1999/2000*. New York: The World Bank and Oxford University Press, pp. 114–129.

—— (2002) 'Small firms, the informal sector, and the devil's deal', *IDS Bulletin*, 33(3), 1–15.

Thibert, J. and Osorio, G. A. (2014) 'Urban segregation and metropolitics in Latin America: The case of Bogotá, Colombia', *International Journal of Urban and Regional Research*, 38(4), 1319–1343.

Thomas, A. (2000a) 'Poverty and the "end of development"', in Tim Allen and Alan Thomas, (eds), *Poverty and Development in the 21st Century*. Oxford: Oxford University Press, pp. 3–22.

—— (2000b) 'Meanings and views of development', in Tim Allen and Alan Thomas, (eds), *Poverty and Development in the 21st Century*. Oxford: Oxford University Press, pp. 23–48.

Tilly, C. (1984) 'Social movements and national politics', in Charles Bright and Susan Harding, (eds), *Statemaking and Social Movements*. Ann Arbor: University of Michigan Press, pp. 297–317.

—— (1985) 'War making and state making as organized crime', in P. B. Evans, D. Rueschemeyer and T. Skocpol, (eds), *Bringing the State Back In*. Cambridge: Cambridge University Press, pp. 169–191.

—— (1992) *Coercion, Capital, and European States, AD 990–1992*. Oxford: Blackwell.

—— (1994) 'Entanglements of European cities and states', in Charles Tilly and Wim P. Blockmans, (eds), *Cities and the Rise of States in Europe: A.D. 1000–1800*. Colorado: Westview Press, pp. 1–27.

Tilly, C. and Tarrow, S. (2007) *Contentious Politics*. Oxford: Oxford University Press.

TMG (2015) 'Tokyo Cap-and-Trade Program achieves reduction after 4th year', Tokyo Metropolitan Government, Bureau of Environment, February 19th, Japan.

Todaro, Michael (2000) *Economic Development* (seventh edition). Harlow, Essex: Pearson Education Limited and New York: Addison-Wesley Longman Inc.

Torfing, J., Peters, B. G., Pierre, J. and Sorensen, E. (2012) *Interactive Governance: Advancing the Paradigm*. Oxford; New York, Oxford University Press.

Townsend, A. M. (2013) *Smart Cities: Big Data, Civic Hackers, and the Quest for a New Utopia*. New York: W.W. Norton & Company.

Townsend, Peter (1993) *The International Analysis of Poverty*. Hemel Hempstead: Harvester Wheatsheaf.

Toye, J. (1987) *Dilemmas of Development*. Oxford: Blackwell.

Treisman, D. (2007) *The Architecture of Government: Rethinking Political Decentralization*. Cambridge: Cambridge University Press.

Türkün, A. (2011) 'Urban regeneration and hegemonic power relationships', *International Planning Studies*, 16(1), 61–72.

Turner, J. F. C. (1969) 'Uncontrolled urban settlement: Problems and policies', in G. Breese, (ed.), *The City in Newly Developing Countries: Readings on Urbanism and Urbanization*. Englewood Cliffs, NJ: Prentice Hall, pp. 507–534.

Turner, John (1972) 'Housing as a verb', in John Turner and Robert Fichter, (eds), *Freedom to Build*. London: MacMillan, pp. 148–176.

—— (1976) *Housing by People: Towards Autonomy in Building*. London: Marion Boyars.

Tyler, S. and Moench, M. (2012) 'A framework for urban climate resilience', *Climate and Development*, 4(4), 311–326.

UN (2014) 'Final Compilation of Amendments to Goals and Targets by Major Groups and other stakeholder including citizen's responses to MY World 6 priorities', available at http://sustainabledevelopment.un.org/content/documents/4438mgscompilationowg13.pdf, accessed 11 November 2014.

UNCHS (1996) *An Urbanizing World: Global Report on Human Settlements*. Oxford: Oxford University Press.

UNDESA (2013) *The World Economic and Social Survey 2013*. New York: United Nations.

UNDP (1994) *Human Development Report 1994*. New York: Oxford University Press.

—— (2014) *Human Development Report United Nations Development Programme* Available at: http://hdr.undp.org/en/2014-report/download, accessed 6 October 2015.

—— (2015) *What is the Multidimensional Poverty Index*. Available at: http://hdr.undp.org/en/content/what-multidimensional-poverty-index, accessed 30 June 2015.

UN-HABITAT (2003a) *Water and Sanitation in the World's Cities*. London and Nairobi: UN-HABITAT and Earthscan Publications.

—— (2003b) *The Challenge of Slums: Global Report on Human Settlements*. Nairobi: UN-Habitat.

—— (2005) *Financing Urban Shelter: Global Report on Human Settlements 2005*. London: Earthscan and UN Habitat.

—— (2006) *State of the World's Cities Report 2006/7*. Nairobi: United Nations Centre for Human Settlements.

—— (2007) *Enhancing Urban Safety and Security: Global Report on Human Settlements*. London, UN-Habitat.

—— (2008) *The State of the World's Cities Report 2006/7: The Millennium Development Goals and Urban Sustainability*. London: Earthscan. UN-HABITAT.

—— (2010) *Report of the Fifth Session of the World Urban Forum*. Rio de Janeiro: UN-HABITAT.

—— (2011a) *Global Report on Human Settlements 2011: Cities and Climate Change*. United Nations Human Settlements Program, Earthscan.

—— (2011b) *Condominium Housing in Ethiopia: The Integrated Housing Development Programme*. Nairobi: UN-HABITAT.

UN-ISDR (2011) *Global Assessment Report on Disaster Risk Reduction: Revealing Risks, Redefining Development*. United Nations, Geneva, Switzerland.

United Nations (1980) 'Patterns of urban and rural population growth', Department of International Economic and Social Affairs, Population Studies, No. 68. New York: United Nations

—— (2015) The Millennium Development Goals Report 2015. New York: United Nations.

United Nations, Department of Economic and Social Affairs (1968) 'Urbanization: Development policies and planning', *International Social Development Review*, No. 1, New York: United Nations.

—— (1999) *The World at Six Billion*. New York: United Nations.

United Nations, Department of Economic and Social Affairs, Population Division (2013) *World Population Policies 2013*. New York: United Nations.

—— (2014) *World Urbanization Prospects: The 2014 Revision*. CD-ROM Edition.

UNODC (2013) *Global Study on Homicide 2013*. Vienna: United Nations Office on Drugs and Crime.

UNPD (2011) *World Urbanization Prospects: The 2011 Revision.* United Nations, Department of Economic and Social Affairs, Population Division. Available at: http://esa.un.org/unpd/wup/index.htm, accessed 29 January 2013.

Ünsal, Ö. and Kuyucu, T. (2010) 'Challenging the neoliberal urban regime: Regeneration and resistance' in Başıbüyük and Tarlabaşı, *Orienting Istanbul: Cultural Capital of Europe*, London: Routledge, 51–70.

Urdal, H. and Hoelscher, K. (2012) 'Explaining urban social disorder and violence: An empirical study of event data from Asian and sub-Saharan African cities', *International Interactions*, 38(4), 512–528.

UTEP (2012) *Borderplex Economic Outlook: 2012–2014.* UTEP Border Region Modeling Project. The University of El Paso Texas.

Uvin, P. (2004) *Human Rights and Development* (Vol. 37). Streling: Kumarian Press.

Vambe, M. T. (2008) *The Hidden Dimensions of Operation Murambatsvina.* African Books Collective.

Van Dijk, Meine Pieter (2003) 'Government policies with respect to an information technology cluster in Bangalore, India,' *European Journal of Development Research*, 15 (2), 93–108.

Vanek, J, Chen, M. A., Carré, F., Heintz, J. and Hussmanns, R. (2014) 'Statistics on the informal economic: Definitions, regional estimates & challenges'. WEIGO Working Paper No 2.

Varshney, Ashutosh (2002) *Ethnic Conflict and Civic Life: Hindus and Muslims in India.* New Haven, CT: Yale University Press.

Viguié, V. and S. Hallegatte (2012) 'Trade-offs and synergies in urban climate policies', *Nature Climate Change*, 2(5), 334–337.

Vlassenroot, K. and Büscher, K. (2013) 'Borderlands, identity and urban development: The case of Goma (Democratic Republic of the Congo)', *Urban Studies*, 50(15), 3168–3184.

Wackernagel, Mathis and Rees, William (1996) 'What Is an ecological footprint?', in Stephen M. Wheeler and Timothy Beatley, (eds), *The Sustainable Urban Development Reader*. London: Routledge, pp. 211–219.

Wade, R. (2004, originally published 1990) *Governing the Market: Economic Theory and the Role of Government in East Asian Industrialization*, second edition. Princeton, NJ: Princeton University Press.

Wagenaar H. (2014) 'The agonistic experience: Informality, hegemony and prospects for democratic governance', in S. Griggs, A. Norval and H. Wagenaar, (eds), *Practices of Freedom: Decentred Governance, Conflict and Democratic Participation.* Cambridge: Cambridge University Press, pp. 217–248.

Wajahat, F. (2012) 'Perceptions of tenure security in a squatter settlement in Lahore, Pakistan', in N. Perera, (ed.), *Transforming Asian Cities: Intellectual Impasse, Asianizing Space, and Emerging Translocalities.* Oxon: Routledge, pp. 137–148.

Walker, G. and King, D. (2008) *The Hot Topic: How to Tackle Global Warming and Still Keep the Lights On.* London: Bloomsbury Publishers.

Walsh, J. and Amponstira, F. (2013) 'Infrastructure development and the repositioning of power in three Mekong Region capital cities', *International Journal of Urban and Regional Research*, 37(3), 879–893.

Walton, John (1998) 'Urban conflict and social movements in poor countries: Theory and evidence of collective action', *International Journal of Urban and Regional Research*, 22(3), 460–481.

Wan Xiaoyuan (2015 forthcoming) 'Governmentalities in everyday practices: The dynamic of urban neighbourhood governance in China', *Urban Studies*. Doi: 10.1177/0042098015589884.

Wang, H., Zhang, R. Liu, M. and Bi, J. (2012) 'The carbon emissions of Chinese cities', *Atmospheric Chemistry and Physics*, 12, 6197–6206.

Wang, L., Ma, A., Zhou, X., Tan, Y. et al. (2008) 'Environment and energy challenge of air conditioners in China'. Paper presented to the 2nd International Conference on Bioinformatics and Biomedical Engineering, Shanghai, May.

Wang, Y. P., Wang, Y. and Wu, J. (2009) 'Urbanization and informal development in China: Urban villages in Shenzhen', *International Journal of Urban and Regional Research*, 33(4), 957–973.

Warner, K. (2010) 'Global environmental change and migration: Governance challenges', *Global Environmental Change*, 20(3), 402–413.

Warren, M. E. (2002) 'What can democratic participation mean today?', *Political Theory*, 30(5), 677–701.

—— (2014) 'Governance-driven democratization', in Steven Griggs, Aletta Norval and Hendrik Wagenaar, (eds), *Practices of Freedom: Democracy, Conflict and Participation in Decentred Governance*. Cambridge: Cambridge University Press, pp. 38–59.

WaterAid (2001) *Looking Back: The Long-term Impacts of Water and Sanitation Projects.* London: WaterAid, June.

Watson, V. (2003) 'Conflicting rationalities: Implications for planning theory and ethics', *Planning Theory & Practice*, 4(4), 395–407.

—— (2009a) '"The planned city sweeps the poor away…": Urban planning and 21st century urbanisation', *Progress in Planning*, 72(3), 151–193.

—— (2009b) 'Seeing from the South: Refocusing urban planning on the globe's central urban issues', *Urban Studies*, 46(11), 2259–2275.

—— (2014) 'African urban fantasies: Dreams or nightmares?', *Environment and Urbanization*, 26(1), 215–231.

Watt, Paul (2000) *Social Investment in Economic Growth: A Strategy to Eradicate Poverty*. Oxford: Oxfam.

Weber, M. (1994) 'The profession and vocation of politics', in M. Weber, *Political Writings*. Cambridge: Cambridge University Press, pp. 309–369.

Weinstein, L. (2013) 'Demolition and dispossession: Toward an understanding of state violence in millennial Mumbai', *Studies in Comparative International Development*, 48(3), 285–307.

Wekwete, Kadmiel H. (1997) 'Urban management: The recent experience', in Carole Rakodi, (ed.), *The Urban Challenge in Africa*. Tokyo and New York: United Nations University Press, pp. 536–555.

Werlin, H. (1999) 'The slum upgrading myth', *Urban Studies*, 36(9), 1523–1534.

Wheeler, Stephen and Beatley, Timothy (eds) (2014) *The Sustainable Urban Development Reader*. London: Routledge.

While, Aidan and Mark Whitehead (2013) 'Cities, urbanisation and climate change', *Urban Studies*, 50(7), 1325–1331.

While, A., Jonas, A. E. G. and Gibbs, D. (2010) 'From sustainable development to carbon control: Eco-state restructuring and the politics of urban and regional development', *Transactions of the Institute of British Geographers*, 35(1), 76–93.

WHO (2004) 'World Health Day: Road Safety is no Accident!', available at: http://www.who.int/mediacentre/news/releases/2004/pr24/en, accessed 6 October 2015.

—— (2008) *Global Burden of Disease: 2004 Update*. Geneva: World Health Organization.

WHO and UN-Habitat (2010) *Hidden Cities: Unmasking and Overcoming Health Inequities in Urban Settings*. Kobe: World Health Organization and United Nations Human Settlements Programme.

WHO and UNICEF (2014) *Progress on Drinking Water and Sanitation: 2014 Update*. Geneva: WHO.

Williams, C. C. and Lansky, M. A. (2013) 'Informal employment in developed and developing economies: Perspectives and policy responses', *International Labour Review*, 152, 355–380.

Williams, G. (2004) 'Evaluating participatory development: Tyranny, power and (re)politicisation', *Third World Quarterly*, 25(3), 557–578.

Williams, G. and Thampi, B. V. (2013) 'Decentralisation and the changing geographies of political marginalisation in Kerala', *Environment and Planning A*, 45(6), 1337–1357.

Williams, G. O., Aandahl, G. and Devika, J. (2015 Forthcoming) 'Making space for women in urban governance? Leadership and claims-making in a Kerala slum', *Environment and Planning A*.

Williams, J. M. (2010) *Chieftaincy, the State and Democracy: Political Legitimacy in Post-Apartheid South Africa*. Bloomington: Indiana University Press.

Williamson, J. G. (1965) 'Regional inequality and the process of national development: A description of the patterns', *Economic Development and Cultural Change*, 13(4), Part 2, 1–84.

Willis, K. (2005) *Theories and Practices of Development*. London: Routledge.

Winton, A. (2004) 'Urban violence: A guide to the literature', *Environment and Urbanization*, 16(2), 165–184.

—— (2014) 'Gangs in global perspective', *Environment and Urbanization*, 26(2), 401–416.

Wirth, L. (1938) 'Urbanism as a way of life', *The American Journal of Sociology*, 44(1), 1–24.

Wisner, B. (2003) 'Disaster risk reduction in megacities: Making the most of human and social capital', in A. Kreimer, M. Arnold and A. Carlin, (eds), *Building Safer Cities: The Future of Disaster Risk*. Washington DC: The World Bank, pp. 181–196.

Wolfe, M. (2016 Forthcoming) *Watering the Revolution: An Environmental and Technological History of Agrarian Reform in La Laguna, Mexico*. Durham: Duke University Press.

Wood, Geoff (2003) 'Staying secure, staying poor: The Faustian bargain', *World Development*, 31(3), pp. 455–471.

World Bank (1990) *World Development Report 1990*. Washington DC: The World Bank.

—— (1997) *World Development Report: The State in a Changing World.* Oxford and New York: Oxford University Press.

—— (2000) *World Development Report 2000.* Washington DC: The World Bank.

—— (2005) *World Development Report 2005: A Better Investment Climate for Everyone.* Washington DC: The World Bank.

—— (2006) *Hazards of Nature, Risks to Development: An IEG Evaluation of World Bank Assistance for Natural Disaster.* Washington DC: The World Bank.

—— (2010a) *Cities and Climate Change: An Urgent Agenda.* The World Bank, Washington DC: The World Bank.

—— (2010b) *Violence in the City: Understanding and Supporting Community Responses to Urban Violence.* Washington DC: The World Bank.

—— (2011) *World Development Report 2011: Conflict, Security, and Development.* Washington DC: The World Bank.

—— (2013) *Global Monitoring Report 2013: Rural–Urban Dynamics and the Millennium Development Goals.* Washington DC: International Bank for Reconstruction and Development/The World Bank.

—— (2015) *World Development Indicators*, Online Database, http://data.worldbank.org/data-catalog/world-development-indicators, accessed July 2015.

World Bank and UNHCS (1989) *Urban Management Program: Overview of Program Activities.* Washington DC: The World Bank.

World Commission on Environment and Development (WCED) (1987) *Our Common Future (Brundtland Commission)*, Oxford: Oxford University Press.

Wratten, Ellen (1995) 'Conceptualising urban poverty', *Environment and Urbanization*, 7(1), 11–38.

Wu, Fulong (2000) 'The global and local dimensions of place-making: Remaking Shanghai as a world city', *Urban Studies*, 37(8), 1359–1377.

Wu, F., Zhang, F. and Webster, C. (2013) 'Informality and the development and demolition of urban villages in the Chinese peri-urban area', *Urban Studies*, 50(10), 1919–1934.

Wu, Kang-Li (2013) 'Sustainable urban development in China', in Stephen Wheeler and Timothy Beatley, (eds), (2014) *The Sustainable Urban Development Reader.* London: Routledge, pp. 447–456.

Yel, A. M. and Nas, A. (2013) 'After Gezi: Moving towards post-hegemonic imagination in Turkey', *Insight Turkey*, 15(4), 177–190.

Young, K. (1997) 'Gender and development', in Nalini Visvanathan, Lynn Duggan, Laurie Nisonoff, Nan Wiegersma, (eds), *The Women, Gender and Development Reader.* London, Zed Books, pp. 51–54.

Zimring, W. D. S. F. E. (2006) *The Great American Crime Decline.* Oxford: Oxford University Press.

Žižek, S. (2009) *Violence: Six Sideways Reflections.* London: Profile Books.

Index

Page numbers in italics and in bold typeface refer to information in figures and tables respectively.